Artificial Ground Freezing Technology

Artificial Ground Freezing Technology

Guest Editors

Jie Zhou

Kai-Qi Li

Jun Hu

MDPI

Basel • Beijing • Wuhan • Barcelona • Belgrade • Novi Sad • Cluj • Manchester

Guest Editors
Jie Zhou
Department of Geotechnical Engineering
School of Civil Engineering
Tongji University
Shanghai
China

Kai-Qi Li
Institute of Engineering Risk and Disaster Prevention
Wuhan University
Wuhan
China

Jun Hu
School of Civil Engineering and Architecture
Hainan University
Haikou
China

Editorial Office
MDPI AG
Grosspeteranlage 5
4052 Basel, Switzerland

This is a reprint of the Special Issue, published open access by the journal *Applied Sciences* (ISSN 2076-3417), freely accessible at: https://www.mdpi.com/journal/applsci/special_issues/AGF_technologies.

For citation purposes, cite each article independently as indicated on the article page online and as indicated below:

Lastname, A.A.; Lastname, B.B. Article Title. *Journal Name* **Year**, *Volume Number*, Page Range.

ISBN 978-3-7258-4279-7 (Hbk)
ISBN 978-3-7258-4280-3 (PDF)
https://doi.org/10.3390/books978-3-7258-4280-3

Contents

About the Editors vii

Preface ix

Changqiang Pang, Haibing Cai, Rongbao Hong, Mengkai Li and Zhe Yang
Evolution Law of Three-Dimensional Non-Uniform Temperature Field of Tunnel Construction Using Local Horizontal Freezing Technique
Reprinted from: *Appl. Sci.* **2022**, *12*, 8093, https://doi.org/10.3390/app12168093 1

Chao Wang, Jing Chen, Lilei Chen, Yue Sun, Zelei Xie, Guoan Yin, et al.
Experimental and Modeling of Residual Deformation of Soil–Rock Mixture under Freeze–Thaw Cycles
Reprinted from: *Appl. Sci.* **2022**, *12*, 8224, https://doi.org/10.3390/app12168224 20

Jun Hu, Ke Li, Yuwei Wu, Dongling Zeng and Zhixin Wang
Optimization of the Cooling Scheme of Artificial Ground Freezing Based on Finite Element Analysis: A Case Study
Reprinted from: *Appl. Sci.* **2022**, *12*, 8618, https://doi.org/10.3390/app12178618 33

Chuan-Xin Rong, Zhi Wang, Yi Cao, Qing Yang and Wei Long
Orthogonal Test on the True Triaxial Mechanical Properties of Frozen Calcareous Clay and Analysis of Influencing Factors
Reprinted from: *Appl. Sci.* **2022**, *12*, 8712, https://doi.org/10.3390/app12178712 51

Jie Zhou, Zhongqiu Guo, Chuanhe Wang, Zeyao Li, Huade Zhou and Wansheng Pei
Analysis of Freeze–Thaw Response and Pore Characteristics of Artificially Frozen Soft Soil under Combined Formation Seepage
Reprinted from: *Appl. Sci.* **2022**, *12*, 10687, https://doi.org/10.3390/app122010687 66

Jie Zhou, Zeyao Li and Wansheng Pei
The Quantification and Evolution of Particle Characteristics of Saturated Silt under Freeze–Thaw Cycles
Reprinted from: *Appl. Sci.* **2022**, *12*, 10703, https://doi.org/10.3390/app122110703 80

Yin Duan, Chuanxin Rong, Xianwen Huang and Wei Long
An Analytical Solution to Steady-State Temperature Field in the FSPR Method Considering Different Soil Freezing Points
Reprinted from: *Appl. Sci.* **2022**, *12*, 11576, https://doi.org/10.3390/app122211576 98

Mengkai Li, Haibing Cai, Zheng Liu, Changqiang Pang and Rongbao Hong
Research on Frost Heaving Distribution of Seepage Stratum in Tunnel Construction Using Horizontal Freezing Technique
Reprinted from: *Appl. Sci.* **2022**, *12*, 11696, https://doi.org/10.3390/app122211696 113

Bin Wang, Shenwei Liang, Yi Cao, Chuanxin Rong and Shengmin Yu
Analysis of Elastoplastic Mechanical Properties of Non-Uniform Frozen Wall Considering Frost Heave
Reprinted from: *Appl. Sci.* **2023**, *13*, 1038, https://doi.org/10.3390/app13021038 132

Natalija Batoćanin, Wojciech Wróblewski, Ivana Carević, Uroš Durlević, Violeta Gajić and Aleksandar Valjarević
Facies and Origin of Tufa Deposits from the Gostilje River Basin and the Sopotnica River Basin (SW Serbia)
Reprinted from: *Appl. Sci.* **2023**, *13*, 3190, https://doi.org/10.3390/app13053190 150

Siyuan Shu, Zhishu Yao, Yongjie Xu, Chen Wang and Kun Hu
Mechanical Properties and Constitutive Relationship of Cretaceous Frozen Sandstone under Low Temperature
Reprinted from: *Appl. Sci.* **2023**, *13*, 4504, https://doi.org/10.3390/app13074504 . . . . **164**

Irina S. Ivanova, Liudmila S. Shirokova, Jean-Luc Rols and Oleg S. Pokrovsky
Partitioning of Dissolved Organic Carbon, Major Elements, and Trace Metals during Laboratory Freezing of Organic Leachates from Permafrost Peatlands
Reprinted from: *Appl. Sci.* **2023**, *13*, 4856, https://doi.org/10.3390/app13084856 . . . . **183**

Shilong Peng, Yuhao Xu, Guangyong Cao and Lei Pei
Research on the Elastoplastic Theory and Evolution Law of Plastic Zone Contours of Horizontal Frozen Walls under Nonuniform Loads
Reprinted from: *Appl. Sci.* **2023**, *13*, 9398, https://doi.org/10.3390/app13169398 . . . . **200**

Sergey Bublik, Mikhail Semin, Lev Levin, Andrey Brovka and Ivan Dedyulya
Experimental and Theoretical Study of the Influence of Saline Soils on Frozen Wall Formation
Reprinted from: *Appl. Sci.* **2023**, *13*, 10016, https://doi.org/10.3390/app131810016 . . . . **219**

Yazhou Ou, Long Wang, Hui Bian, Hua Chen, Shaole Yu, Tao Chen, et al.
Numerical Analyses of the Effect of the Freezing Wall on Ground Movement in the Artificial Ground Freezing Method
Reprinted from: *Appl. Sci.* **2024**, *14*, 4220, https://doi.org/10.3390/app14104220 . . . . **245**

Tao Wang, Huixi Lin, Kexiong Ren, Jian Gao and Di Wang
Sensitivity Analysis of Different Hydrothermal Characteristics in the Variable Thermodynamic Processes of Soft Clay Rock
Reprinted from: *Appl. Sci.* **2024**, *14*, 10253, https://doi.org/10.3390/app142210253 . . . . **262**

About the Editors

Jie Zhou

Jie Zhou is a professor and MEng/PhD supervisor at the Department of Geotechnical Engineering, School of Civil Engineering, Tongji University. She was honored as a 2022 Young Yangtze Scholar by the Ministry of Education and has decades of expertise in urban engineering geology, disaster mitigation, and AGF technology. At present, she also serves as the Vice Secretary-General of the International Affairs Office, Chinese Society for Rock Mechanics and Engineering (CSRME). She has published 60 SCI papers, authored 5 books in both Chinese and English, and certified 18 patents, including an international innovation patent. She has led several key projects funded by the National Natural Science Foundation and the Shanghai Young Talent Sail Program and sub-projects as part of the National Key Research and Development Program. She has always been unwavering in her determination to combine innovation in fundamental theory with engineering practice. She has earned widespread recognition and numerous awards, including second prize in Natural Science from the Ministry of Education, second prize in Scientific and Technological Progress from the Shanghai Civil Engineering Society, second prize in Technological Invention in Shanghai, and first prize in Geological Achievements in Shanghai.

Kai-Qi Li

Kai-Qi Li is a Postdoctoral Fellow in the Department of Civil and Environmental Engineering at the Hong Kong Polytechnic University (PolyU). She is particularly recognized in the field of artificial ground freezing (AGF), in which she has developed advanced thermo-hydro-mechanical (THM) models for freezing processes in complex geological environments. Her research resulted in her being awarded the 2023 Hubei Provincial Science & Technology Progress Award. She has published more than 30 papers in leading international journals and also serves as an Early-Career Editorial Panel Member for the *European Journal of Soil Science* and regularly completes reviews for top-tier journals such as the *Journal of Geotechnical and Geoenvironmental Engineering* and *Computers and Geotechnics*.

Jun Hu

Jun Hu is a Professor and MEng/PhD supervisor in the School of Civil Engineering and Architecture at Hainan University and the Director of the Institute of Marine Geotechnical Research. He was appointed as the first batch of Hainan Province's young "South China Sea Masters" and "Double Hundred" talent team. He was also honored as an outstanding talent in Hainan province. With more than 200 publications to his name in international refereed journals, more than 120 national patents, 18 software copyrights, 5 monographs, and 5 textbooks, he is currently conducting interdisciplinary research relating to marine geotechnical engineering, tunnel and underground engineering, artificial ground freezing technology, and intelligent construction.

Preface

Artificial ground freezing (AGF) has emerged as a vital geotechnical engineering technique, offering innovative solutions for ground stabilization, groundwater control, and underground construction in challenging environments. With increasing urbanization and infrastructure development in complex geological conditions, AGF plays a crucial role in enabling safe and sustainable engineering practices. The following Special Issue brings together cutting-edge research to advance the understanding and application of AGF technology, addressing critical challenges in thermal process optimization, the mechanical behavior of frozen soils, environmental interactions, and field applications.

The collection features 16 peer-reviewed papers that can be broadly categorized into four research themes: Pang et al. and Duan et al. investigate temperature field evolution and analytical solutions for freezing processes, while Hu et al. present optimization approaches for cooling schemes using advanced numerical simulations. Several papers, including those by Wang et al., Rong et al., and Shu et al., examine the mechanical properties and constitutive relationships of various frozen soils and rocks under different loading and environmental conditions. Ivanova et al. and Batoćanin et al. explore the environmental implications of freezing processes and the unique geological applications of AGF technology. Contributions from Li et al., Ou et al., and others demonstrate practical applications of AGF in tunnel construction and ground improvement, providing valuable insights for engineering practice.

This Special Issue highlights significant advancements in AGF technology while identifying key areas for future research, including energy-efficient freezing systems, long-term performance monitoring, and climate-adaptive solutions. We are incredibly grateful to all authors, reviewers, and the editorial team for their valuable contributions to this collection, which we believe will serve as an important reference for researchers and practitioners in the field. The presented research was funded by the National Natural Science Foundation of China (42477182 and 52471281) and the Hainan Provincial Natural Science Foundation Innovation Research Team Project (522CXTD511).

It is our hope that this Special Issue will stimulate further innovation and promote the continued development of artificial ground freezing technology to meet the growing demands of modern geotechnical engineering.

Jie Zhou, Kai-Qi Li, and Jun Hu
Guest Editors

applied sciences

MDPI

Article

Evolution Law of Three-Dimensional Non-Uniform Temperature Field of Tunnel Construction Using Local Horizontal Freezing Technique

Changqiang Pang [1], Haibing Cai [1,*], Rongbao Hong [1,2], Mengkai Li [1] and Zhe Yang [1]

1 School of Civil Engineering and Architecture, Anhui University of Science and Technology, Huainan 232001, China
2 School of Architectural Engineering, Nantong Vocational University, Nantong 226007, China
* Correspondence: author: haibingcai@163.com; Tel.: +861-39-6645-0807

Abstract: The formation quality of a frozen wall is one of the prerequisites for tunnel excavation using artificial ground freezing techniques. However, the non-uniformity of temperature distribution along the length direction of the freezing pipe is often ignored in the actual freezing engineering, which leads to a thin frozen wall at a local position that does not meet the design requirements. Therefore, exploring the evolution law of three-dimensional non-uniform freezing temperature fields is necessary. In this paper, a tunnel horizontal freezing model test system was established based on the similarity criterion of hydro–heat coupling, and the temperatures at three sections were tested using thermocouple temperature sensors. The results show that the temperature drop curves of measurement points suffer from three periods: steep drop, slow drop and tending to be stable. The temperature curves on the main and vice planes of the frozen wall all present a "V" type; specifically, the temperature on the axis plane is the lowest, while the temperature away from the axis plane is higher, and the temperature gradient outside the axis plane is greater than that inside. The frozen wall develops from frozen soil columns to a sector ring, and the average thickness of the frozen wall at three sections is 50.6, 40.7 and 75.1 mm after freezing for 60 min, respectively, which shows an obvious non-uniformity. The temperature distribution along the length of the freezing pipe is $T = -0.000045z^2 + 0.0205z - 13.5125$. The freezing temperature contours calculated by ABAQUS are basically consistent with those calculated by the model test after calling the temperature function of the freezing pipe wall.

Keywords: tunnel; horizontal freezing; non-uniformity; freezing temperature field

1. Introduction

Artificial ground freezing techniques, used as ground reinforcement methods that freeze pore water into ice to form an impermeable layer, have the advantages of good water sealing, high strength and stiffness, strong adaptability, green environmental protection and so on, so they are widely used in underground engineering. In freezing construction, it is very important to master the thickness of the frozen wall, which depends on the distribution of the freezing temperature field. Thus, exploring the evolution law of the freezing temperature field is the foundation of tunnel construction using artificial ground freezing techniques [1,2]. Scholars' research achievements through the following three methods are mainly as follows:

(1) Theoretical analysis: Trupak [3] and Bakholdin [4] presented the calculation methods of single-piped, single-circle-piped and double-circle-piped steady-state freezing temperature fields. Subsequently, Sanger and Sayles [5,6] optimized the analytical solution of a single-circle-piped freezing temperature field. On this basis, Tobe [7] derived an analytical solution for a multi-circle-piped freezing temperature field, and then

Hu [8–12] optimized these analytical solutions. In addition, Aziz [13], Hosseini [14], Jiji [15], Jiang [16] and Cai [17] derived the analytical solution of a single-piped transient freezing temperature field.

(2) Numerical simulation: Yang [18], Yu [19,20], Fu [21] and Cai [22] studied the distribution law of the freezing temperature field of a subway connecting passage by different finite element software. Hong [23] studied the evolution law of the local horizontal freezing temperature field of an underground tunnel with a shallow depth by ABAQUS. Hu [24] studied the distribution law of the cup-type freezing temperature field of the tunnel port and found that the closure of outer-circle pipes was earlier than that of inner-circle pipes.

(3) Model tests: Shang [25] established a rectangular tunnel construction model and found that the outer edge of the frozen wall developed slowly due to the heat dissipation of the model surface. Shi [26] established a shield docking freezing model and determined the positive freezing time. Cai [27] and Duan [28] established a pipe-roof freezing model with different types of freezing pipe and found that the freezing effect of an empty pipe with a double circular freezer was the best. Zhang [29] established a tunnel vertical freezing model and found that the thickness of the basin-type frozen wall upstream was smaller than that downstream.

None of the above studies consider the loss of cooling capacity along the length direction of the freezing pipe, which leads to the three-dimensional non-uniformity of the freezing temperature field in actual engineering. In this paper, a tunnel horizontal freezing model test system was established based on the similarity theory to explore the evolution law of the three-dimensional non-uniform transient temperature field of construction of an underground tunnel using the local horizontal freezing method.

2. Model Test Design

2.1. Project Overview

A local horizontal freezing project is located south of the section between the Dabeiyao and thermal power plant section of the Beijing Metro Fu-ba line [23]. The buried depth of this tunnel is 10 m, and the maximum excavation diameter is 3.0 m. A 1.2 m-thick freezing mild clay layer at the arch of the tunnel was formed through 8 freezing pipes with an outer diameter of 108 mm. There are some long-term disrepaired sewage pipelines over the tunnel, and the downward seepage is serious. The mild clay layer is in a saturated state all year round, and soil mass moisture content can reach 21.72%.

2.2. Derivation of Similarity Criteria

The heat conduction differential equation of soil freezing is as follows:

$$\frac{\partial \theta_n}{\partial \tau} = a_n\left(\frac{\partial^2 \theta_n}{\partial r^2} + \frac{1}{r}\frac{\partial \theta_n}{\partial r}\right) \ (0 < r_0 < r < \infty,\ \tau > 0,\ n = 1,\ 2) \tag{1}$$

where r is radial coordinate, e is time; θ is soil temperature, which is the temperature of unfrozen area when $n = 1$ and that of frozen area when $n = 2$, r_0 is the outer radius of freezing pipes, and a is the thermal diffusivity of soil.

The initial and boundary conditions of soil freezing temperature field [30] are

$$\begin{cases} \theta(r,0) = \theta_0 \\ \theta(\infty,\tau) = \theta_0 \\ \theta(\rho,\tau) = \theta_f \\ \theta(r_0,\tau) = \theta_y \end{cases} \tag{2}$$

where θ_0 is the initial temperature of soil, ρ is the external surface coordinates of frozen wall, θ_f is the freezing temperature of soil, θ_y is the wall temperature of freezing pipes.

The heat balance equation at the freezing front ($r = \rho$) is

$$\lambda_2 \frac{\partial \theta_2}{\partial r}|_{r=\rho} - \lambda_1 \frac{\partial \theta_1}{\partial r}|_{r=\rho} = L\frac{d\rho}{d\tau} \tag{3}$$

where λ is the thermal conductivity of soil, L is the latent heat of soil.

According to Equations (1)–(3), the similarity criterion equation of soil freezing temperature field can be obtained by equation analysis method.

$$F(F_0, K_0, R, \theta) = 0 \tag{4}$$

where $F_0 = \frac{a\tau}{r^2}$ is the Fourier criterion of temperature field; $K_0 = \frac{L}{c\theta}$ is Kosović criterion and c is the specific heat of soil; R is geometric criterion, and θ is temperature criterion.

In the model test, the soil was taken from the site, $C_\rho = 1$, $C_a = 1$, $C_c = 1$, $C_L = 1$. The water content of soil is the same, so the latent heat released during freezing is equal, which is substituted into Equation (4) to obtain:

$$C_\tau = {C_r}^2 \tag{5}$$

$$C_\theta = 1 \tag{6}$$

Equation (5) shows that the time similarity ratio in the model test is the square of the geometric similarity ratio; according to Equation (6), the temperature of each point in the model test is the same as that of each corresponding point in the prototype.

Water migration occurs in the soil freezing process, and its essence is the humidity field problem in the freezing process. The mathematical model is:

$$\frac{\partial h}{\partial \tau} = b\left(\frac{\partial^2 h}{\partial r^2} + \frac{1}{r}\frac{\partial h}{\partial r}\right) \tag{7}$$

The boundary condition is:

$$\begin{cases} h(r,0) = h_0 \\ h(\infty,\tau) = h_0 \\ h(\rho,\tau) = 0 \end{cases} \tag{8}$$

where h is humidity, and b is moisture conductance.

According to Equations (7) and (8), the similarity criterion equation of soil moisture field can be obtained by the equation analysis method as follows:

$$F(F_h, R, H) = 0 \tag{9}$$

where $F_h = \frac{b\tau}{r^2}$ is the Fourier criterion of humidity field, H is humidity criterion.

It shows that the water transfer process and soil freezing process are similar in the mathematical model, and they all obey the Fourier criterion. Therefore, under the condition of geometric similarity, as long as the temperature field is similar, the humidity field is also similar to the temperature field.

In this paper, the similarity ratio of geometry (C_l) in the model test was determined to be 30; this ratio can be used to calculate the similarity ratio of other parameters (Table 1) according to the aforementioned similarity criteria.

In Table 1, the similarity ratio is the ratio of the value of the same physical quantity in the prototype to that in the model; that is, the size will be reduced by 30 times and time will be reduced by 900 times in the model test, while temperature, humidity (soil moisture), density, thermal conductivity, specific heat and latent heat of phase change remain consistent with those in the prototype.

Table 1. Similarity ratios.

Parameter	Similarity Ratio
Geometry (m)	30
Time (s)	900
Temperature (°C)	1
Humidity (%)	1
Density (kg/m^3)	1
Thermal conductivity ($kcal{\cdot}m^{-1}{\cdot}d^{-1}{\cdot}°C^{-1}$)	1
Specific heat ($kcal{\cdot}kg^{-1}{\cdot}°C^{-1}$)	1
Latent heat of phase change ($kcal{\cdot}kg^{-1}$)	1

2.3. Model Test System

Figure 1 shows the tunnel horizontal freezing model test system, which includes a model box, freezing system and temperature measurement system.

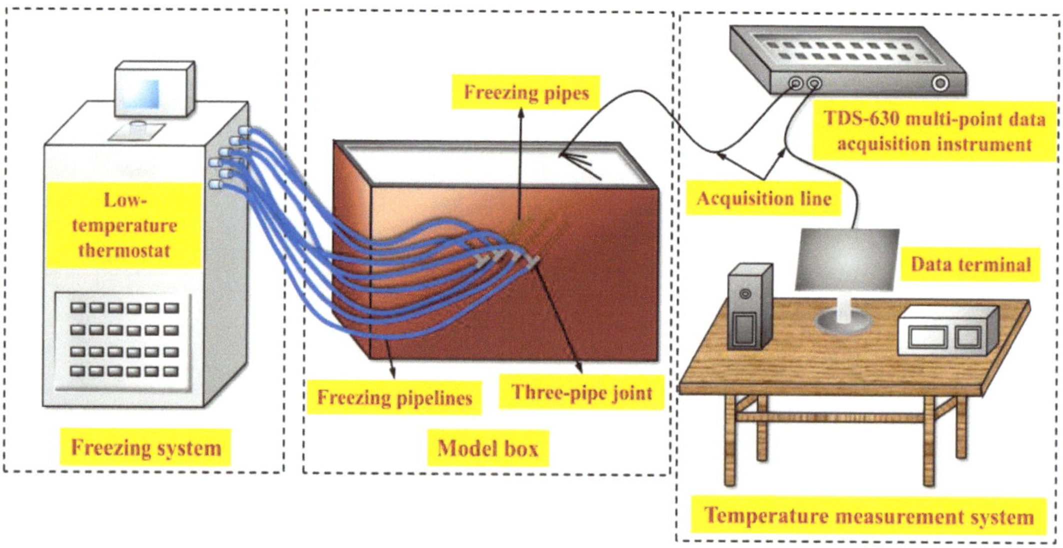

Figure 1. Three-dimensional diagram of model test system.

(1) Model box

The size of the model box in Figure 2e was 1000 mm × 800 mm × 700 mm, and freezing pipes in Figure 2f were set horizontally along the width direction of the box. The parameters of mild clay are shown in Table 2.

Table 2. Soil parameters.

Parameter		Density /(kg/m^3)	Thermal Conductivity /(kcal/(m °C·d))	Specific Heat /(kcal/(kg·°C))	Latent Heat of Phase Change /(kcal/kg)	Freezing Temperature /(°C)
Mild clay	Unfrozen	2100	23.00	0.357	8.93	−1
	Frozen		38.06	0.240		

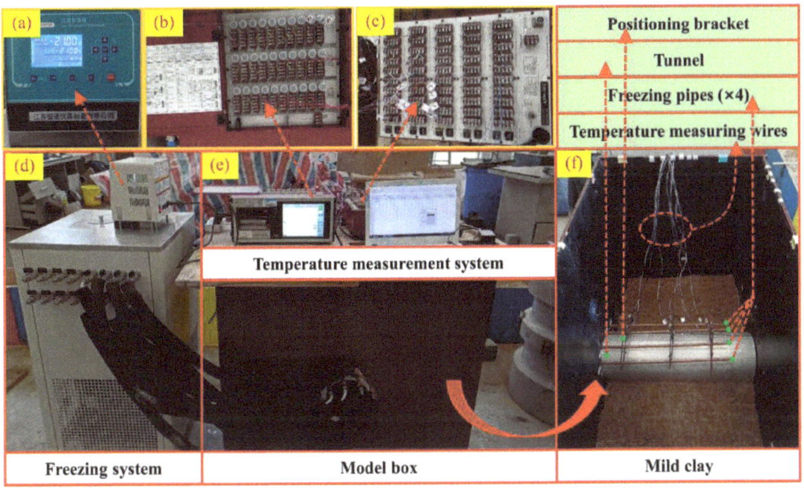

Figure 2. Tunnel horizontal freezing model test system: (**a**) the control panel of low-temperature constant temperature bath; (**b**,**c**): TDS-630 multi-point data acquisition instrument; (**d**) low-temperature thermostat; (**e**) model box; (**f**) mild clay and freezing pipes in the model box.

(2) Freezing system

As shown in Figure 2a,d, the freezing equipment adopted the low-temperature thermostat produced by Jiangsu Hengnuo Instrument Manufacturing Co., Ltd. (Nanjing, China), which allowed alcohol to maintain a temperature of −21 °C for cycle operation. The test time was set as 72 min, where the positive time was 60 min, and the design thickness of the frozen wall was set as 40 mm. In this model test, 4 freezing pipes with an outer diameter of 8 mm were determined by the equivalent heat transfer and similarity criteria, and these pipes were symmetrically distributed above the tunnel at an angle of 28°. The structure of the freezing pipe is a coaxial sleeve, and the circulation mode of alcohol is entering from the inner pipe and leaving from the outer pipe, and its flow rate is 1.2 m^3/h, as shown in Figure 3.

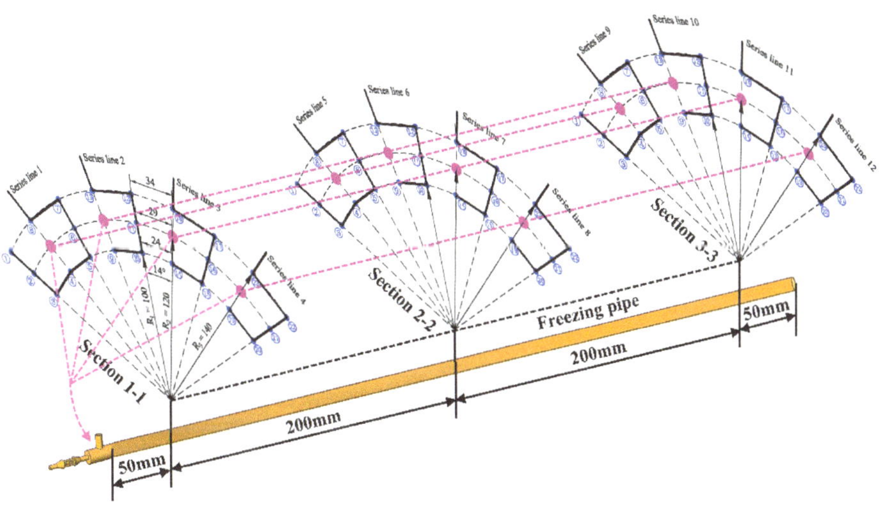

Figure 3. Temperature measurement points.

(3) Temperature measurement system

A TDS-630 multi-point data acquisition instrument in Figure 2b,c was selected to record temperature, and the acquisition frequency was set as 1 min/time. In this model test, three temperature measurement sections were set along the length direction of the freezing pipe, that was 50 mm, 250 mm and 450 mm from the orifice of the freezing pipe. At each temperature measurement section, there were 23 measurement points in the soil and 4 measurement points on the freezing pipe wall, as shown in Figure 3.

3. Model Test Results

3.1. Temperature Variation Analysis

The temperature duration curves of each measurement point at three sections in the model test are shown in Figure 4.

Figure 4 shows that the temperature drop curves of the measurement points suffer from three periods: steep drop, slow drop and tending to be stable. The temperature of the measurement point between the adjacent two freezing pipes decreases faster, while that outside the freezing pipe decreases slower. Figure 4a shows that the temperatures of measurement points are all below −1 °C after freezing for 55 min. The temperature drop rates of measurement point ⑪ are 0.365, 0.182 and 0.024 °C/min at three periods; Figure 4b shows that the temperatures of measurement points are all below −1 °C after freezing for 70 min. The temperature drop rates of measurement point ⑪ are 0.318, 0.120 and 0.011 °C/min at three periods; Figure 4c shows that the temperatures of measurement points are all below −1 °C after freezing for 60 min. The temperature drop rates of measurement point ⑥ are 0.334, 0.148 and 0.019 °C/min at three periods.

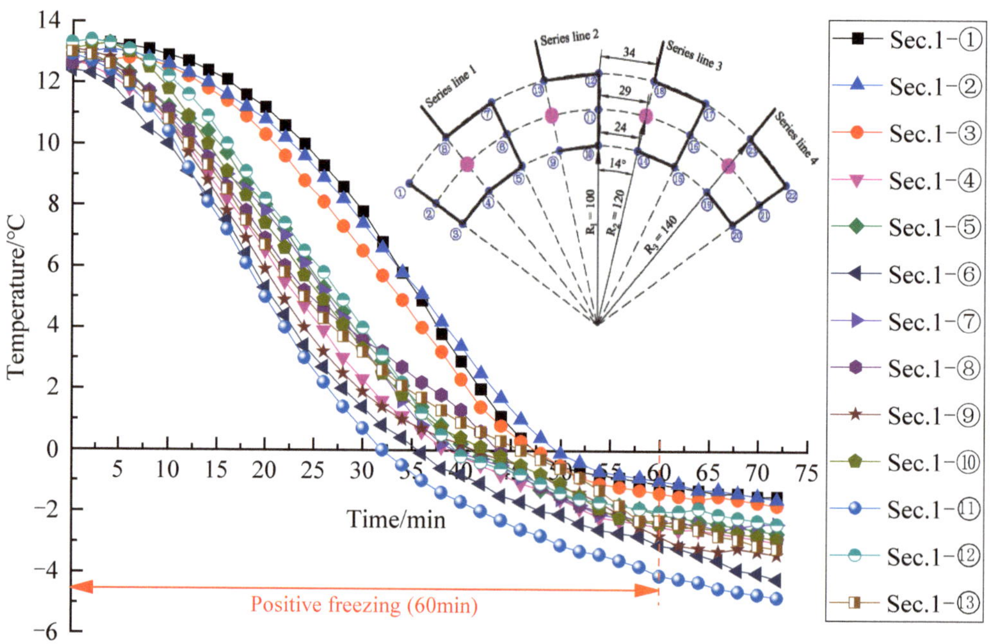

(a)

Figure 4. *Cont.*

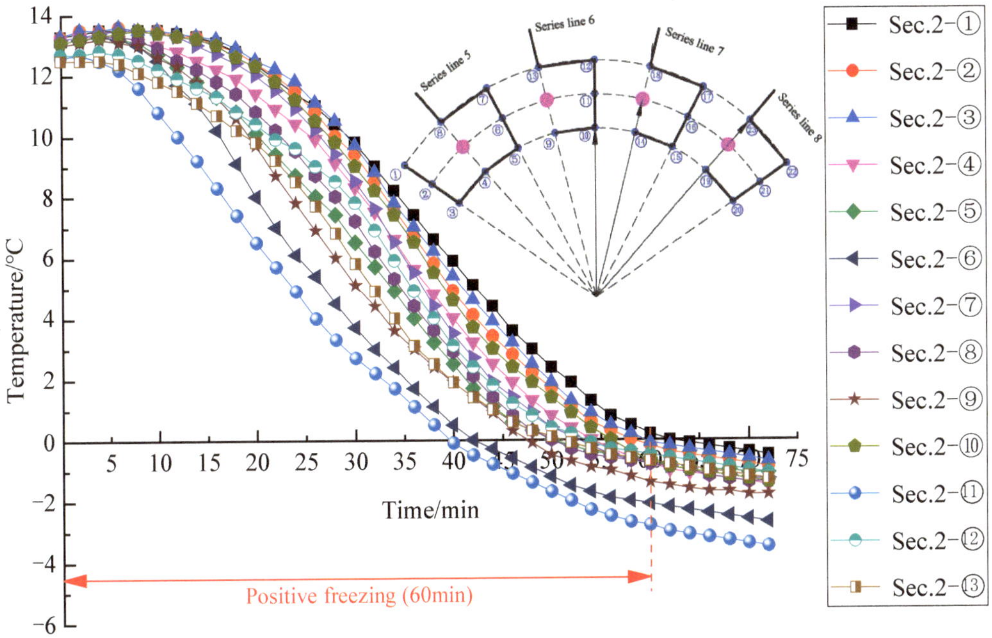

(b)

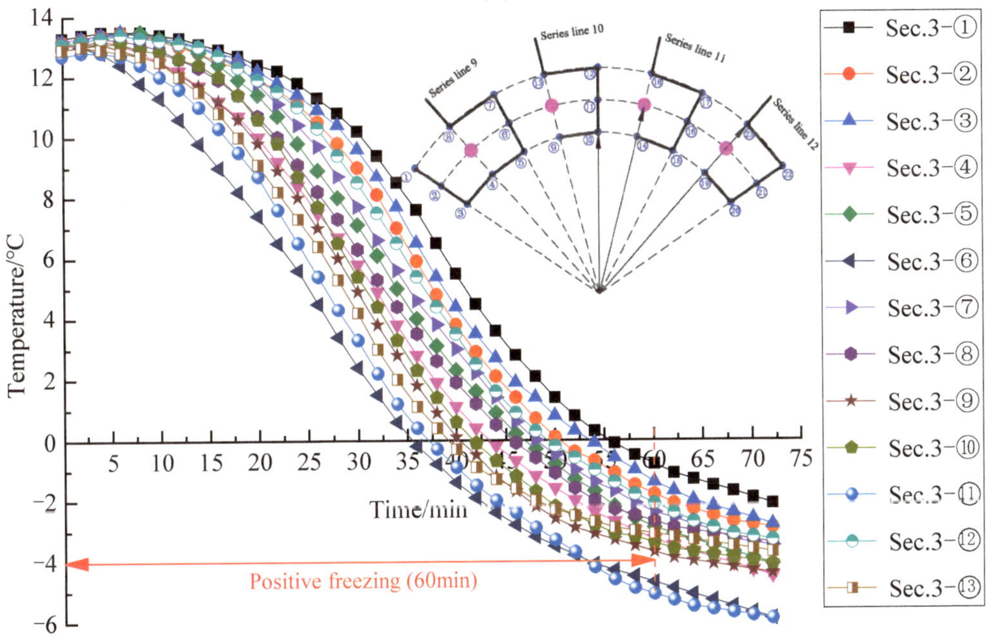

(c)

Figure 4. Temperature variation curves of all measurement points at each section: (**a**) Section 1-1; (**b**) Section 2-2; (**c**) Section 3-3.

3.2. Temperature Field Evolution Perpendicular to Freezing Pipes

The freezing temperature fields at three sections are shown in Figures 5–7 after freezing for 4 min, 28 min, 44 min and 60 min.

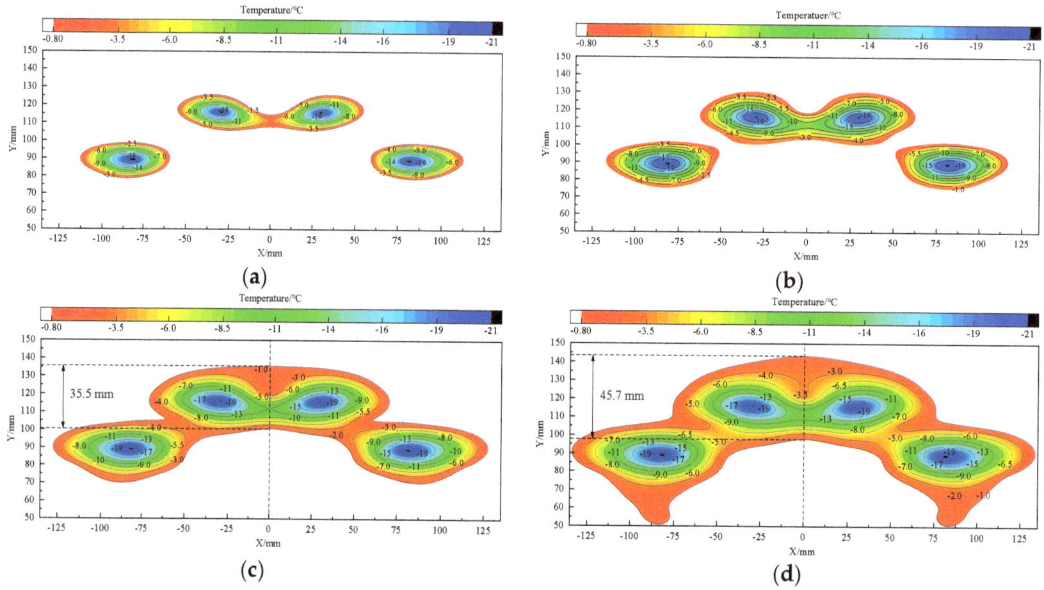

Figure 5. Temperature field at Section 1-1: (**a**) 4 min; (**b**) 28 min; (**c**) 44 min; (**d**) 60 min.

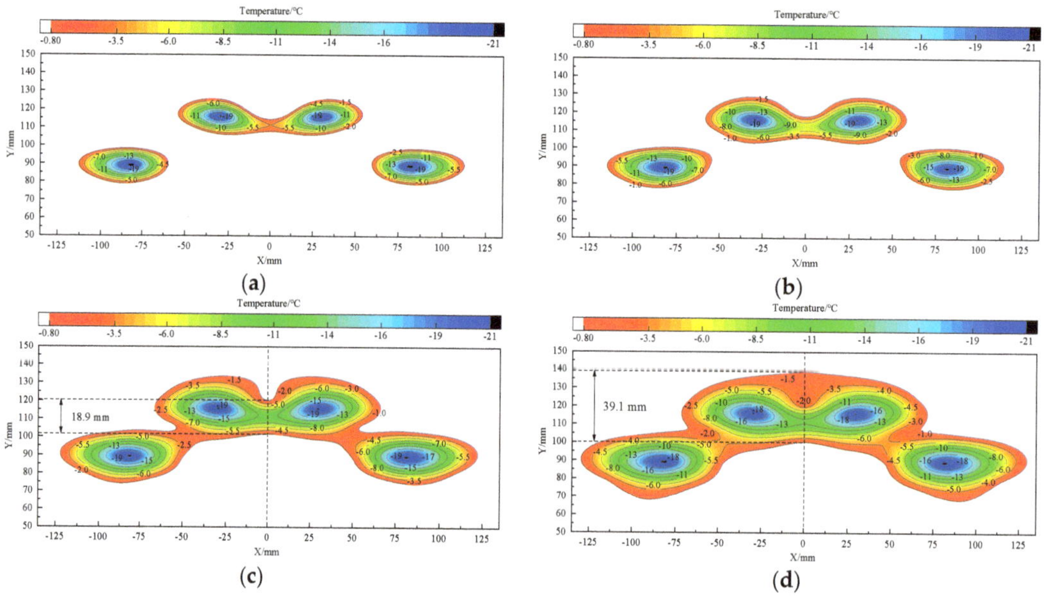

Figure 6. Temperature field at Section 2-2: (**a**) 4 min; (**b**) 28 min; (**c**) 44 min; (**d**) 60 min.

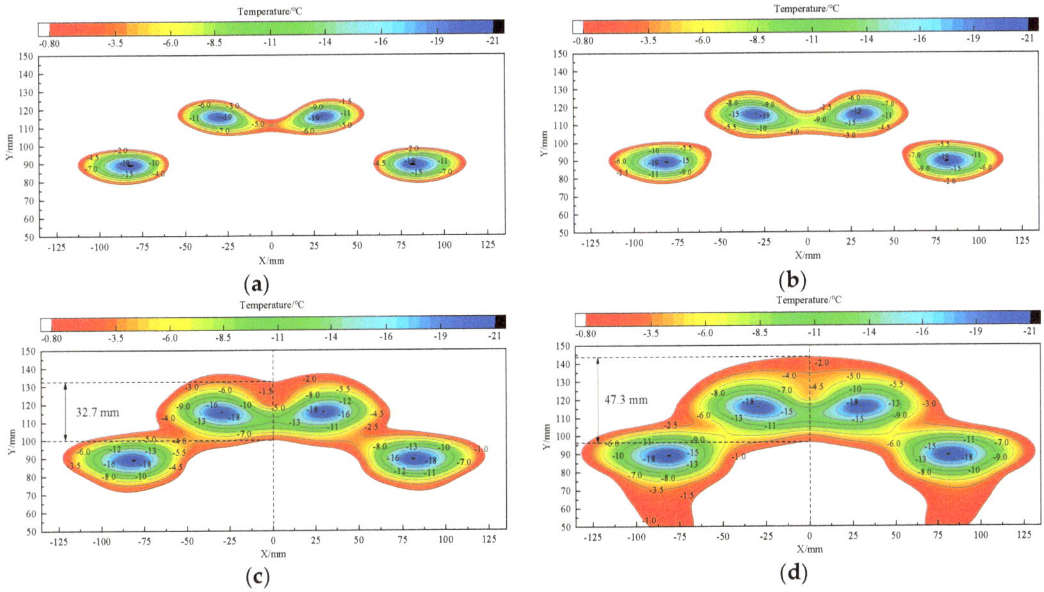

Figure 7. Temperature field at Section 3-3: (**a**) 4 min; (**b**) 28 min; (**c**) 44 min; (**d**) 60 min.

Figures 5–7 show that after freezing for 4 min, a circular frozen soil column is formed around the freezing pipes. After freezing for 28 min, the diameter of the frozen soil columns further expands, the two frozen soil columns in the middle intersect first. After freezing for 44 min, the frozen wall basically intersects, the thickness of the frozen wall at the tunnel arch in three sections is 35.5 mm, 18.9 mm and 32.7 mm, respectively. After freezing for 60 min, the frozen wall completely intersects, and the thickness of the frozen wall at the tunnel arch in three sections is 45.7 mm, 39.1 mm and 47.3 mm, respectively. Moreover, the freezing temperature field can be calculated by the single pipe freezing theory within freezing for 28 min, while it can be calculated by the flat-panel freezing theory [31] after freezing for 28 min. After freezing for 60 min, the order of the freezing effect is Section 3-3 > Section 1-1 > Section 2-2.

To quantitatively analyze the non-uniformity of temperature distribution at three sections, the temperature on two characteristic planes (the main and vice planes) of the horizontal frozen wall are selected to draw curves. Where the main plane refers to the characteristic plane passing through the center of the tunnel and each freezing pipe, the vice plane refers to the characteristic plane passing through the center of the tunnel and the midpoint of the connecting line between adjacent freezing pipes, while the axis plane refers to the arrangement circle of freezing pipes, as shown in Figure 8.

Figures 9–11 show that the temperature curves on the main and vice planes of the frozen wall present a "V" type; specifically, the temperature on the axis plane is the lowest, while it gradually rises with the increase in distance to the axis plane. Moreover, the temperature curve near the inner and outer edges (R = 100 mm and R = 140 mm) of the designed frozen wall tends to be flat, and the temperature at the inner edge (R = 100 mm) is always lower than that at the outer edge (R = 140 mm) during freezing for 60 min. It indicates that the dissipation capacity of the cooling capacity inside the axis plane of the frozen wall is less than that outside the axis plane.

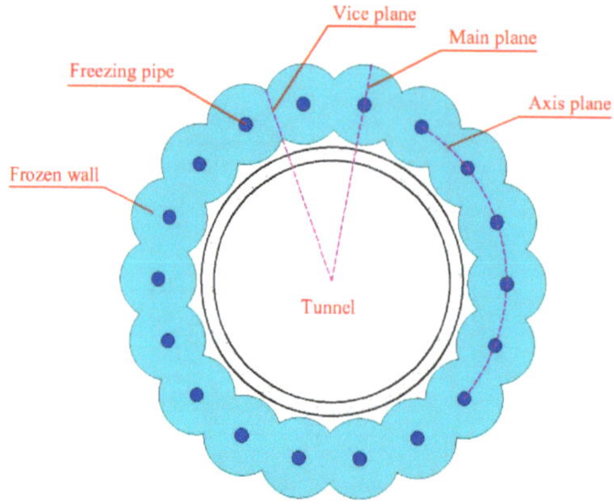

Figure 8. Characteristic planes of frozen wall.

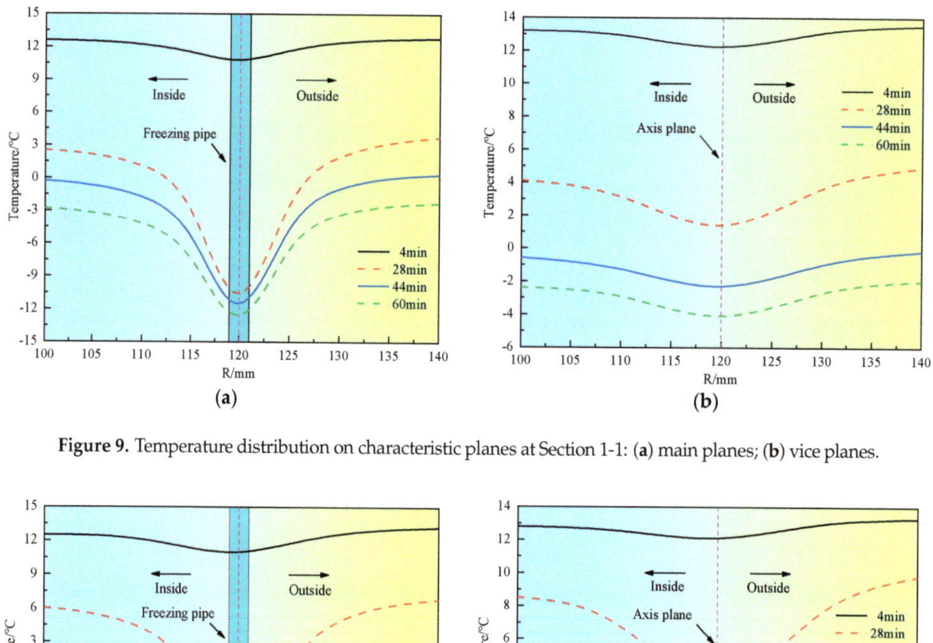

Figure 9. Temperature distribution on characteristic planes at Section 1-1: (**a**) main planes; (**b**) vice planes.

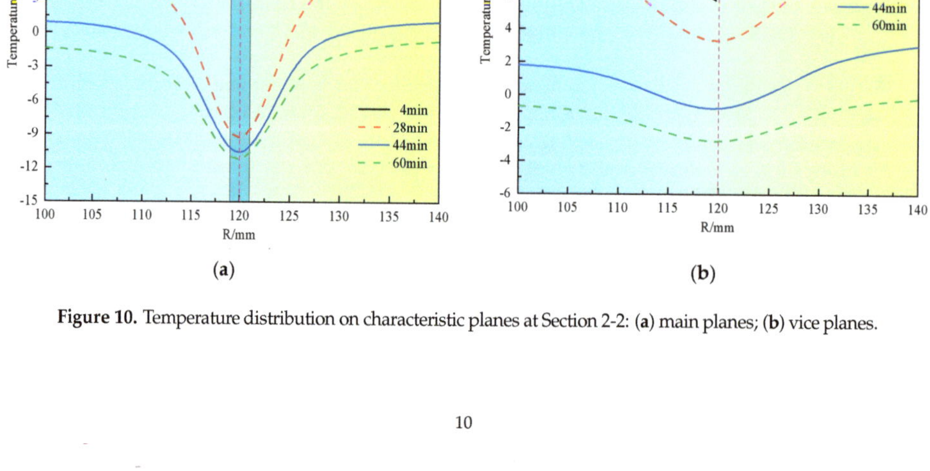

Figure 10. Temperature distribution on characteristic planes at Section 2-2: (**a**) main planes; (**b**) vice planes.

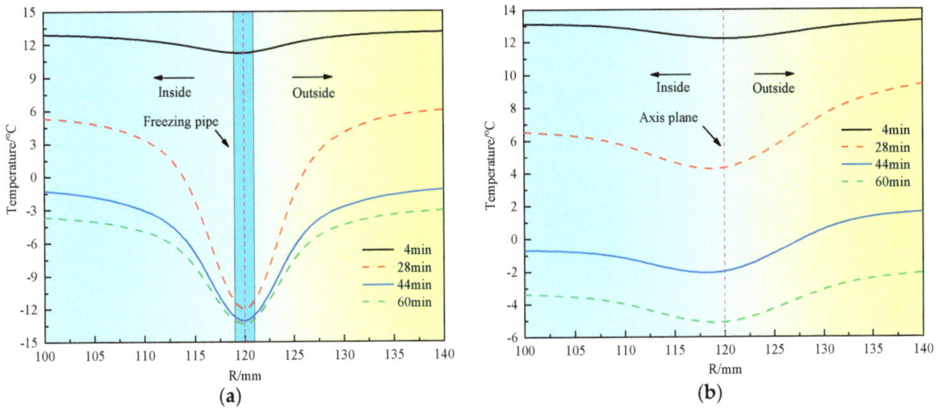

Figure 11. Temperature distribution on characteristic planes at Section 3-3: (**a**) main planes; (**b**) vice planes.

In addition, at the early stage of freezing, the temperature drop rate inside the axis plane is bigger than that outside the axis plane. At Section 1-1, along the main plane, the average temperature drop rate outside the axis plane is approximately 0.254 °C/min, while it is approximately 0.259 °C/min inside the axis plane. Meanwhile, along the vice plane, the average temperature drop rate outside the axis plane is approximately 0.252 °C/min, while it is approximately 0.254 °C/min inside the axis plane. At the later stage of freezing, the temperature drop rate inside and outside the axis plane is roughly the same.

The duration curves of the average temperature gradient of the frozen wall at three sections are shown in Figure 12.

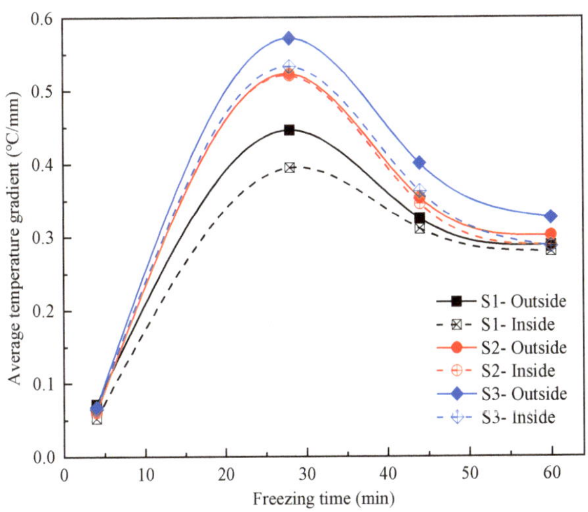

Figure 12. Duration curve of average temperature gradient of frozen wall at three sections.

Figure 12 shows that the average temperature gradient at the inside and outside of the frozen wall first increases, then decreases and finally tends to be stable with the increase in freezing time. At Section 1-1, after freezing for 4 min, the average temperature gradient outside the axis plane is only 0.07 °C/mm, while it is 0.05 °C/mm inside the axis plane. After freezing for 28 min, the average temperature gradient outside the axis plane is only

0.446 °C/mm, while it is 0.395 °C/mm inside the axis plane. After freezing for 44 min, the average temperature gradient outside the axis plane is only 0.325 °C/mm, while it is 0.311 °C/mm inside the axis plane. After freezing for 60 min, the average temperature gradient outside the axis plane is only 0.288 °C/mm, while it is 0.280 °C/mm inside the axis plane. It is obvious that the average temperature gradient on the outside of the frozen wall is more than that on the inside of the frozen wall. Moreover, the order of the value of the average temperature gradient at three sections is Section 3-3 > Section 2-2 > Section 1-1.

3.3. Temperature Field Evolution Parallel to Freezing Pipes

The temperature field parallels to the freezing pipe at the tunnel vertical center line after positive freezing for 60 min is shown in Figure 13.

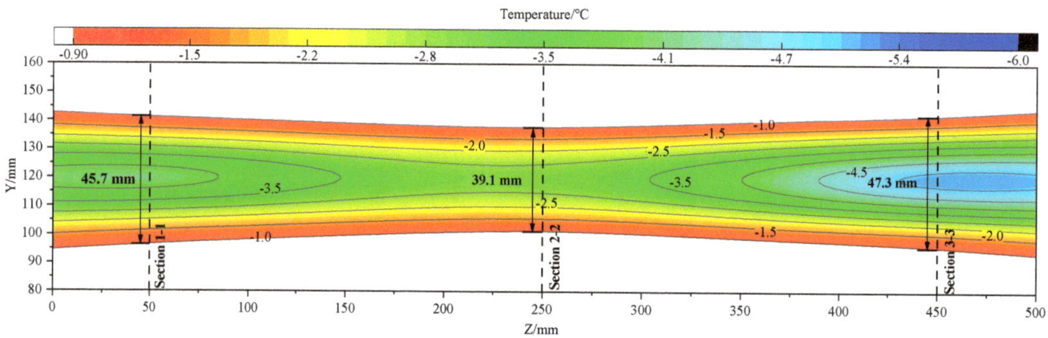

Figure 13. The temperature field parallels to the freezing pipe at tunnel vertical center line.

Figure 13 shows the morphology of the frozen wall, which is thicker at the ends and thinner in the middle; it is intuitive and obvious that the frozen wall is non-uniform. The average thickness of the frozen wall at three sections after positive freezing for 44 and 60 min is shown in Figure 14.

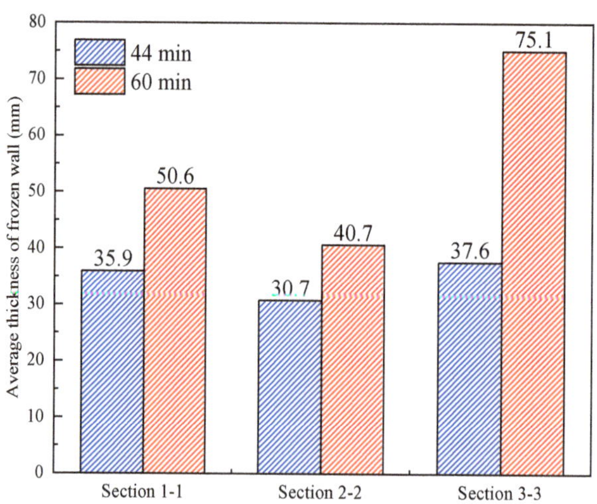

Figure 14. The average thickness of frozen wall at three sections.

Figure 14 shows that the average thickness of the frozen wall at three sections is not all equal after freezing for 44 min and 60 min. After freezing for 44 min, the average thickness of the frozen wall at Sections 1-1, 2-2 and 3-3 is 35.9 mm, 30.7 mm and

37.6 mm, respectively; after freezing for 60 min, the average thickness of the frozen wall at Sections 1-1, 2-2 and 3-3 is 50.6 mm, 40.7 mm and 75.1 mm, respectively. The order of the average thickness of the frozen wall at three sections is Section 3-3 > Section 1-1 > Section 2-2, which presents the non-uniformity of a frozen wall along the length direction of the freezing pipe.

The pipe wall temperature fitting curves along the length direction of the freezing pipe are shown in Figure 15 after freezing for 4 min, 28 min, 44 min and 60 min.

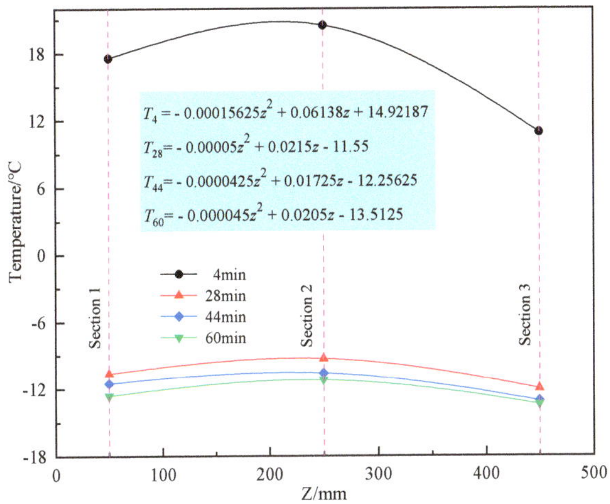

Figure 15. Pipe wall temperature fitting curves along the length direction of the freezing pipe.

Figure 15 shows that the pipe wall temperature first increases and then decreases along the length direction of the freezing pipe, where it reaches the maximum at Section 2-2. After freezing for 4 min, 28 min, 44 min and 60 min, the temperature fitting functions are as follows:

$$\begin{cases} T = -0.00015625z^2 + 0.06138z + 14.92187 & t = 4\text{min} \\ T = -0.00005z^2 + 0.0215z - 11.55 & t = 28\text{min} \\ T = -0.0000425z^2 + 0.01725z - 12.25625 & t = 44\text{min} \\ T = -0.000045z^2 + 0.0205z - 13.5125 & t = 60\text{min} \end{cases} \quad (10)$$

The conclusion obtained from the model test is inconsistent with the common sense that the alcohol temperature in the outer pipe gradually rises from the distal end to the proximal end of the freezing pipe; the main reason lies in the structure of the freezing pipe. As shown in Figure 16, the freezing pipe used in this model test is composed of an outer pipe, an inner pipe and a three-pipe joint; it is also commonly used in practical engineering. The low-temperature alcohol enters from the inner pipe, flows through the inner pipe to the end of the freezing pipe, and then flows back to the three-pipe joint through the outer pipe. When the alcohol flows in the outer pipe, it indirectly cools the soil by absorbing the heat transferred by the soil to the pipe wall, so as to achieve the purpose of freezing soil. Usually, the temperature of the pipe wall should gradually rise along the flow direction of the alcohol in the outer pipe, and the temperature of the pipe wall on each section should be shown as Section 3-3 < Section 2-2 < Section 1-1. However, the alcohol flowing in the outer pipe not only has heat exchange with the outer pipe wall, but also has heat exchange with the inner pipe wall, and there is also heat exchange between the inner pipe wall and the alcohol flowing in the inner pipe. The complex heat exchange process eventually leads to the non-uniform distribution of the temperature of the outer pipe wall of the freezing pipe, and the specific thermodynamic theory needs further study.

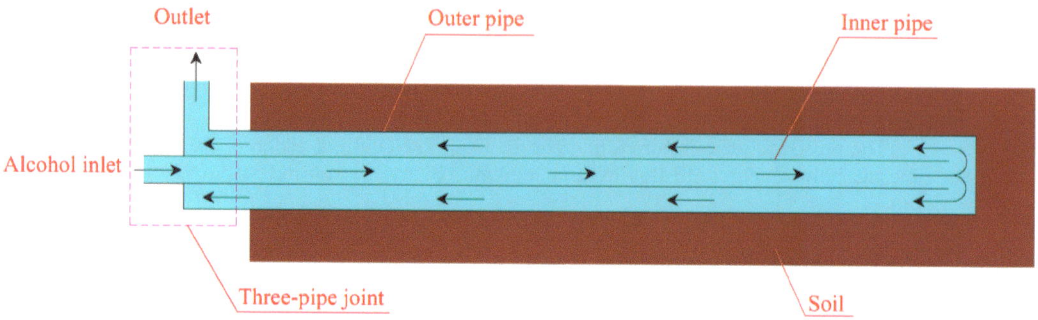

Figure 16. Diagram of freezing pipe structure.

4. Numerical Simulation Design

4.1. Establishment of Model

The numerical simulation can more intuitively unfold the development process of the frozen wall. Based on the pipe wall temperature fitting function calculated by the model test, a three-dimensional finite element model is established to simulate the freezing temperature field, as shown in Figure 17.

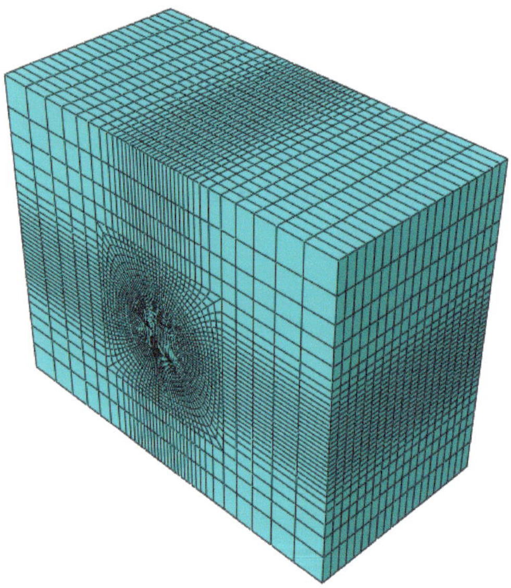

Figure 17. Three-dimensional finite element model.

Figure 17 shows that the model size is 1000 mm × 800 mm × 500 mm. The element type is DC3D8 with good heat-transfer capability, and the number of elements is 59120, where the four freezing pipes are simplified as linear heat sources, and each freezing pipe consists of 25 linear elements.

4.2. Boundary Conditions

The initial temperature of soil elements is set as 20 °C, while the temperature of line elements of the freezing pipe is set as the temperature fitting function.

4.3. Material Parameters

The soil elements are given the material parameters of mild clay, as shown in Table 2.

5. Numerical Simulation Results

The validation of the numerical simulation results with model test results at temperature measuring point ⑪ at Section 3-3, which has a good representative, is shown in Figure 18.

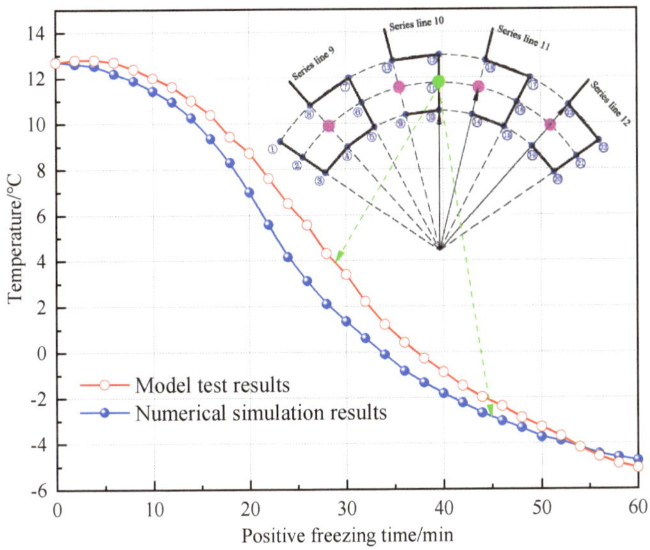

Figure 18. The validation of numerical simulation results with model test results.

As shown in Figure 18, the numerical simulation results are basically consistent with the model test results; the two curves both show a change process of steep drop first, then slow drop and finally stabilized. The difference between them mainly occurs at the steep drop stage, where the model test results are always lower than the numerical simulation results, and the maximum reaches 2.5 °C. The difference in refrigerating capacity may be due to the unstable ambient temperature in the laboratory or the influence of some unpredictable factors; that is, the model test results may have errors in reflecting the development of freezing temperature field. In contrast, the numerical simulation results can reflect the development of the freezing temperature field more reasonably.

The evolution law of the three-dimensional freezing temperature field is shown in Figure 19 after freezing for 4 min, 28 min, 44 min and 60 min.

Figure 19 shows that the three-dimensional freezing temperature field has obvious non-uniformity after freezing for 4 min, 28 min, 44 min and 60 min. The frozen walls at both ends of the freezing pipe connect at first, and the frozen walls in the middle of the freezing pipe connect subsequently. After freezing for 4 min, four independent frozen soil columns are formed with the characteristic that both ends are coarser and the middle is thinner. After freezing for 28 min, the two frozen soil columns in the middle connect in turn at Section 3-3 and Section 1-1, then expand along with the negative and positive directions of freezing depth, respectively, and eventually converge at Section 2-2. During this period, the frozen soil columns on both sides also gradually develop following this law. After freezing for 48 min, the frozen wall basically intersects, and the whole frozen wall still presents the law that is thin in the middle and thick at both ends. After freezing for 60min, the frozen wall completely intersects, and the thickness of the frozen wall meets the design requirements.

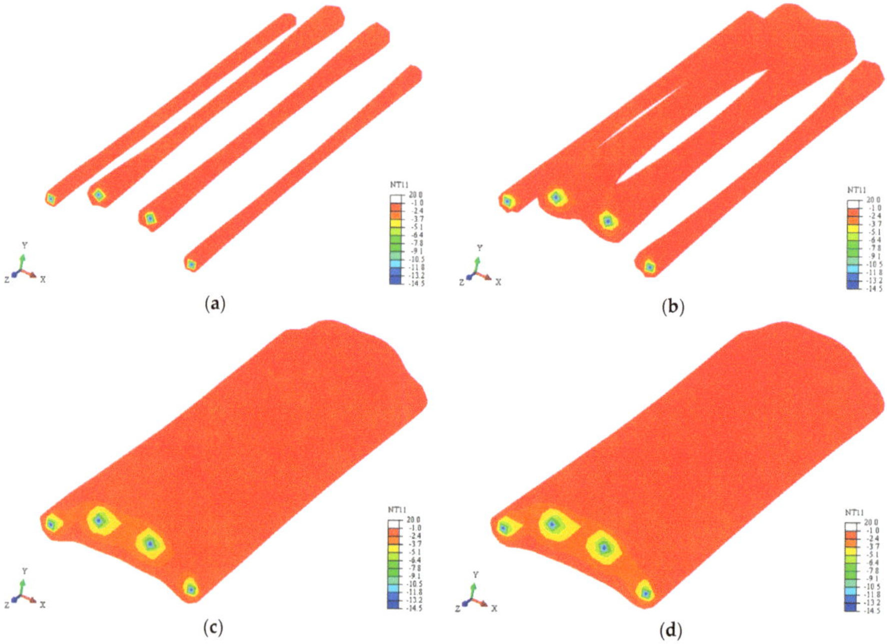

Figure 19. Evolution of three-dimensional freezing temperature field: (**a**) 4 min; (**b**) 28 min; (**c**) 44 min; (**d**) 60 min.

The comparison of the freezing temperature field calculated by numerical simulation and model test is shown in Figure 20 after freezing for 60 min.

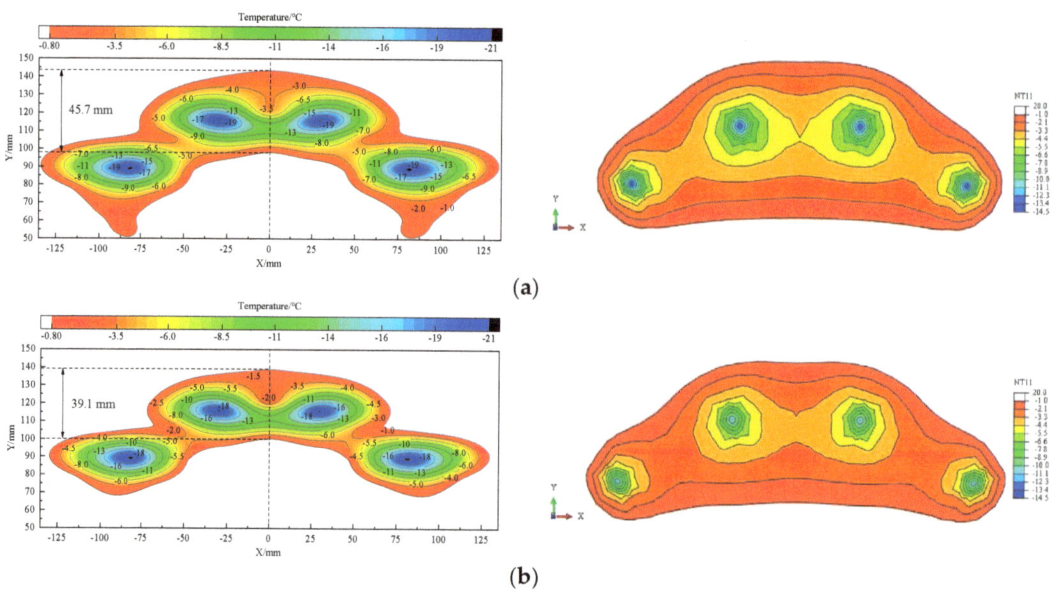

Figure 20. *Cont.*

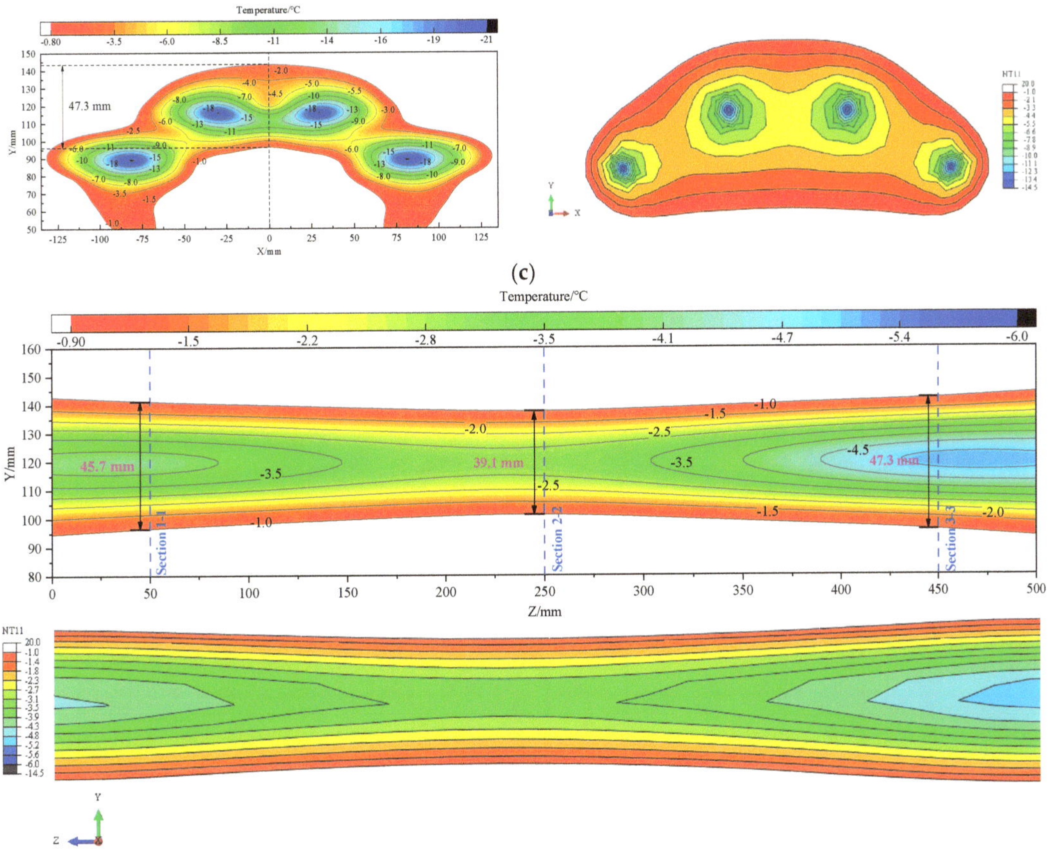

Figure 20. Temperature contours of model test and numerical simulation after freezing for 60 min: (**a**) Section 1-1; (**b**) Section 2-2; (**c**) Section 3-3; (**d**) the section parallels the freezing pipe at tunnel vertical center line.

Figure 20 shows that after freezing for 60 min, the thickness of the frozen wall at three sections is different; that is, the thickness of Section 1-1, 2-2 and 3-3 at tunnel arch is 45.7 mm, 39.1 mm and 47.3 mm in the model test, while it is 46.3 mm, 40.9 mm and 49.4 mm in the numerical simulation, respectively. The difference between them is 0.6 mm, 1.8 mm and 2.1 mm, and no matter the sections perpendicular or parallel to the length direction of the freezing pipes, the shapes of the frozen wall at each section are all basically consistent in the numerical simulation and the model test, which indicates that the numerical simulation method has high prediction accuracy after introducing the temperature fitting function of the freezing pipe wall.

6. Conclusions

In this paper, the evolution law of a three-dimensional non-uniform temperature field is analyzed by combining the physical model test and numerical simulation during the tunnel construction using a local horizontal freezing technique. The main conclusions are as follows:

The temperature drop curves of the measurement points suffer from three periods: steep drop, slow drop and tending to be stable, and the temperature of the measurement point between the adjacent two freezing pipes decreases faster, while that located outside the freezing pipes decreases slower.

The temperature curves on the main and vice planes of the frozen wall present a "V" type, specifically, and the temperature on the axis plane is the lowest, while it gradually rises with the increase in distance to the axis plane. The average temperature gradient at the inside and outside of the frozen wall first increases, then decreases and finally tends to be stable with the increase in freezing time.

After freezing for 60 min, the frozen wall presents an obvious non-uniformity along the length direction of the freezing pipe, and the thickness at Section 1-1, 2-2 and 3-3 at the tunnel arch is 45.7 mm, 39.1 mm and 47.3 mm, respectively, and the difference with the numerical simulation results is 0.6 mm, 1.8 mm and 2.1 mm, respectively, which indicates that the new numerical simulation method introducing the temperature fitting function of the freezing pipe wall has high accuracy in predicting the evolution law of the non-uniform temperature field.

The results of this study are helpful to understand the temperature field development of the soil around sleeve-type freezing pipes and can provide a certain reference for the design and construction of such freezing projects and other aspects of research. Next, research on the evolution law of non-uniform temperature fields in the whole process of freezing and thawing under complex formation conditions will be carried out, as well as its frost heave and thaw settlement effects.

Author Contributions: Data curation, R.H., M.L. and Z.Y.; Funding acquisition, H.C.; Writing—original draft, C.P. All authors have read and agreed to the published version of the manuscript.

Funding: This research was funded by the National Natural Science Foundation of China, grant number 51778004; Academic Funding for Top-notch Talents in University Disciplines (Majors) of Anhui Province, China, grant number gxbjZD10; Research on Graduate Science Project in Anhui Province, China, grant number YJS20210383; Graduate Innovation Fund Project of Anhui University of Science and Technology, grant number 2020CX2024; and Graduate Innovation Fund Project of Anhui University of Science and Technology, grant number 2022CX2047.

Data Availability Statement: The datasets generated and analyzed during the current study are available from the corresponding author upon reasonable request.

Acknowledgments: The authors would like to thank Fangxing Yao, Longfei Yang, Heyin Wang for the field data measurements.

Conflicts of Interest: The authors declare no conflict of interest.

References

1. Li, S.Q.; Gao, L.X.; Chai, S.X. Significance and interaction of factors on mechanical properties of frozen soil. *Rock Soil Mech.* **2012**, *33*, 1173–1177.
2. Hu, X.D.; Wu, Y.H.; Li, X.Y. A field study on the freezing characteristics of freeze-sealing pipe roof used in ultra-shallow buried tunnel. *Appl. Sci.* **2019**, *9*, 1532. [CrossRef]
3. Trupak, N.G. *The Freezing of Rock during Drilling*; Ugliet Hizdat: Moscow, Russia, 1954.
4. Bakholdin, B.V. *Select the Best Ore Freezing Mode for Construction Purposes*; State Press: Moscow, Russia, 1963.
5. Sanger, F.J. Ground freezing in construction. *J. Soil Mech. Found. Div.* **1968**, *94*, 923–950. [CrossRef]
6. Sanger, F.J.; Sayles, F.H. Thermal and rheological computations for artificially frozen ground construction. *Dev. Geotech. Eng.* **1979**, *26*, 311–337.
7. Tobe, N.; Akimata, O. Temperature distribution formula in frozen soil and its application. *Refrigeration* **1979**, *54*, 3–11.
8. Hu, X.D.; Chen, J.; Wang, Y.; Li, W.P. Analytical solution to steady-state temperature field of single-circle-pipe freezing. *Rock Soil Mech.* **2013**, *34*, 874–880.
9. Hu, X.D.; Yu, J.Z.; Ren, H.; Wang, Y.; Wang, J.T. Analytical solution to steady-state temperature field for straight-row-piped freezing based on superposition of thermal potential. *Appl. Therm. Eng.* **2017**, *111*, 223–231. [CrossRef]
10. Hu, X.D.; Fang, T.; Han, Y.T. Generalized analytical solution to steady-state temperature field of double-circle-piped freezing. *J. China Coal Soc.* **2017**, *42*, 2287–2294.
11. Hu, X.D.; Fang, T.; Zhang, L.Y. Analytical solution to temperature distribution in frozen soil wall with wavy boundaries by single-row- and double-row-piped freezing. *Cold Reg. Sci. Technol.* **2018**, *148*, 208–228. [CrossRef]

12. Hong, Z.Q.; Hu, X.D.; Fang, T. Analytical solution to steady-state temperature field of Freeze-Sealing Pipe Roof applied to Gongbei tunnel considering operation of limiting tubes. *Tunn. Undergr. Space Technol.* **2020**, *105*, 103571. [CrossRef]
13. Aziz, A.; Lunardini, V.J. Perturbation techniques in phase change heat transfer. *Appl. Mech. Rev.* **1993**, *46*, 29–68. [CrossRef]
14. Hosseini, S.M.; Akhlaghi, M.; Shaken, M. Transient heat conduction in functionally graded thick hollow cylinders by analytical method. *Heat Mass Transfer.* **2007**, *43*, 669–675. [CrossRef]
15. Jiji, L.M.; Ganatos, P. Approximate analytical solution for one-dimensional tissue freezing around cylindrical cryoprobes. *Int. J. Therm. Sci.* **2009**, *48*, 547–553. [CrossRef]
16. Jiang, B.S.; Wang, J.G.; Zhou, G.Q. Analytical calculation of temperature field around a single freezing pipe. *J. China Univ. Min. Technol.* **2009**, *38*, 463–466.
17. Cai, H.B.; Xu, L.X.; Yang, Y.G.; Li, L.Q. Analytical solution and numerical simulation of the liquid nitrogen freezing-temperature field of a single pipe. *AIP Adv.* **2018**, *8*, 055119. [CrossRef]
18. Yang, P.; Ke, J.M.; Wang, J.G.; Chow, Y.K.; Zhu, F.B. Numerical simulation of frost heave with coupled water freezing, temperature and stress fields in tunnel excavation. *Comput. Geotech.* **2006**, *33*, 330–340. [CrossRef]
19. Yu, C.Y.; Lu, M.Y. Study on development characteristics of horizontal freezing temperature field in subway connecting passage. *IOP Conf. Ser. Earth Environ. Sci.* **2020**, *619*, 012086. [CrossRef]
20. Yu, C.Y. Research on development law of horizontal freezing temperature field of complex curtain. *IOP Conf. Ser. Mater. Sci. Eng.* **2020**, *780*, 032029. [CrossRef]
21. Fu, Y.; Hu, J.; Wu, Y.W. Finite element study on temperature field of subway connection aisle construction via artificial ground freezing method. *Cold Reg. Sci. Technol.* **2021**, *189*, 103327. [CrossRef]
22. Cai, H.B.; Huang, Y.C.; Pang, T. Finite element analysis on 3D freezing temperature field in metro connected aisle construction. *J. Railw. Sci. Eng.* **2015**, *12*, 1436–1443.
23. Hong, R.B.; Cai, H.B.; Li, M.K. Integrated prediction model of ground surface deformation during tunnel construction using local horizontal freezing technology. *Arab. J. Sci. Eng.* **2022**, *47*, 4657–4679. [CrossRef]
24. Hu, J.; Yang, P. Numerical analysis of temperature field within large-diameter cup-shaped frozen soil wall. *Rock Soil Mech.* **2015**, *36*, 523–531.
25. Shang, H.S.; Yue, F.T.; Shi, R.J. Model test of artificial ground freezing in shallow-buried rectangular cemented soil. *Rock Soil Mech.* **2014**, *35* (Suppl. S2), 149–155+161.
26. Shi, R.J.; Yue, F.T.; Zhang, Y.; Lu, L. Model test on freezing reinforcement for shield junction Part 1: Distribution characteristics of temperature field in soil stratum during freezing process. *Rock Soil Mech.* **2017**, *38*, 368–376.
27. Cai, H.B.; Liu, Y.J.; Hong, R.B.; Li, M.K.; Wang, Z.J.; Ding, H.L. Model test and numerical simulation analysis on freezing effect of different freezers in freeze-sealing pipe-roof method. *Geofluids* **2022**, *2022*, 5350650. [CrossRef]
28. Duan, Y.; Rong, C.X.; Cheng, H.; Cai, H.B.; Xie, D.Z.; Ding, Y.L. Model test of freezing temperature field of the freeze-sealing pipe roof method under different pipe arrangements. *J. Glaciol. Geocryol.* **2020**, *42*, 479–490.
29. Zhang, J.X.; Qi, Y.; Yang, H.; Song, Y.W. Temperature field expansion of basin-shaped freezing technology in sandy pebble stratum of Beijing. *Rock Soil Mech.* **2020**, *41*, 2796–2813.
30. Cai, H.B.; Li, P.; Wu, Z. Model test of liquid nitrogen freezing-temperature field of improved plastic freezing pipe. *J. Cold Reg. Eng.* **2020**, *34*, 04020001. [CrossRef]
31. Cai, H.B.; Liu, Z.; Li, S.; Zheng, T.L. Improved analytical prediction of ground frost heave during tunnel construction using artificial ground freezing technique. *Tunn. Undergr. Space Technol.* **2019**, *92*, 103050. [CrossRef]

applied sciences

MDPI

Article

Experimental and Modeling of Residual Deformation of Soil–Rock Mixture under Freeze–Thaw Cycles

Chao Wang [1], Jing Chen [2], Lilei Chen [1], Yue Sun [1], Zelei Xie [1], Guoan Yin [3], Minghao Liu [3] and Anyuan Li [1,*]

1 Key Laboratory of Rock Mechanics and Geohazards of Zhejiang Province, Shaoxing University, Shaoxing 31200, China
2 School of Ecological and Environmental Sciences, East China Normal University, Shanghai 200241, China
3 State Key Laboratory of Frozen Soil Engineering, Northwest Institute of Eco-Environmental and Resources, CAS, Lanzhou 730000, China
* Correspondence: anyuanli1013@lzb.ac.cn

Abstract: Projects in seasonal frozen soil areas are often faced with frost heaving and thawing subsidence failure, and the foundation fill of most projects is a mixture of soil and rock. Therefore, taking soil–rock mixture with different rock contents as research objects, the residual deformation of soil–rock mixture under multiple freezing–thawing cycles is studied. In addition, the deep learning method based on the artificial neural network was pioneered combined with the freezing–thawing test of the soil–rock mixture, and the Long short-term memory (LSTM) model was established to predict the results of the freezing–thawing test. The LSTM model has been verified to be feasible in the exploration of the freeze–thaw cycle law of a soil–rock mixture, which can not only greatly reduce the period of the freeze–thaw test, but also maintain a high prediction accuracy to a certain extent. The study found that the soil–rock mixture will repeatedly produce frost heave and thaw subsidence under the action of freeze–thaw cycles, and the initial frost heave and thaw subsidence changes hugely. With the increase of the number of freeze–thaw cycles, the residual deformation decreases and then becomes steady. Under the condition that the content of block rock in the soil–rock mixture is not more than 80%, with the increase of block rock content, the residual deformation caused by the freeze–thaw cycle will gradually decrease due to the skeleton function of block rock, while the block rock content's further increase will increase the residual deformation. Furthermore, the LSTM model based on an artificial neural network can effectively predict the freezing and thawing changes of soil–rock mixture in the short term, which can greatly shorten the time required for the freezing and thawing test and improve the efficiency of the freezing and thawing test to a certain extent.

Keywords: soil–rock mixture; freeze–thaw cycle; residual deformation; rock content; long short-term memory network

1. Introduction

In recent years, for the sake of adapting to the transition of our country's economic development model from fast to stable, construction of the transportation system, especially the high-speed railway, is increasing. However, the complex geographical conditions also bring more challenges to the project's construction. In particular, projects in seasonal permafrost areas also need to take into account the freezing and thawing deformation of foundation fill composed of soil–rock mixture. Therefore, it is necessary to fully explore the change law of the volume and strength of soil–rock mixture under the action of freeze–thaw cycles, and to grasp the influence of rock content on the residual deformation of the soil–rock mixture, so as to ensure the engineering quality and operation safety to the greatest extent.

Many scholars at home and abroad have done plenty of research on the freeze–thaw cycle mechanism and theoretical model, and they usually focus on the basic theory of the soil frost heave effect. There is still no systematic reference model for the research on the freeze–thaw cycle mechanism of soil–rock mixture. Foreign research on the mechanism of the freeze–thaw cycle roughly started in the 1960s. Corte [1,2] observed that the rocks in the soil were exposed to the surface due to repeated freezing and thawing in the cold winter area. After research, it was found that under the action of repeated freezing and thawing, when the soil contains enough water, the non-uniform soil particles will undergo vertical sorting of particle sizes. Further laboratory experiments determined the basic conditions for vertical sorting to occur, that is, the effects of the freezing rate range, water content on the frozen surface, and particle size on soil particle migration. Inglis and Corte [3] proposed another explanation for this phenomenon, which was, assuming that the soil particles are spherical and are of sufficient mass, embedded in a cold mud composed of fine particles frozen from top to bottom, then the direction of the particles' movement is decided by the orientation of freezing and thawing, which is not determined by the direction of gravity. Jackson and Uhlmann [4] believe that, in addition to the freezing expansion of soil water, the frost heave effect of soil water on soil during the freezing process should also be taken into consideration. Due to the repulsion between ice and particles and the tension created by unfrozen water, larger particles tend to be sorted to the surface layer, while smaller particles are migrated deep into the soil. Vliet-Lanoë [5] expounded the frost heave theory from a geological point of view. That is to say, the separation of ice occurs during the frost heave process of soil. Different forms of separated ice such as ice layers and ice wedges will cause the soil to bulge with the continuous growth. With the seasonal freezing and thawing, the soil particles are sorted and the rocks in the soil are migrated. The study by Viklander [6] pointed out the soil conditions necessary for the uplift of the rock under the action of freeze–thaw cycles. The test results show that the void ratio of the soil is the key parameter that determines the upward or downward movement of the rock. In addition, the soil must contain enough water for ice to form and to cause frost heave. During the freeze–thaw cycle, vertical cracks are induced in the fine-grained soil, and the water in the soil continuously replenishes the water for the formation of ice lenses. This process causes the rock and coarse particles to continuously move towards the cold end, that is, generally towards the surface. An artificial neural network model was proposed and validated by Ren et al. [7], who tested the modulus of elasticity of pavement subgrade soils in seasonal permafrost zones. Thus, the elastic modulus of the soil can be estimated under freeze–thaw cycling conditions by combining the initial moisture content of the specimen and the soil type. The introduction of the artificial neural network model provides a new idea for the investigation of the freeze–thaw cycle mechanism.

The research on the freeze–thaw cycle mechanism of the soil–rock mixture by domestic scholars started a little later, but it has developed rapidly due to the increase in the demand for engineering construction in seasonally frozen land areas. Xu et al. [8–10] defined a soil–rock mixture as an extremely inhomogeneous and loose geotechnical material composed of pore space, soil particles, and rocks with large strength and a certain engineering scale. To explore the freeze–thaw cycle mechanism of soil–rock mixture, it is necessary to figure out the freeze–thaw cycle mechanism of soil first. Xu et al. [11], Peng et al. [12], Yan et al. [13], Zhao et al. [14], Wang et al. [15], Ma et al. [16], and Fang et al. [17] researched the freeze–thaw characteristics of loess by indoor freeze–thaw tests and obtained the relationships between freezing rate, freeze–thaw volume, freezing temperature, recharge conditions and water content. Secondly, it is essential to find out the moisture condition in the soil. Li et al. [18] believed that repeated freezing and thawing would increase the moisture content of soil samples and cause water to accumulate in the upper part. The research of Wang et al. [19] proposed that the moisture content of the samples after multiple freeze–thaw cycles under the condition of replenishment will be much greater than the initial moisture content, and the replenishment amount during the freezing process is much greater than that during the thawing process. Yu et al. [20] proposed the conditions that ice lens formation must meet during the frost heave process of soil, which is stress failure and the necessary water field. The sufficient water supply will cause partial condensation and frost heave, but if the partial condensation temperature is too low, the migration of unfrozen water will be interrupted, so it needs to be controlled within a certain temperature range. Ming et al. [21] supplemented the ice lens judgment criteria. The formation position of the ice lens is significantly affected by the temperature boundary, while the generation of segregated ice is restricted by the amount of water migration. However, as the load increases, it will show an exponential decay. At the same time, according to the research of Zhang et al. [22], freeze–thaw cycles will destroy the original structure between soil particles, resulting in an increase in soil permeability and a decrease in the plasticity index. It can even increase the large pores in the loess-like soil without collapsibility and produce collapsibility. Zhao et al. [23] believed that under the repeated freezing and thawing of the slope soil in the seasonal frozen soil area, the soil compactness decreased and the pores increased, and proposed the relationship between the frost heave amount and the freezing depth, and the residual deformation amount and the freezing depth. The residual deformation of soil in freeze–thaw cycles is formed by the superposition of frost heave and thaw subsidence. Qi et al. [24] believes that the research on soil frost heave is more in-depth, and the research on thaw settlement is mostly based on empirical methods, and the influence of soil mechanical properties is not paid enough attention. The laboratory test and actual monitoring should be combined to build a model. Liang et al. [25] proposed the thawing characteristics of different soils during repeated freezing and thawing under different moisture content, compactness, and load conditions, and believed that the increase in load would inhibit the frost heave deformation, but at the same time, increase the thawing deformation. Qiu et al. [26] proposed the influence of soil compaction and saturation on the law of freeze–thaw deformation characteristics. That is to say, under the condition of water replenishment, the frost-heave rate, thaw coefficient, and volume change rate of the sample are proportional to the saturation, but the volume change rate tends to be the same with the increase in the number of freeze–thaw cycles. Saturation only affects the number of freeze–thaw cycles at which the volume variability reaches a stable value, and the degree of compaction also affects the size of the stable value. With the in–depth investigation of the mechanism of freeze–thaw cycles, the tests on soil–rock mixture have been carried out gradually.Li et al. [27], Zhou et al. [28], Yao et al. [29], Zhao et al. [30], Gao et al. [31], Zhang et al. [32], Hu et al. [33], Shen et al. [34], Li et al. [35], Hu et al. [36] conducted uniaxial compression strength testing and other indoor tests on the soil–rock mixture under freeze–thaw cycles, and obtained its mechanical properties and damage mechanisms, which can provide valuable references for the engineering in cold regions.

In summary, the research and practice of the freeze–thaw cycle mechanism of soil–rock mixtures have their strengths. Studies by foreign scholars focus on the formation of ice lenses and the movement of soil particles during the freeze–thaw process, and are dominated by indoor experiments. Although domestic scholars' research on this topic started relatively late, it was gradually carried out due to the support of a large number of engineering construction data. In the research direction, not only the stress, water, and other conditions of soil freezing and thawing are discussed, but also factors such as changes in the physical properties of soil through freezing and thawing cycles are considered. Nevertheless, a relatively systematic theoretical model has not been formed for the frost heave, thaw settlement, and residual deformation during the freeze–thaw cycle. Therefore, the test takes the residual deformation of the soil–rock mixture after freeze–thaw cycles as the research object and analyzes the frost heave and thaw settlement after the freeze–thaw cycles. It can fit the actual engineering conditions in the frozen soil area to a certain extent so that the exploration of the freezing–thawing cycle mechanism of the soil–rock mixture has certain practical significance.

2. Materials and Methods

2.1. The Basic Physical Index of the Test Sample

The test sample is a saturated remodeled soil–rock mixture, of which the soil sample is taken from Lvliang, Shanxi Province, and belongs to Q3 Malan loess. According to the

"Standards for Geotechnical Test Methods" (GB/T50123-2019), the soil mechanics basic test was carried out on the soil samples. The particle size distribution curve of the soil samples is shown in the Figure 1.

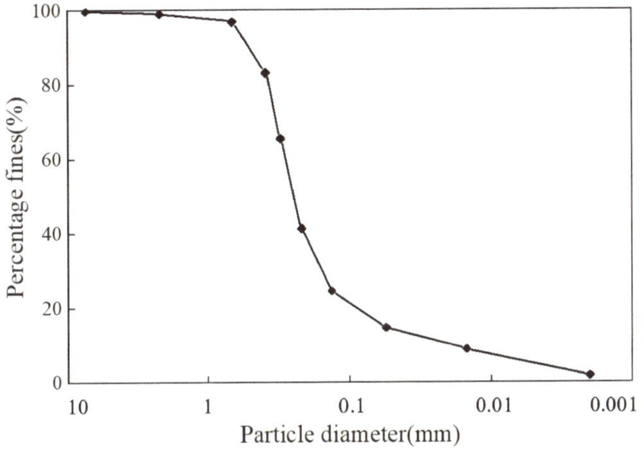

Figure 1. Particle size distribution curve of soil.

After calculation, the uniformity coefficient is 20.500 and the coefficient of curvature is 5.378, so the test soil sample has poor gradation and is not easy to be compacted. In addition, the results obtained by testing and calculating the basic physical properties of the soil samples are shown in Table 1.

Table 1. Basic physical parameters of soil.

Material	Density (g/cm^3)	Dry Density (g/cm^3)	Water Content (%)	Grain Specific Gravity
Soil	1.65	1.44	14.20	2.72

The block stone is gray-white marble with rough surface, as shown in Figure 2. The sample diameter is 60 mm and the height is 100 mm. According to the "Geotechnical Test Method Standard" (GB/T50123-2019), it can be seen that the maximum particle size of the block stone in the soil–rock mixture does not exceed the diameter of the sample, 1/8 of the height, or 1/4 of the height, so the particle size of the stone selected for this test was 10 mm.

Figure 2. The main constituent materials of soil–rock mixture.

2.2. Specimen Preparation and Test Protocol

2.2.1. Preparation of Soil–Rock Mixture Samples

The size of the soil–rock mixture sample was 60 mm in diameter and 100 mm in height. The undisturbed loess soil sample was air-dried, crushed, and passed through a 2 mm sieve, then dried, mixed with distilled water, configured into a saturated state, and sealed for 12 h to make the water uniform. After that, the block stone particles were rinsed with distilled water to remove surface impurities, saturated in distilled water by vacuum saturation method, and then fully mixed with soil samples with rock contents of 60%, 70%, 80%, and 90%, stirred, and put into Layered static compaction in the mold. Since the thickness of each layer needed to be greater than 1.5 times the maximum particle size of the rock, the design thickness of the layered compaction in this test was 20 mm. After compaction, the samples were sealed at an ambient temperature of 0 °C for 36 h.

2.2.2. Equipment and Programs for the Test

The test used the precision triaxial frost heave tester of the State Key Laboratory of Frozen Soil Engineering to carry out the freeze–thaw cycle test. The main instruments used in the freeze–thaw cycle system are shown in Figure 3 and the schematic diagram is shown in Figure 4. In the test, EYELA's NCB–3100 precision low-temperature constant temperature water tank was connected to the top plate to provide a ±20 °C circulating cold bath for one-way freezing test, freezing for 6 h, thawing for 6 h, and circulating for 20 times, and the Martens bottle was used for free water replenishment. The top of the sample was equipped with an infrared displacement sensor, which could monitor the deformation of the sample in the vertical direction. The temperature and relative displacement changes were collected by Keysight's 34972A data acquisition instrument. The acquisition accuracy was 0.01 mm, and the acquisition interval was 5 min. The effects of different boulder contents on the residual deformation of saturated soil under the action of freeze–thaw cycles were systematically studied through the data system of frost heave, thaw settlement, and residual deformation of samples after freeze–thaw cycles.

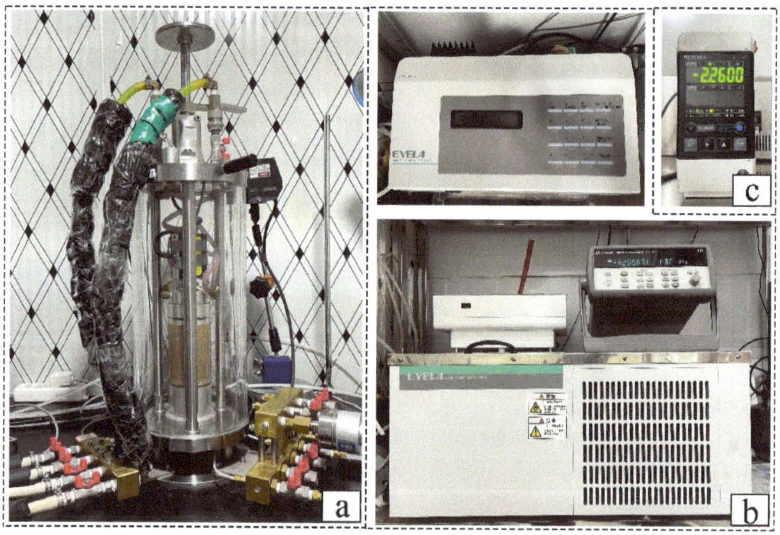

Figure 3. The main instrument used in the experiment: (**a**) Precision frost heave triaxial instrument; (**b**) Constant temperature water tank; (**c**) Data collector.

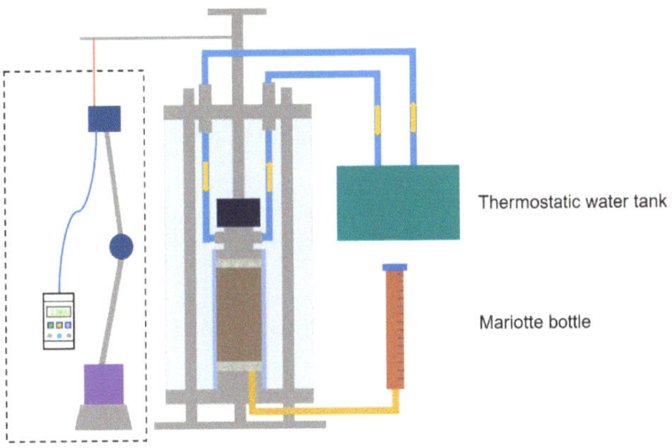

Figure 4. Freeze–thaw cycle device schematic diagram.

3. Experimental Results

3.1. Freeze–Thaw Deformation Analysis

According to the available experimental data, the skeletal effect of the blocks will gradually increase when the rock content in the soil–rock mixture is greater than 40%. Generally speaking, the amount of soil freezing and swelling is positively correlated with the water content, so four sets of tests were conducted with specimens saturated and under open recharge conditions at 60%, 70%, 80%, and 90% rock contents to obtain the freeze–thaw changes of soil–rock mixture with different rock contents. The freeze–thaw cycle process of the soil–rock mixture is shown in Figure 5 (taking group S1 as an example).

Figure 5. Freeze–thaw cycle process of soil–rock mixture samples.

The temperature of the top plate of each group of specimens varied cyclically within ±20 °C during the freeze–thaw cycle, and the data recorded by the temperature sensors at the top and bottom of the specimens were observed to vary regularly within the predetermined range (as shown in Figure 6). It was found that the soil–rock mixture specimens showed more obvious and different degrees of freezing and thawing changes in each group during the freeze–thaw cycle. In the S1 group specimens with a higher rock content, as the number of freeze–thawing cycles increases, it can be observed that the blocks become closer to the inner wall of the transparent container. That is, the specimens are subject to significant radial displacement by freeze–swelling during the freeze–thawing cycle.

According to the change of frost heave of each group of samples, the soil rock mixture samples have different degrees of frost heave during the freeze–thaw cycles (as shown in Figure 7). With the increase in the freeze–thaw cycles, the frost heave of the sample gradually decreases and tends to be stable. However, with the increase of the content of block rock in the sample, the number of freeze–thaw cycles required for the frost heave of the sample to reach a stable state also increases correspondingly. With the increase of the rock content, the frost heave of the sample first decreases at 60–80% and increases at

80–90%. The S2 group sample with the rock content of 80% at the extreme value even has the phenomenon of frost shrinkage.

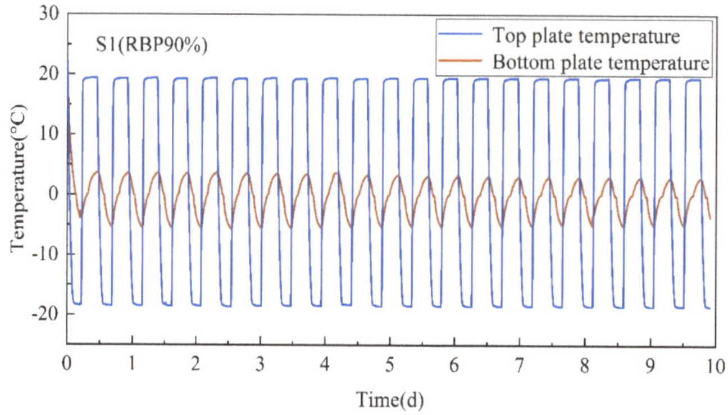

Figure 6. Freeze–thaw cycle temperature of samples in group S1.

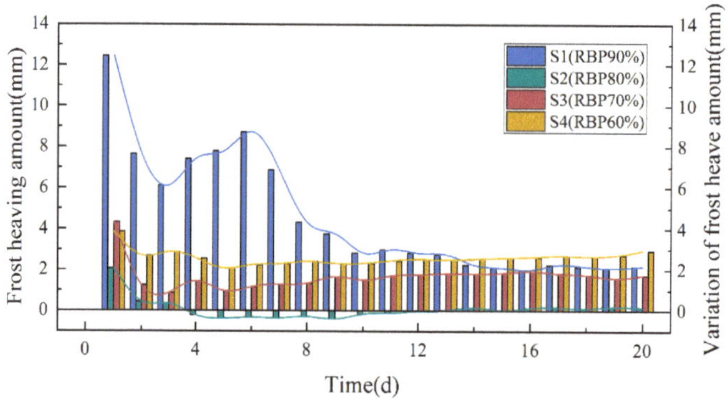

Figure 7. Frost heave amount of each group of samples.

3.2. Residual Deformation Analysis

The soil–rock mixture specimens will alternately produce freezing expansion and thawing settlement during the freeze–thaw cycle, and the superposition of the two is the residual deformation of the specimens. The residual deformation of each group of specimens is shown in the Figure 8.

As can be seen from the Figure 8, almost all of the soil–rock mixture specimens produced thawing and sinking after the freeze–thaw cycles. The initial residual deformation of the specimens is larger than the later residual deformation, but with the increase of the number of freeze–thaw cycles, the residual deformation gradually stabilizes. The residual deformation decreases and then increases with the increase in the rock content in the specimens, and reaches the minimum residual deformation in the S2 group specimen with 80% rock content.

According to the freeze–thaw change curves of the soil–rock mixture specimens as shown in the Figure 9, the residual deformation of each specimen is shown as settlement deformation. When the rock content of the specimens is between 70% and 80%, the freeze-swelling and thawing changes of the specimens in the S2 and S3 groups are smaller than those in the S1 group with 90% rock content and the S4 group with 60% rock content, or the blocks in the specimens in this rock content range can provide the best skeletal support. The freeze-swelling and thawing changes of the specimens in the other two groups are larger, with a maximum deformation of about 12%, and thus eventually also produce larger residual changes.

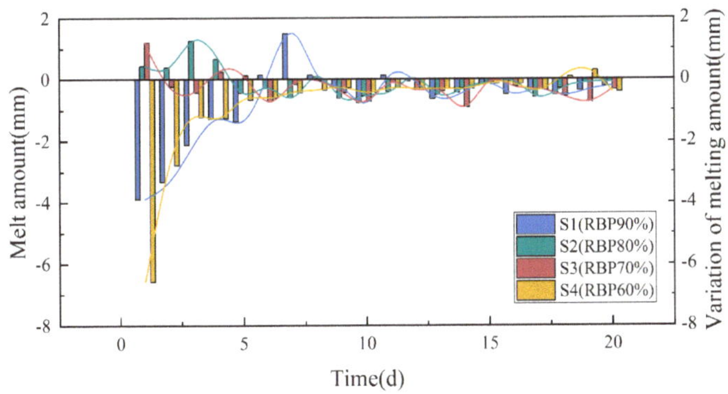

Figure 8. The residual deformation of each group of samples.

The final residual deformation produced by each specimen is shown in Table 2. When the rock content of the soil–rock mixture specimen is 60~80%, the residual deformation is all settlement deformation. With the increase of the lumpy rock content in the specimens, the settlement amount gradually decreased and reached the minimum settlement amount of 2.54 mm in the specimens of group S2 with 80% rock content, which was only 48.29% and 25.81% of the settlement amount in the specimens of groups S3 and S4, while the specimens of group S1 produced the maximum settlement amount, which was almost six times of the settlement amount in group S2.

Table 2. The final residual deformation of each sample.

Number	S1	S2	S3	S4
Residual Deformation (mm)	−14.62	−2.54	−5.26	−9.84

It can be inferred that there is an optimal rock content in the range of 80% to 90%, and until this content is reached, the increase of the boulder content can make the skeleton effect of the boulders in the soil–rock mixture specimens more and more obvious, so that they can resist the deformation caused by the freeze–thaw cycles. However, the S1 group specimens with the highest rock content showed the largest settlement, either because the boulder content was too high or the soil could not completely fill the pores between them, thus producing a larger amount of melting and settlement under the freeze–thaw cycles.

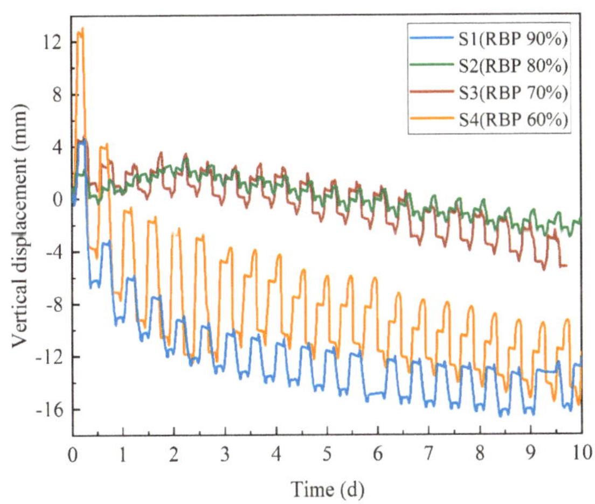

Figure 9. The distribution of freezing and thawing.

4. Model Building and Prediction

The deep learning method based on an artificial neural network can quickly extract effective data features, according to a large number of existing experimental data, and analyze the evolution law of experimental data. In this experiment, the applicability of the freeze–thaw test data prediction of frost heave and thaw deposition was studied by establishing the LSTM model, and the actual test data and the results of the model output were compared, thus confirming the feasibility of the model prediction.

The LSTM model is a neural network model proposed by Hochreiter in the 1990s, and its model structure is shown in the Figure 10. LSTM models contain a special unit, a memory block in the recursive hidden layer. The memory block consists of a cell state and three gating mechanisms (input gate, forget gate, and output gate) to control the flow of information. The gating mechanism is implemented by the Sigmoid function. After the data flow in, they first enters the forget gate. After discarding some redundant information to obtain key feature data, they enter the input gate to control the update of data. Finally, they go through the output gate for output.

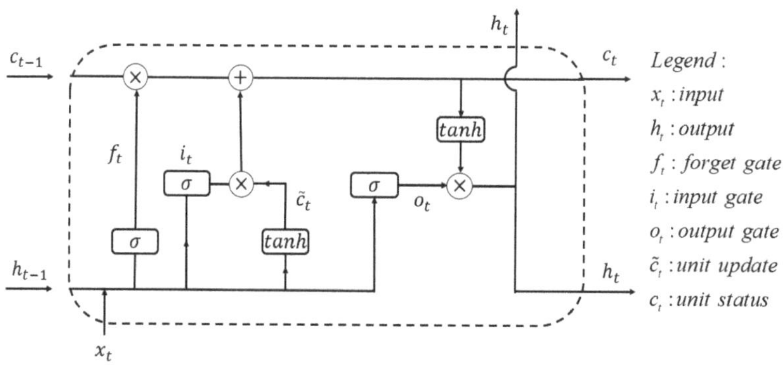

Figure 10. Schematic diagram of LSTM unit structure.

Each LSTM unit updates six parameters (see the legend of the Figure 10) in each time step. The LSTM unit extracts the mapping between input sequence x_t and output sequence h_t, and the update of the LSTM unit can be represented by the following formula:

$$i_t = \sigma(W_i \times [h_{t-1}, x_t] + b_i)$$

$$f_t = \sigma\left(W_f \times [h_{t-1}, x_t] + b_f\right)$$

$$o_t = \sigma(W_o \times [h_{t-1}, x_t] + b_o)$$

$$\tilde{c}_t = tanh \cdot (W_c \times [h_{t-1}, x_t] + b_c)$$

$$c_t = f_t * c_{t-1} + i_t * \tilde{c}_t$$

$$h_t = o_t * tanh(c_t)$$

where W is the matrix of input weights, b is a vector of bias, and σ represents a logistic sigmoid function. The four groups of soil–rock mixture samples were all frozen and thawed 20 times, and each group of experiments lasted 10 days. The LSTM model took the data of the first 7 days of each group of experiments as the training set, and the last 3 days of experiments as the test set. The deep learning method of the neural network predicted the effect in the freeze–thaw cycle test of the soil–rock mixture.

It can be seen from the Figure 11 that the results of the learning and prediction of the LSTM model have achieved good expected results, which are highly consistent with the data obtained from the actual test. Therefore, the long and short-term memory network model in deep learning can effectively and accurately predict the soil–rock mixture. The development law of body freezing and thawing can greatly reduce the time span of the soil–rock mixture freezing and thawing test.

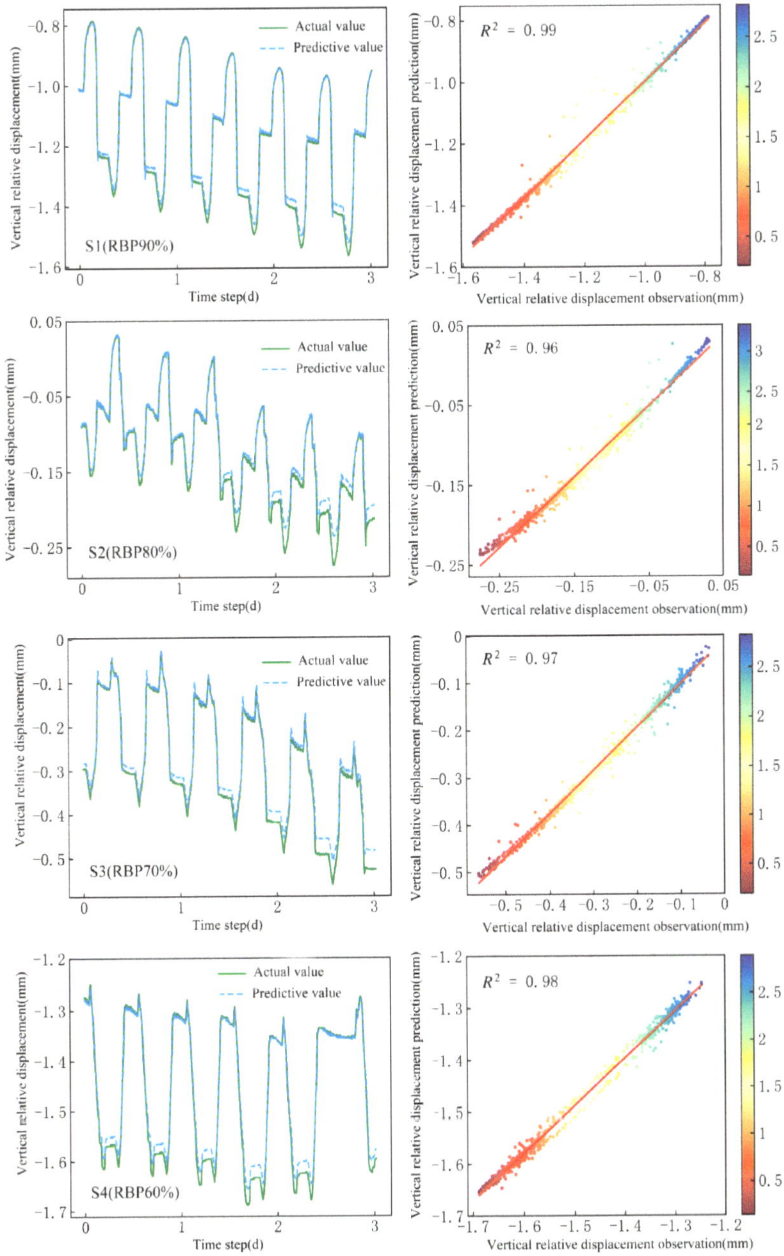

Figure 11. LSTM model predictions.

5. Discussion

In this study, we investigated the residual deformation of soil–rock mixture under the action of freeze–thaw cycles with the rock content as a variable. It was found that the boulders in the soil–rock mixture assumed the main skeletal role. With the increase in the rock content, the residual deformation after the freeze–thaw cycle gradually decreased, and a smaller amount of residual deformation was produced when the rock content was about 80%. This conclusion is similar to the results of Jin et al. [37]. However, when the rock content of the soil–rock mixture reached 90%, a larger residual deformation was produced again. This phenomenon may be related to the "lubricating" effect of the soil filling between the boulders and destroying the skeletal structure, as mentioned in Liao et al. [38]. In addition, the residual deformation of all groups of specimens was settling. According to Zeinali et al. [39], the water migrates upward when the soil thaws, and thus the excess pore space generated leads to the redistribution of fine particles in the soil and the appearance of softening of the soil, which fits with the experimental results. We also tried to build an LSTM model using deep learning methods. The model was validated based on the results of freeze–thaw cycling tests on soil–rock mixture, and the numerical simulation results were relatively small in error with the actual test results, which proved the feasibility of the LSTM model in simulating freeze–thaw tests.

However, there are still some limitations in this study on the freeze–thaw cycle mechanism of soil–rock mixture. First, due to the test equipment, the changes inside the soil–rock mixture during the freeze–thaw process cannot be clearly and directly observed. Secondly, the deep learning method has been little involved in the study of the freeze–thaw cycle mechanism, and the constructed LSTM model is not mature yet. Given the above problems, it is hoped that future research can be combined with high-precision CT scanning equipment to visualize and monitor the stages of freeze–thaw of the soil–rock mixtures so that more specific experimental data can be obtained. It can also provide more references for deep learning models.

6. Conclusions

Through the experimental exploration of the residual deformation of soil–rock mixtures with different rock contents under the action of freeze–thaw cycles, and the simulation prediction and verification of the LSTM model, the following conclusions can be drawn:

(1) Under the action of freeze–thaw cycles, the soil–rock mixture will repeatedly produce frost heave and thaw subsidence, and the initial frost heave and thaw subsidence change greatly and become steady.
(2) Under the condition that the content of boulders in the soil–rock mixture is not more than 80%, as the content of boulders increases, the residual deformation caused by the freeze–thaw cycle will gradually decrease due to the skeleton effect of boulders, while the increase in the content of the boulders will increase the residual deformation.
(3) The LSTM model based on the artificial neural network can effectively predict the freezing and thawing change law of the soil–rock mixture in the short term, thereby reducing the period of the freezing and thawing test and improving the fault tolerance rate.

Author Contributions: Conceptualization, C.W.; Data curation, Y.S.; Formal analysis, J.C.; Methodology, C.W. and A.L.; Project administration, A.L.; Resources, L.C. and Z.X.; Software, J.C.; Supervision, G.Y., M.L. and A.L.; Validation, G.Y. and M.L.; Visualization, J.C.; Writing—original draft, C.W.; Writing—review and editing, A.L. All authors have read and agreed to the published version of the manuscript.

Funding: This research was supported by Zhejiang Collaborative Innovation Center for Prevention and Control of Mountain Geological Hazards (No. PCMGH-2017-Y03), the National Key Research and development of China (Grant No. 2017YFA0603101), the funding of the State Key Laboratory of Frozen Soil Engineering (Grant No. SKLFSE201712), the Program of the State Key Laboratory of Frozen Soil Engineering (Grant No. SKLFSE-ZT-202110).

Conflicts of Interest: The authors declare no conflict of interest.

References

1. Corte, A.E. *The Frost Behavior of Soils. I. Vertical Sorting*; Highway Research Board: Washington DC, USA, 1962.
2. Corte, A.E. Particle sorting by repeated freezing and thawing. *Science* **1963**, *142*, 499–501. [CrossRef] [PubMed]
3. Inglis, D.R.; Corte, A.E. Particle sorting and stone migration by freezing and thawing. *Science* **1965**, *148*, 1616–1617. [CrossRef] [PubMed]
4. Jackson, K.A.; Uhlmann, D.R. Particle sorting and stone migration due to frost heave. *Science* **1966**, *152*, 545–546. [CrossRef] [PubMed]
5. Vliet-Lanoë, B.V. Frost effects in soils. *Soils Quat. Landsc. Evopution* **1985**, *1*, 117–158.
6. Viklander, P. Laboratory study of stone heave in till exposed to freezing and thawing. *Cold Reg. Sci. Technol.* **1998**, *27*, 141–152. [CrossRef]
7. Ren, J.; Vanapalli, S.K.; Han, Z.; Omenogor, K.O.; Bai, Y. The resilient moduli of five Canadian soils under wetting and freeze-thaw conditions and their estimation by using an artificial neural network model. *Cold Reg. Sci. Technol.* **2019**, *168*, 102894. [CrossRef]
8. Xu, W.J.; Hu, R.L.; Yue, Z.Q.; Tan, R.J. Mesostructral character and numerical simulation of mechanical properties of soil-rock mixture. *Chin. J. Rock Mech. Eng.* **2007**, *26*, 300–311. (In Chinese)
9. Xu, W.J.; Hu, R.L.; Yue, Z.Q.; Zhang, R.; Wang, G.L. Research on relationship between rock block proportion and shear strength of soil-rock mixture based on digital image analysis and large direct shear test. *Chin. J. Rock Mech. Eng.* **2008**, *27*, 996–1007. (In Chinese)
10. Xu, W.J.; Hu, R.L. Conception, classification and significations of soil-rock mixture. *Hydrogeol. Eng. Geol.* **2009**, *36*, 50–56. (In Chinese)
11. Xu, J.; Niu, F.J.; Niu, Y.H.; Lin, Z.J.; Xu, Z.Y. Experimental analysis of frost susceptibility for remolded clayey loess. *J. Civ. Archit. Environ. Eng.* **2010**, *32*, 24–30. (In Chinese)
12. Peng, L.Y.; Liu, J.K.; Tian, Y.H. Study on frost heaving property of silty caly. *Hydrogeol. Eng. Geol.* **2009**, *36*, 62–67. (In Chinese)
13. Yan, H.; Wang, T.L.; Liu, J.K. Experimental study of repeated frost heave and thaw settlement properties of silty sand. *Rock Soil Mech.* **2013**, *34*, 3159–3165. (In Chinese)
14. Zhao, Z.H.; Yang, S.S.; Hao, R.Q.; Niu, S.Q.; Yang, H.H.; Pang, X. Experimental and field measure of frozen heave properties of different soil sample. *Sci. Technol. Eng.* **2019**, *19*, 245–252. (In Chinese)
15. Wang, Z.J.; Pan, J.Y.; Zhou, P.; Wu, G.H. Influence of freezing and thawing on collapsibility of loess. *Chin. J. Undergr. Space Eng.* **2016**, *12*, 1710–1716. (In Chinese)
16. Ma, H.Y.; Zhang, F.; Feng, D.C.; Tang, K.W. Experimental study on frost heave of saturated silty clay under single side freezing. *J. Build. Mater.* **2016**, *19*, 926–932. (In Chinese)
17. Fang, L.L.; Qi, J.L.; Ma, W. Freeze-thaw induced changes in soil structure and its relationship with variations in strength. *J. Glaciol. Geocryol.* **2012**, *34*, 435–440. (In Chinese)
18. Li, G.Y.; Ma, W.; Mu, Y.H.; Zhou, C.L.; Mao, Y.C. Process and mechanism of impact of freezing and thawing cycle on collapse deformation of compacted loess. *China J. Highw. Transp.* **2011**, *24*, 1–5+10. (In Chinese) [CrossRef]
19. Wang, T.L.; Bu, J.Q.; Wang, Y.; Xu, L.; Yan, H. Thaw subsidence properties of soils under repeated freeze-thaw cycles. *Chin. J. Geotech. Eng.* **2014**, *36*, 625–632. (In Chinese)
20. Yu, J.N.; Tan, F.Y.; Fu, W. Analysis of frozen heave deformation in process of frozen soil. *Rock Soil Mech.* **2006**, *27*, 203–206. (In Chinese)
21. Ming, F.; Li, D.Q.; Huang, X.; Zhang, Y. Study of the ice lens growth during the freezing process. *China Civ. Eng. J.* **2015**, *48*, 346–350. (In Chinese)
22. Zhang, Z.; Ma, W.; Qi, J.L. Structure evolution and mecheanism of engineering properties change of soils under effect of freeze-thaw cylcle. *J. Jilin Univ. (Earth Sci. Ed.)* **2013**, *43*, 1904–1914. (In Chinese)
23. Zhao, W.; Wang, X.J.; Xu, Z.X.; Fu, W.L.; Jiang, S.Y. Experimental study on natural evolution characteristics of coarse-grained soil slope in seasonal frozen soil region. *J. Eng. Geol.* **2021**, *29*, 1497–1506. (In Chinese)
24. Qi, J.L.; Ma, W. State of art of research on mechanical properties of frozen soil. *Rock Soil Mech.* **2010**, *31*, 133–143. (In Chinese)
25. Liang, B.; Zhang, G.S.; Liu, D.R. Experimental study on thawing subsidence characters of permafrost under frost heaving and thawing circulation. *Chin. J. Geotech. Eng.* **2006**, *28*, 1213–1217. (In Chinese)
26. Qiu, N.; Cui, Z.Z.; Zhan, S.Y.; Hu, Y. Two-dimensional freeze-thaw deformation characteristics of compacted loess under makeup water condition. *J. Xi'an Univ. Technol.* **2020**, *36*, 115–121. (In Chinese)
27. Li, Z.Q.; Hu, F.; Qi, S.W.; Hu, R.L. Strain-softening failure mode after the post-peak as a unique mechanism of ruptures in a frozen soil-rock mixture. *Eng. Geol.* **2020**, *274*, 105725. [CrossRef]
28. Zhou, Z.; Xing, K.; Yang, H.; Wang, H. Damage mechanism of soil-rock mixture after freeze-thaw cycles. *J. Cent. South Univ.* **2019**, *26*, 13–24. [CrossRef]
29. Yao, Y.S.; Li, J.; Ni, J.J.; Liang, C.H.; Zhang, A.S. Effects of gravel content and shape on shear behaviour of soil-rock mixture: Experiment and DEM modelling. *Comput. Geotech.* **2022**, *141*, 104476. [CrossRef]
30. Zhao, Y.L.; Gao, Y.; Zhang, Y.L.; Wang, Y. Effect of fines on the mechanical properties of composite soil stabilizer-stabilized gravel soil. *Constr. Build. Mater.* **2016**, *126*, 701–710. [CrossRef]
31. Gao, W.; Hu, R.L.; Ibrahim, O.A.; Li, Z.Q.; Zhang, X.Y. Geomechanical Characterization of Zhangmu Soil-rock Mixture Deposit. *Geotech. Geol. Eng.* **2014**, *32*, 1329. [CrossRef]

32. Zhang, Z.P.; Sheng, Q.; Fu, X.D.; Zhou, Y.Q.; Huang, J.H.; Du, Y.X. An approach to predicting the shear strength of soil-rock mixture based on rock block proportion. *Bull. Eng. Geol. Environ.* **2020**, *79*, 2423–2437. [CrossRef]
33. Hu, R.L.; Li, X.; Wang, Y.; Gao, W.; Xia, J.G.; Li, Z.Q.; Gao, W.W.; Sun, Y.S. Research on engineering geomechanics and structural effect of soil-rock mixture. *J. Eng. Geol.* **2020**, *28*, 255–281. (In Chinese)
34. Shen, Y.P.; Liu, Y.; Tai, B.W.; Tang, T.X.; Tian, Y.H. Experimental study on thawing settlement coefficient and frost heave rate of sandy silt under freeze-thaw cycle. *J. China Railw. Soc.* **2021**, *43*, 118–126. (In Chinese)
35. Li, G.; Tang, L.Y.; Jin, L.; Cui, Y.P.; Xi, J.M. Study of strength degradation characteristics of soil-rock mixture under freeze-thaw cycles. *J. Henan Univ. Sci. Technol. (Nat. Sci.)* **2022**, *43*, 67–75. (In Chinese)
36. Hu, F.; Li, Z.; Tian, Y.; Hu, R. Failure patterns and morphological soil–rock interface characteristics of frozen soil–rock mixture under compression and tension. *Appl. Sci.* **2021**, *11*, 461. [CrossRef]
37. Jin, L.; Zeng, Y.W.; Li, H.; Li, J.J. Numerical simulation of large-scale triaxial tests on soil-rock mixture based on DEM of irregularly shaped particles. *Chin. J. Geotech. Eng.* **2015**, *37*, 829–838. (In Chinese)
38. Liao, Q.L.; Li, X.; Li, S.D. Sample remodeling, compactness characteristic and mechanical behaviors of rock-soil mixture. *J. Eng. Geol.* **2010**, *18*, 385–391. (In Chinese)
39. Zeinali, A.; Edeskär, T.; Laue, J. Mechanism of thawing. *Cogent Eng.* **2020**, *7*, 1716438. [CrossRef]

applied sciences

Article

Optimization of the Cooling Scheme of Artificial Ground Freezing Based on Finite Element Analysis: A Case Study

Jun Hu [1], Ke Li [1], Yuwei Wu [1,*], Dongling Zeng [2] and Zhixin Wang [2]

1 School of Civil Engineering and Architecture, Hainan University, Haikou 570228, China
2 Hainan Investigation Institute of Hydrogeology and Engineering Geology, Haikou 570206, China
* Correspondence: wuyuwei@eis.hokudai.ac.jp

Abstract: The present study was envisaged to evaluate the influence of different brine cooling schemes on the freezing process in the formation of sand-cobble strata in an underground connection aisle in Hohhot, China. The brine cooling schemes were set up by modifying the starting and ending brine temperatures in the construction of an underground connection aisle. Using ADINA finite element software, the simulation of the temperature field during the freezing process of the sand and pebble strata under three different schemes was performed. It was found that the freezing process was accelerated by lowering the freezing start temperature during the cooling process when the starting and ending brine temperatures remained unchanged. Furthermore, if the initial freezing temperature was changed, keeping the same freezing time at constant soil thermophysical parameters, the final effective thickness of the frozen wall was almost identical. Considering the same location of the temperature measurement points, the measured temperature of the inner and outer holes of the freezing curtain was found to be consistent with the numerical simulation, demonstrating the rationality of the numerical model. On the basis of this study, a brine cooling plan is proposed, which could serve as a reference for future construction.

Keywords: artificial ground freezing; thermal field; finite element method

1. Introduction

Diebe Gorman & Co. developed the first artificial ground freezing (AGF) in Swansea, South Wales, UK, in 1862. The technique was mainly used to sink underground coal mines [1]. In 1883, the German mining engineer Friedrich Poetsch improved the system and patented it [2]. Due to its reliability, the AGF method was chosen in the construction of a 50-meter-deep shaft in a fully saturated sandstone structure [3]. Since then, the system has been used extensively as a temporary ground support system. In contrast to other geotechnical support methods, AGF is not restricted to specific project sizes or ground types. For example, it can be used in small-scale projects such as the in-situ sampling of Pleistocene sands [1,4] as well as in large-scale projects such as underground uranium mines [5,6]. Apart from the use of AGF in fine-grained sands such as slit sand and clay sand [7], the method can also be applied to fractured sedimentary rocks such as sandstones [8]. Additionally, as AGF is a temporary process, it has a minimal environmental impact, and no foreign substances, such as cement or chemical grouting, are added to the ground. Once the AGF process is completed, ground thawing occurs, thereby returning the formation to its original state. The main disadvantage of AGF systems is the substantial operational costs and energy consumption [9,10]. As a consequence, several studies have been conducted to evaluate and enhance the AGF process [5,11–14]. At the same time, many scholars have conducted a lot of research on the use of AGF as a method for reinforcing contact channels. Zhang et al. [15] established a three-dimensional thermos lid coupling model using ANSYS to study the construction development characteristics of the artificial freezing method in water-rich sand layers and derived the distribution law of the temperature diffusion in the ground. Qin et al. [16] analyzed the construction monitoring results in conjunction with the construction of an extra-long connection aisle in Nanjing. They achieved the development of a frozen wall when two shield tunnels were frozen at the same time and obtained the timing of its excavation. Sun et al. [17] introduced the construction process with thawing and sinking grouting of the No. 6 contact channel of a Jinan metro interval and analyzed the surface settlement monitoring results of the freezing method during the construction. It was found that the critical period for the surface and underground pipeline settlement was 5 days before thawing, and the forced thawing measures could effectively control the thawing and sinking phenomenon within a short period of time. Fu et al. [18] analyzed the evolution of a permafrost curtain according to a three-dimensional finite element numerical model of a section of the Nanning metro and observed that the distribution of temperature diffusion in the stratum was regular. Many scholars have studied the laws of temperature field distribution during the freezing process [18–24].

Hohhot is the capital of the Inner Mongolia Autonomous Region and is located at the intersection of the Eurasian Continental Bridge, the Yellow River Basin Economic Belt, and the Bohai Sea Economic Circle. With complex geological conditions and soil types, it is an important city in northern China's frontier region. Presently, numerical simulation methods are used to investigate the development of the temperature field of the connection aisle freezing method during construction. Most of the research content is focused on the physical properties of the soil, freezing tube setting parameters, and external factors of the connection aisle; however, there have been limited studies on artificial freezing methods and the impact of the freezing process on brine cooling plans. It was observed that during the design and construction of the freezing method, the brine cooling plan has a significant impact on the freezing process of the soil, thereby affecting both the time needed to reach the design freezing requirements in the frosted area and the smooth implementation of the construction phase. During the construction of specific buildings, it is often necessary to mix different levels of cement slurry into the soil to reinforce and improve the local soil to meet engineering needs. However, it may affect the initial brine temperature in the soil due to changes in the heat of hydration of the cement slurry after mixing. On the other hand, there are differences between the initial ground temperature and the ambient temperature due to the different construction seasons and the geographical location of the site of construction. Therefore, it is necessary to control the termination temperature and set up four common different starting temperature cooling scenarios to simulate and discuss the variability of the freezing effect due to differences in the starting brine temperature of the soil.

Based on the construction of the No. 2 connection aisle between the Gongzhongfu Station and the Inner Mongolia Stadium Station of Hohhot Metro Line 2, the present study envisages the effect of the brine cooling plan on the spatial and temporal distribution characteristics of the temperature field. Based on the conventional refrigeration system temperature range of -10 to -35 °C, four different termination temperature cooling schemes were devised to investigate the effect of different termination temperatures on the effective thickness of the final formed freezing wall. The construction requirements were such that at the end of active freezing, the average temperature of the frozen curtain in the area was below -10 °C, the average temperature at the interface between the frozen wall and the pipe sheet was below -5 °C, and the frozen wall reached the specified design thickness before subsequent construction, i.e., excavation of the frozen soil could take place. If the construction requirements were not met, the length of active freezing was extended. Furthermore, the changes in the positive freezing time are specifically discussed in order to develop a rational brine cooling plan in accordance with the construction requirements. The study aims to determine the optimal freezing solution in brine freezing so as to ensure successful artificial freezing during construction and guide construction activities accordingly.

2. Project Overview

The traffic network of the first phase of Hohhot Metro Line 2 is an "L"-shaped backbone line running north–south, with a total length of 27.33 km and an average station distance of 1.166 km, including 24 stations that are underground. Among them, the right line of the shield tunnel starts and ends at DK15 + 497.653 to DK16 + 986.740, with a total length of 1490.226 m, including a long chain of 1.139 m. The left line starts and ends at DK15 + 497.653 to DK16 + 986.740, including a long chain of 1.047 m and a short chain of 2.005 m, with a total length of 14.33 km. There are two liaison tunnels constructed using the mining method. The No. 6 liaison tunnels are located at the mileage of DK16 + 500/left DK16 + 495.535, with an overburden of about 16.26 m at the top of the vault and a buried depth of about 21.91 m at the bottom of the water collection pit; the structure is located in the 3–4 chalk and 3–9 round sand layers. The distance between the two tunnel centers is 10 m, the tunnel radius is 2.75 m, and the thickness of the pipe sheet is 0.35 m. The structure has two layers of chalk and rounded sand. Due to its special geological structure, the surface water at the proposed site is slightly corrosive to the concrete structure and the reinforcing steel in the reinforced concrete structure.

The area runs along the west lane of the Meteorological Bureau through Princess House Street and the Zadagai River to the north of the Genghis Khan Primary School in the new city. The terrain is relatively gentle, with the ground elevation ranging from 1062.776 to 1067.348 m, and the geomorphological unit is a pre-hill alluvial sloping plain. According to the borehole data and indoor geotechnical test results, the soil layers in the exploration area of the project are classified into the Quaternary Holocene Artificial Fill Layer (Q4 ml), the Quaternary Holocene to Upper Pleistocene Alluvium Layer (Q3–4 al + l), and the Quaternary Pleistocene Lake Layer (Q21) according to the deposition age and genesis type of the strata. According to the drilling survey, the groundwater at the site is submerged, and the depth of the stable water table measured by the borehole is 8.5–12.2 m, corresponding to an elevation of 1054.211–1055.166 m, with an elevation difference of 0.955 m and annual water level change of 1.5–3.0 m. The aquifers in this area are mostly strongly permeable, and the lower water barrier is mostly discontinuous and incomplete. This offers a major construction challenge. Figure 1 shows the schematic diagram of the geological stratigraphy, freeze tube construction around the contact channel, and the arrangement of the frozen tube on both sides of the contact channel.

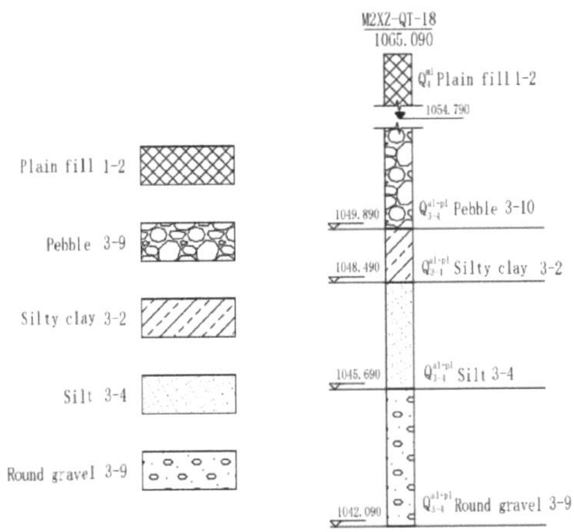

(a)

Figure 1. *Cont.*

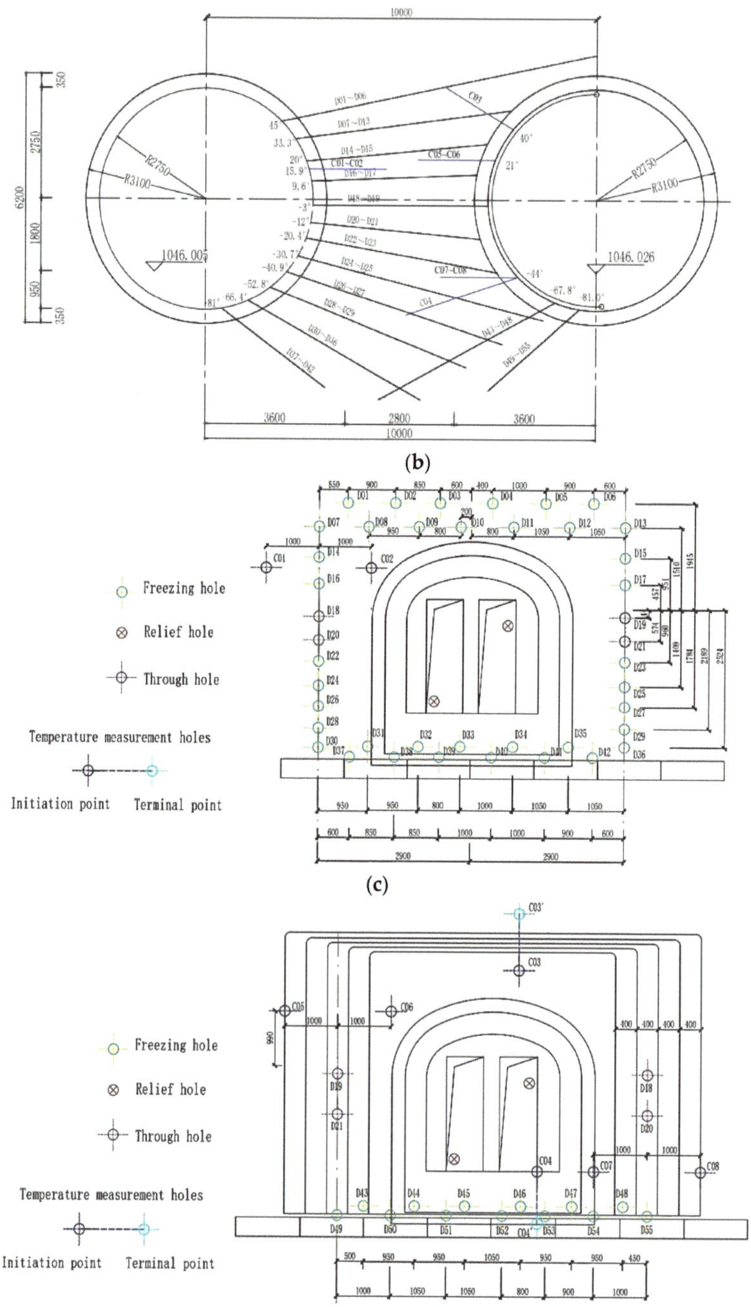

Figure 1. Freeze tube and left and right line tunnel borehole layout of the No. 2 contact channel. (**a**) Geological stratigraphy, (**b**) section of the freeze hole arrangement, (**c**) plan of the freezing hole on the side of the freezing station, and (**d**) plan of the freezing hole on the opposite side of the freezing station.

The specific arrangement of the contact channel freeze pipe is shown in Table 1.

Table 1. Summary of depth and pitch (elevation) angle of temperature measurement and pressure relief holes.

Hole Type	Drill Hole Number	Hole Depth (m)	Positioning Angle (°)	Elevation Angle of Perforation (°)	Perforated Horizontal Angle (°)	Total Hole Depth (m)
Temperature measurement holes	C1~C2	2.0	0	0	0	4.0
	C3	2.0	40	31	0	2.0
	C4	3.0	−44	−18.4	0	3.0
	C5~C6	2.0	−21	0	0	4.0
	C7~C8	2.0	−44	0	0	4.0
Pressure relief hole	X1	2.0		0	0	2.0
	X2	2.0		0	0	2.0
	X3	2.0		0	0	2.0
	X4	2.0		0	0	2.0
Total						25.0

The No. 2 contact channel consisted of 55 freeze holes, 8 temperature measurement holes, and 4 pressure relief holes. Of these, 42 freezing holes, 2 temperature measurement holes, and 2 pressure relief holes were arranged on the side of the machine room; 13 freezing holes, 6 temperature measurement holes, and 2 pressure relief holes were set on the opposite side. The specific layout characteristics of the temperature measurement tubes are shown in Table 1. In order to keep abreast of the changes in soil temperature with the freezing time, three to five measurement points were generally set up in each temperature measurement hole to monitor the soil temperature in real-time. The temperature measurement tubes were installed at the left and right ends of the connection channel, and the points were set at 0.5, 1.25, and 2 m, respectively, to monitor the development of temperature changes.

3. Finite Element Modeling

Based on the Thermal module of ADINA, a numerical model was developed and its geometry size was based on the actual dimensions of the tunnel. This model simulates the real freezing situation by aligning the freezing tubes with the actual layout of the freezing tubes. In order to reasonably describe the evolution of the temperature field in the soil, the model employs a transient thermal conductivity double-tunnel model with phase changes since the temperature changes with time during the freezing process are accompanied by phase changes of water and ice at the same time. The energy generated by latent heat during the ice-water phase change is simplified to the change in heat capacity and thermal conductivity of the soil in the frozen and unfrozen regions. Therefore, this item is not written in the governing equations. The governing equation used in this study is shown below.

For the frozen soil

$$C_f \frac{\partial T_f}{\partial t} = \frac{\partial}{\partial_x}\left(k_f \frac{\partial T_f}{\partial_x}\right) + \frac{\partial}{\partial_y}\left(k_f \frac{\partial T_f}{\partial_y}\right) + \frac{\partial}{\partial_z}\left(k_f \frac{\partial T_f}{\partial_z}\right) \tag{1}$$

For the unfrozen soil

$$C_u \frac{\partial T_u}{\partial t} = \frac{\partial}{\partial_x}\left(k_u \frac{\partial T_u}{\partial_x}\right) + \frac{\partial}{\partial_y}\left(k_u \frac{\partial T_u}{\partial_y}\right) + \frac{\partial}{\partial_z}\left(k_u \frac{\partial T_u}{\partial_z}\right) \tag{2}$$

where C_u is the volumetric heat capacity of the soil; T is temperature; k_f is the thermal conductivity of the soil.

A densely meshed model can accommodate more cells per unit interval, which can improve the accuracy of the calculations. Thus, in this numerical model, the mesh density was set at 0.5 m near the frozen tube area, with complex shape and severe temperature changes and 1.5 m at the boundary of the frozen tube, in order to reduce the number of calculations and ensure higher accuracy. After free division, the model consisted of 1,455,790 cells after meshing in a 4-node meshing mode with a merging tolerance of 0.0001. As the temperature is time-dependent, in the finite element calculation, the temperature-time curve was discretized into load steps. Each load step consisted of multiple sub-steps, and each sub-step used an iterative algorithm. In this study, 1 day was used as a time step to divide, resulting in a period of 40 days of active freezing. Based on the strata survey report and the actual brine cooling scheme that was employed in the project, the model calculations were performed by entering the soil material parameters into the corresponding module of the software. Previous studies have also proven that ADINA performs well in simulating temperature development and distribution. Fu [18] used a three-dimensional finite element method to study the development and spatial distribution of the temperature field during the construction of an artificial ground freezing technique in the context of the reinforcement construction of a liaison channel of the Nanning Metro. The finite element results were first verified against the measured results to verify the feasibility of the numerical method. The formation of the permafrost curtain around the connection channel with time was then discussed, and a series of influencing factors, such as thermophysical parameters and soil condition factors, were systematically and rationally investigated in depth with the help of numerical models. This assisted us in understanding the behavior of the temperature field in the permafrost around the metro connection channel during artificial freezing. Chen [25] developed a three-dimensional finite element model based on ADINA-TMC that considers both thermal coupling and seismic loading. In this model, the laws of heat transfer to the temperature field, seismic loading, and fault movement of the soil are elaborated. Based on the numerical results, the stress–strain under temperature loads, gravity, and seismic loads were compared, providing a theoretical approach to the failure analysis of buried thermal pipes. Nisar A [26] built a numerical model based on ADINA to investigate the temperature distribution patterns generated during the grout sealing of tiles by a high-power diode laser (HPDL). The analysis involved a simulation of a three-dimensional transient temperature field generated by a laser beam that was scanned with constant power at a constant speed across a glazed enamel surface. Latent heat effects due to the melting and solidification of the glaze were considered in the finite element model, thus facilitating a more realistic thermal analysis.

3.1. Basic Assumptions

In this study, ADINA finite element software was used to simulate the changes in the temperature field of the contact channel in the metro tunnel interval. To simplify the calculations, the following assumptions were made without affecting the results of the calculation and analysis.

1. As the variation of temperature in the soil layer was small and negligible, it was assumed that the soil was homogeneous, continuous, and isotropic and that the soil layer was horizontally distributed from top to bottom. Additionally, it was assumed that the initial temperature of the soil was 10 °C, with a uniform initial temperature field.
2. According to the survey report and the construction log of the project, when the freezing construction started on 18 May 2015, all the measured temperatures at different depths of C1~C8 temperature measurement holes were around 10 °C, with a maximum difference of 1.4 °C. Under the premise of ensuring the feasibility of the numerical simulation calculation of the contact channel for basic assumptions, the ground cooling process was simplified to a uniform initial temperature field. The original ground temperature of the soil was set to 10 °C, and it was set as the initial temperature during the freezing period.
3. Temperature-dependent loads were applied to the freeze hole to simulate the temperature of the outer surface of the freeze tube during freezing, thereby ignoring the

complex heat exchange process inside and outside the freeze tube during refrigerant circulation [6].

4. The effects of freeze hole deflection and the mechanical properties of the material of the frozen tube were ignored.
5. The heat exchange in the frozen curtain was more complex due to the presence of groundwater, which also shares part of the cooling capacity, thus causing errors in the numerical calculations. Therefore, the calculation was performed only for the simulation of the freezing process under simulated hydrostatic conditions, without considering the influence of groundwater seepage and the migration of water molecules.
6. From the actual measurement report, it was found that freezing started when the soil temperature dropped to −1 °C, and the frozen soil curtain formed stably when it dropped to −10 °C. The development of the permafrost curtain was observed more easily through the −1 and −10 °C isotherms, wherein the envelope area of the −10 °C isotherm was the minimum freezing area and the envelope area of the −1 °C isotherm was the maximum freezing area.

3.2. Geometric Models

Based on ADINA finite element analysis software, a realistic three-dimensional transient thermal conductivity double-tunnel temperature field model was established for the connection aisle to dynamically simulate the evolution of the permafrost curtain. The distance between the centers of the two tunnels was 10 m, and the tunnel radius was 2.75 m along the tunnel boring direction (longitudinal Z). The dimension was determined as 23 m along the tunnel section lateral, and the distance between the central axes of the twin tunnels was 10 m. The lateral (X-direction) dimension was determined as 16 m, and along the vertical direction (Y-direction), the dimension was 10 m. In this way, the whole model dimension was set in the form of X × Y × Z as a 16 m × 10 m × 23 m freezing tube. In this study, the frozen tube radius was set at 0.045 m. The geometry size and meshing of the model used in this study are shown in Figure 2. In order to ensure the accuracy of the calculation and reduce the amount of calculation, a 4-node meshing method was chosen to mesh the model. When the model was encrypted, the freezing time was set to 40 d, the calculation time step was 40, and each step was for 24 h. The freezing tube arrangement and the grid model are shown in Figure 3.

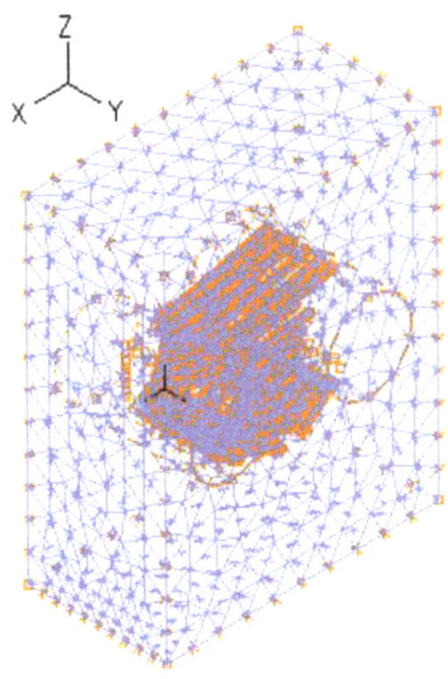

Figure 2. Geometric model and meshes of soil and tunnel (unit: mm).

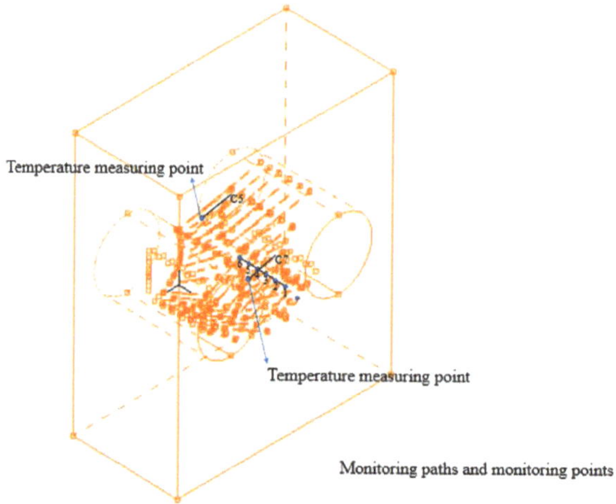

Figure 3. Monitoring points on monitoring paths C5 and C7 and monitoring paths along the centerline of the Y-axis of the tunnel.

3.3. Calculation Parameters

The boundary temperature of the calculation area was assumed to be the temperature of the brine inside the frozen tube, i.e., the load acting directly on the soil surface in contact with it. The heat flux density at the outer boundary of said calculation area was always zero and was an adiabatic boundary [7].

According to the measured data, the initial ground temperature on the numerical model of the soil was set as 10 °C according to the stratigraphic investigation report. The soil thermophysical parameters are shown in Table 2.

Table 2. Soil parameters.

Density /(kg-m^{-3})	Thermal Conductivity/(kJ-m^{-1} °C^{-1})		Specific Heat (kJ-m^{-1} °C^{-1})		Latent Heat of Phase Change/(108 J-m^{-3})
	Unfrozen Soil	Frost Soil	Unfrozen Soil	Frost Soil	
2.010	129	155	1.15	1.29	1.2

3.4. Setting of Observation Sections, Points, and Observation Paths

For better visual observation of the overall development of the temperature field during the construction of the contact channel, the analysis was carried out in the X and Y directions, respectively. Due to the inclined radial arrangement of the freezing tubes, the sparsity of the freezing tubes in the upper and lower rows was different. To facilitate the analysis, several typical cross-sections were selected in the X and Y directions, respectively. The freezing conditions close to the tunnel excavation, where the soil freezes last and the temperature is the highest, were the most difficult part of the construction, and the worst freezing was expected in the weakest part. In particular, the section at X = −4.989 m was a flare with the largest spacing between freeze tubes and, thus, the most unfavorable position in the whole freezing process. Moreover, the areas at X = −2.75 m and X = −6.9 m were the ends of the freezing tubes, which were in a weak position during the freezing process. Therefore, the temperature distribution on the three cross-sections of X = −2.75 m, X = −4.989 m, and X = −6.9 m was selected for specific analysis, as shown in Figure 3. The observation path used in this study is shown in Figure 4, and one observation point was set at every 1 m along the observation path. A total of six observation points were used to monitor the temperature variation at different observation points.

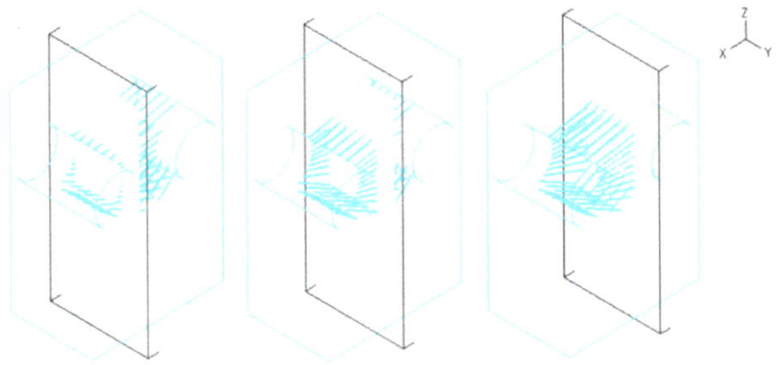

Figure 4. Schematic representation of the observation section.

3.5. Calculation of Thickness and the Average Temperature of Freezing Wall

On 18 May 2015, the temperature measurements were started at the temperature measurement points on the measured path of the connecting passage. By 9 June, the temperature at all the measured points dropped below 0 °C. A temperature-measured point on the C8 path with the highest temperature, No. 3, was selected as an example to analyze the development rate of the frozen wall. On 9 June, the temperature at this measured point fell below 0 °C, which indicated that this place had been frozen since then. The measured point was 0.7 m away from the freezing hole. Thus, the freezing wall development rate was 30.43 mm/d. Table 3 shows the lowest freezing wall development rate at the freezing paths.

Table 3. Slowest freeze wall development rate summary at the measured point.

Sr. No.	Distance From the Freezing Tube (m)	Days Needed to Freeze to 0 °C (Days)	Freezing Wall Development Rate (mm/Day)
C1	0.7	16	43.75
C2	0.85	17	50
C3	0.2	13	15.38
C4	0.38	23	16.52
C5	0.7	17	41.18
C6	0.85	16	53.13
C7	0.85	19	44.74
C8	0.7	23	30.43

It can be seen from Table 3 that the slowest freezing wall development rate was 15.38 mm/d and the fastest freezing wall development rate was 44.74 mm/d. The average speed of all measurement points was 36.59 mm/d. The radius of frozen wall development in 40 days was 1463.75 mm. The thinnest thickness of the frozen curtain was 2825 mm, according to the calculation of a cylindrical intersection circle in the frozen wall. The effective frost wall thickness after the connecting passage freezing process was 2.0 m; hence, the freezing curtain thickness could reach the design requirements.

$$t_{oc} = t_b \bullet \left(1.135 - 0.352\sqrt{L}\right) - 0.875\frac{1}{\sqrt[3]{E}} + 0.266\sqrt{\frac{L}{E}}) - 0.466$$

where t_{oc} is the average temperature of the frozen wall; t_b is the temperature of the brine, L is the distance between the freezing tube, and E is the thickness of the frozen wall.

The average temperature of the frozen wall was obtained as −9.859 °C, which was slightly lower than the average temperature of the frozen wall, which was −10 °C. The reason for the slightly higher temperature at each measured point on the C3 path was that the measured path was far away from the freezing tube or/and the installation of the measured path was affected by a serious deviation caused by the poor quality of the borehole during construction [20].

4. Results and Discussion

4.1. Validity of the Numerical Model

The measured result was compared with the test data to illustrate the validity of the FEM model. The temperature variation during the freezing point in C7 and C5 is shown in Figure 5.

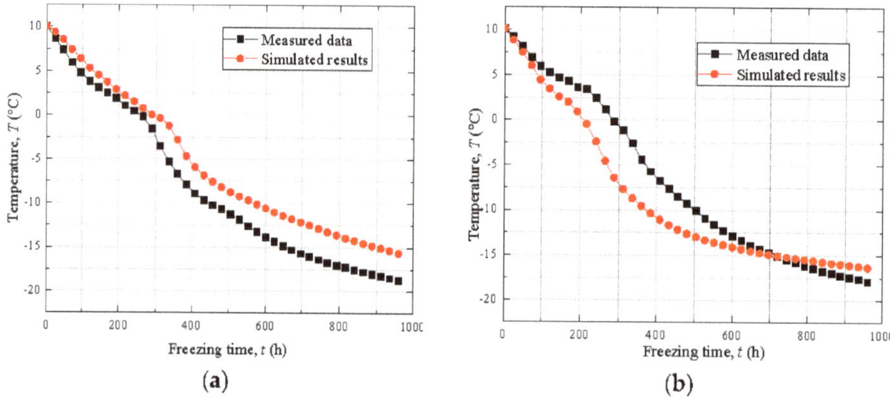

Figure 5. Comparison of test results and simulated temperature at the observation points. (**a**) Comparison on C5; (**b**) comparison on C7.

As shown in Figure 5, the simulation results agreed well with the test data. It can be seen that the temperature decreased faster in the early stage of the freezing development but was slower in the later stages of the freezing development. The temperature field model, established based on the finite element method, could accurately describe the temperature development trend and values at the observation points during freezing.

4.2. Effect of Different Initial Brine Temperatures on Freezing Effect

This section examines the effect of initial brine temperature on the freezing effect by simulation. Four sets of controlled simulation experiments with experimental conditions are shown in Table 4.

Table 4. Cooling schedule for different initial temperatures.

Time (Days)	0	1	5	10	15	20	30	40
Case 1 temperature (°C)	10	2	−16	−23	−25	−27	−29	−30
Case 2 temperature (°C)	0	−6	−20	−25	−26	−28	−30	−30
Case 3 temperature (°C)	−10	−14	−24	−27	−28	−30	−30	−30
Case 4 temperature (°C)	−20	−22	−26	−28	−29	−30	−30	−30

The simulated results are presented in Figure 6.

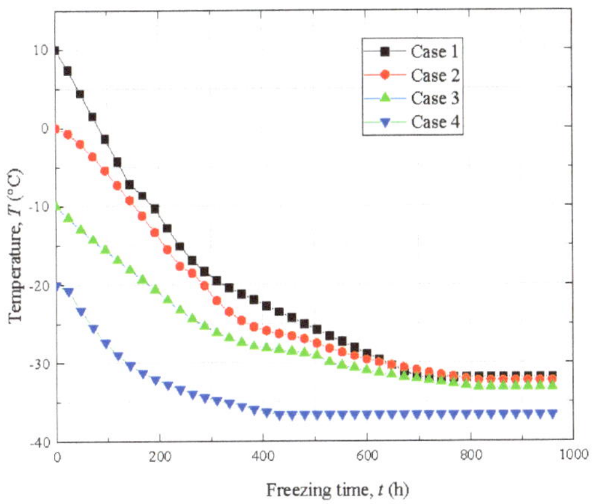

Figure 6. Temperature development at the observation point during freezing at different initial temperatures.

It can be seen from Figure 6 that the different starting brine temperatures led to variability in temperature development at the beginning of freezing. However, the effect of the different initial temperatures on the final freezing effect of the soil layer was not significant. The reasons for the different initial temperatures of the soils are described below. In order to meet different engineering needs, it is often necessary to mix different levels of cement slurry into the soil to strengthen and improve the local soil, and the initial temperature of the soil may be too high due to the effect of the heat of hydration. Alternatively, the initial ground temperature is affected by the ambient temperature differently due to seasonal variations during the construction. The initial freezing temperature will only affect the freezing effect at the early stage of freezing, but the freezing temperature field development change pattern will not change and will exhibit a minor effect on the final freezing effect in the frozen area [11–13].

The temperature development at the observed cross-section for different cases is shown in Figure 7.

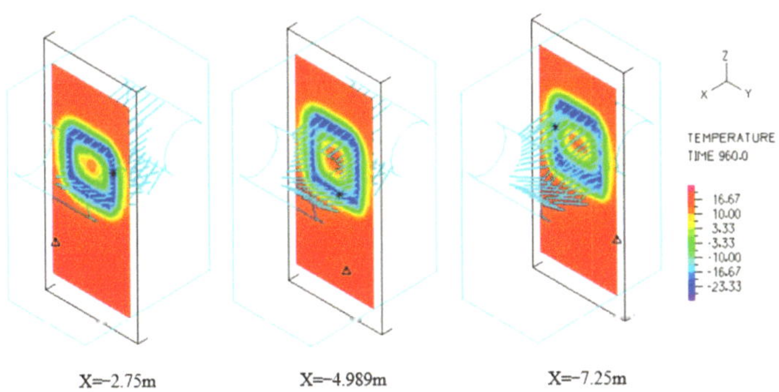

(**a**) Case 1

Figure 7. *Cont.*

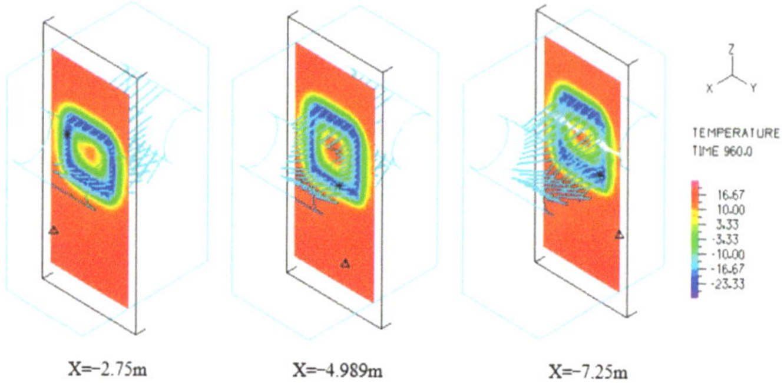

(**b**) Case 2

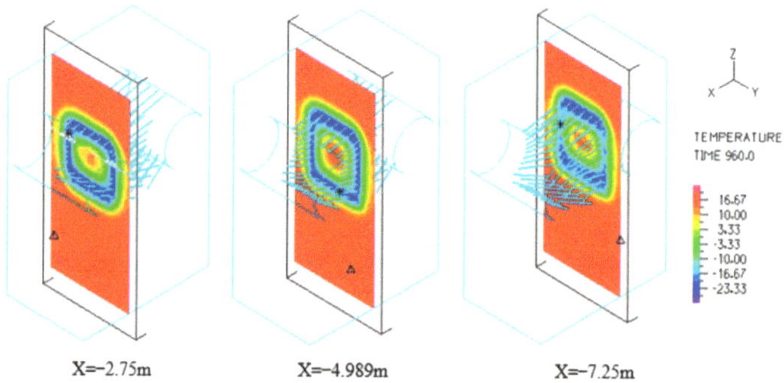

(**c**) Case 3

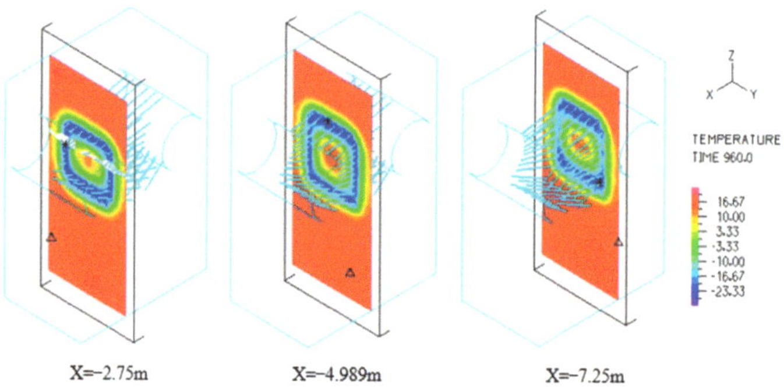

(**d**) Case 4

Figure 7. Temperature distribution clouds on each cross-section under different working conditions.

In Figure 7, it can be seen that due to the different starting brine temperatures, the variability of temperature development at the beginning of freezing at the observation points was large, but the effect of different initial temperatures on the final freezing effect of the soil layer was not significant. The difference in freezing effect caused by the difference in the starting brine temperature of the soil mainly occurred before the soil froze to 0 °C. In an actual project, during the construction of the building, the local soil needs to be reinforced and improved by mixing different degrees of cement slurry into the soil to meet the different engineering needs. The initial temperature of the soil during freezing is influenced by the heat of the hydration of the cement, and its initial ground temperature varies from the ambient temperature due to seasonal variations during the construction. Different initial temperatures affect the development of the freezing process at the beginning of active freezing, but the freezing temperature field development change pattern does not change and exhibits a negligible effect on the final freezing effect of the frozen zone.

4.3. Effect of Different Final Brine Temperatures on the Freezing Effect

The final freezing temperature also affects the pattern of temperature development during the freezing process. This section discusses the effect of the final freezing temperature on temperature change. The changes in temperature development at the temperature measurement points when the final freezing temperature was −10, −20, −30, and −35 are discussed. The different brine cooling schedules are shown in Table 5.

Table 5. Cooling schedule for different initial temperatures.

Time (Days)	0	1	5	10	15	20	30	40
Case 1 temperature (°C)	10	0	−10	−10	−10	−10	−10	−10
Case 2 temperature (°C)	10	0	−10	−20	−20	−20	−20	−20
Case 3 temperature (°C)	10	0	−10	−25	−25	−25	−30	−30
Case 4 temperature (°C)	10	0	−10	−30	−30	−30	−30	−35

In order to observe the temperature distribution of each section at different freezing times under different cases, the temperature distribution clouds at different freezing times were intercepted, as shown in Figure 8.

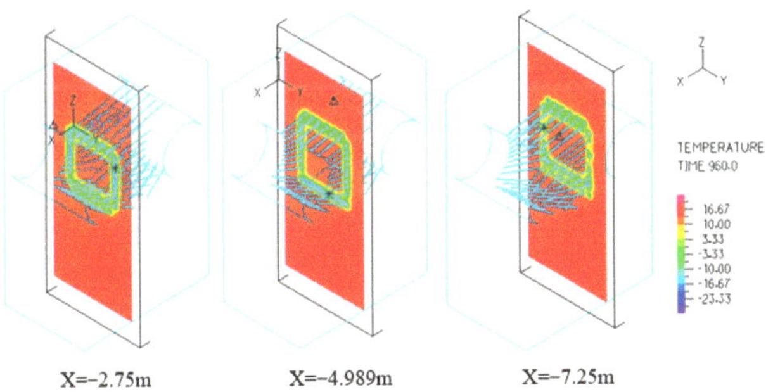

(**a**) Case 1

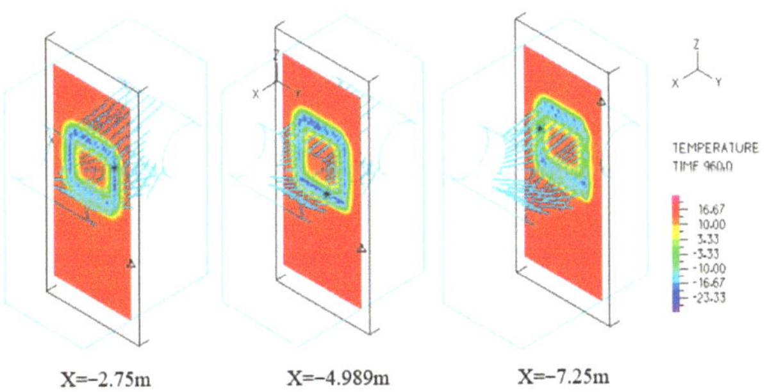

(**b**) Case 2

Figure 8. *Cont.*

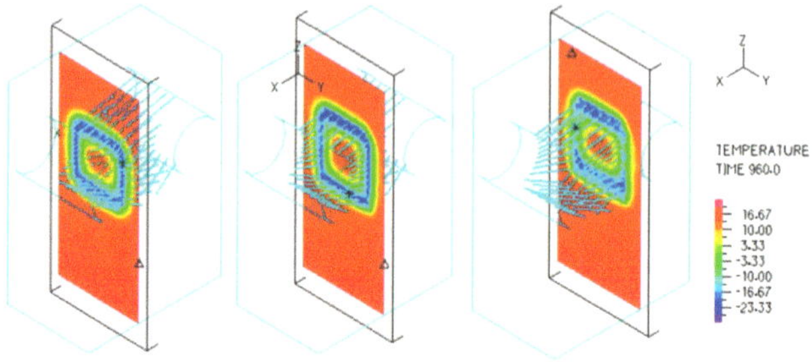

X=-2.75m X=-4.989m X=-7.25m

(**c**) Case 3

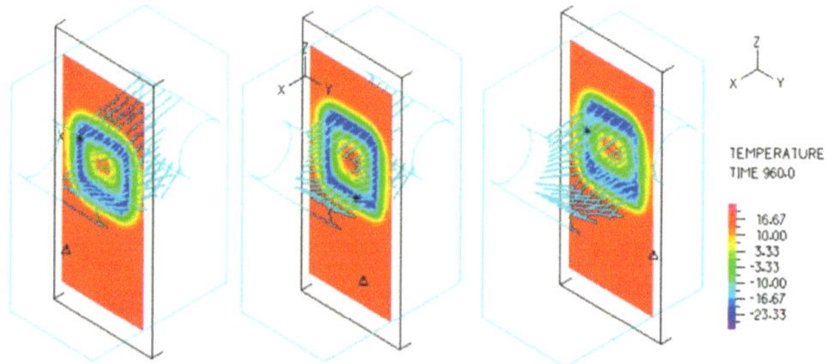

X=-2.75m X=-4.989m X=-7.25m

(**d**) Case 4

Figure 8. Temperature distribution clouds at each cross-section under different working conditions.

The temperature variation curves at the observation points of the C7 observation path under different termination temperature cooling plans are shown in Figure 9.

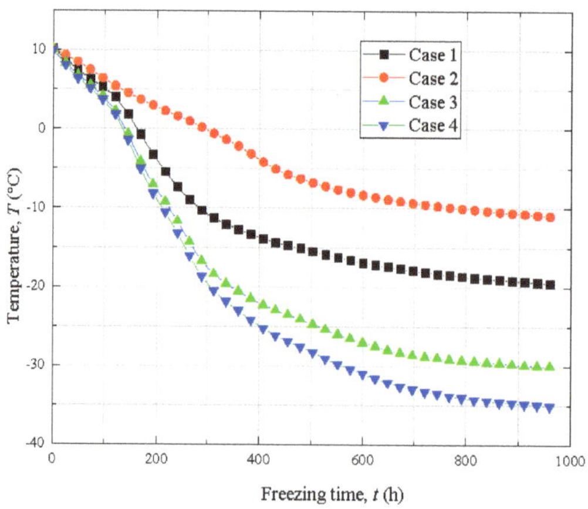

Figure 9. Temperature variation curves at the observation points during freezing at different final temperatures.

It can be seen from Figure 9 that the temperature development trend at the observation points under different brine cooling schedules was basically the same during the first 240 h after the initiation of freezing. When the freezing was carried out to 360 h, the temperature at the observation points under different experimental conditions dropped to below 0 °C, indicating that the observation points were completely frozen under various experimental conditions. After the freezing proceeded to 360 h, the temperature change at the temperature measurement points showed a large variability with the change of the final brine cooling plan. That is, for the experimental case with a lower final freezing temperature, the temperature measurement point presented a lower temperature when freezing at the same temperature measurement point up to the same moment. The general trend of soil cooling under different brine cooling schemes was that the cooling rate of the soil was faster at the beginning of freezing, slowed down after reaching the freezing temperature, and then gradually stabilized until stable permafrost was formed. The temperature development pattern of the soil showed large variability in the maintenance freezing stage and directly affected the final freezing effect.

The project requirements were that the average temperature of the frozen soil curtain in the frozen area should not be higher than −10°C and the average temperature of the interface between the frozen wall and the pipe sheet should not be higher than −5 °C. The final frozen soil temperature was stable at around −28 °C for 40 d, and the frozen wall reached the design thickness, which ensured a smooth excavation of the catchment well and met the soil stability requirements in the excavation process. When the brine temperature and brine flow rate do not meet the construction requirements, the active freezing time should be extended.

Under the different initial temperature conditions, Case 4 (initial temperature of −20 °C) exhibited the best freezing effect, with a cooling rate of 0.925 °C/d and an extreme difference of 16.65 °C/d. The soil temperature was the first to drop to −28 °C in 3 d and remained stable. Case 1 (initial temperature of 10 °C) had the worst freezing effect, with a cooling rate of 0.754 °C/d and an extreme difference of 21.8644 °C/d. Thus, a 30 °C reduction in the initial temperature shortened the freezing time by 24 d and increased the cooling rate by 0.171 °C/d, thereby presenting a significant gain in the freezing process.

Under different termination temperature conditions, the freezing walls of Case 3 (final temperature of −30 °C) and Case 4 (final temperature of −35 °C) reached the design thickness at the late stage of freezing, and the final frozen soil temperature was below −28 °C for 40 d. The freezing effect achieved the strength and stability required by the engineering design. In Case 1 (final temperature of −10 °C) and Case 2 (final temperature of −20 °C), the final freezing temperature was too high and, therefore, did not meet the requirements for subsequent excavation. The average cooling rate of Case 4 was the fastest, at 0.656 °C/d, with an extreme difference of 25.0367 °C/d. The soil temperature dropped to −28 °C at 22 d and remained stable. The average cooling rate for Case 1 was the slowest at 0.492 °C/d, with an extreme difference of 9.5408 °C/d. The soil temperature eventually stabilized at around −10.5408 °C after 36 d. Therefore, a 5 °C reduction in the termination temperature shortened the freezing time by 7 d and increased the cooling rate by 0.144 °C/d, while a 25 °C reduction in the termination temperature shortened the freezing time by 14 d and increased the cooling rate by 0.164 °C/d. Therefore, at a constant initial temperature, the greater the reduction in termination temperature, the more obvious the gain observed in the freezing process.

5. Conclusions and Future Prospects

Addressing the influence of the brine cooling plan on the construction process of the freezing method, this study was based on the background of the construction of a contact channel and pump room in the interval between Gongzhongfu Station and Inner Mongolia Stadium Station by the Hohhot City Railway Line 2 Phase I Project using the freezing method. Subsequently, the numerical model pair of the stratum-freezing tube was established, and the temperature field distribution of the contact channel control section under different brine cooling plans and soil freezing wall morphology was analyzed and compared with the actual measured temperatures. The measured data and the cooling curve obtained from the numerical calculation of the cooling law were basically consistent, and the values were relatively close. This indicates that the numerical calculation method based on the 3D finite element model can simulate field results more accurately. At the early stage of freezing, the discrepancy between the numerical calculation results and the measured data in the field was relatively small, and a deviation gradually appeared with an increase in time. The conclusions of the study can be drawn as follows:

(1) The measured average rate of temperature drop in the C5 temperature measurement hole was about 0. 67 °C/d, with a very high difference of 20.7 °C. The numerical simulation results were about 0.57 °C/d, with a very high difference of 20.1 °C. After 40 d of active freezing, the measured temperature in the C5 temperature measurement hole was −8.1 °C, and the numerical simulation temperature was −7.3 °C.
(2) The measured average cooling rate of the C7 temperature measurement hole was about 0. 852 °C/d, and the extreme difference was 21.3 °C; the numerical simulation result was about 0.908 °C/d, and the extreme difference was 22.7 °C. After 40 d of active freezing, the measured temperature of the C7 temperature measurement hole was −5.2 °C, and the numerical simulation temperature was −6.2 °C.
(3) The trend of the numerically simulated temperature and the temperature obtained from the actual measurement was essentially the same at the same location of the temperature measurement point. As the numerical calculation did not take into account the influence of groundwater, there can be a situation where the numerically simulated temperature is slightly lower than the actual measured temperature. This means that in the actual project, the heat exchange in the permafrost curtain is more complicated due to the presence of groundwater, which also shares part of the cooling. Thus, the measured temperature is slightly lower than the numerical simulation because there is a certain error between the physical parameters and boundary conditions obtained from the test and the actual project.
(4) The overall trends of the development change of the numerical simulation calculation results and the actual measured data in the field were basically the same, and the cooling law was similar. For the temperature measurement points at the same location, the measured temperature data were close to the values obtained from the numerical simulation.
(5) Keeping the freezing time constant, any change in the starting brine temperature in the brine cooling plan showed almost no effect on the final formation of the effective thickness of the freezing wall, and the final effective thickness of the freezing wall was not related to the starting brine temperature. In the case of the termination brine temperature, the freezing curtain form did not change by only changing the freezing starting brine temperature. However, within the tolerable range of the freezing pipe, lowering the starting brine temperature increased the cooling rate of the soil and accelerated the freezing process.
(6) Keeping the freezing time constant, changes in the terminating brine temperature in the brine cooling plan showed a great direct effect on the final freezing temperature. In general, the final soil freezing temperature decreased with a decrease in the final brine temperature.

In summary, due to the influence of unmeasurable factors at the project site, the actual measured data had some deviations in reflecting the development pattern of the permafrost curtain temperature field. Nevertheless, the numerical calculation results could still simulate the temperature field evolution in the contact channel and surrounding soil layers during the construction process in a more accurate and dynamic way. The instantaneous freezing temperature field obtained by numerical simulation with the finite element software ADINA reflected the actual situation of the project realistically. However, seepage may also greatly influence the development of temperature during freezing, so this should also be considered in future studies.

In this study, numerical simulations were carried out using finite element software. For a long time, the construction method of building engineering was limited by traditional construction tools and technical means. On the one hand, it is difficult to put the imagination and creativity of architects for 3D building forms into practice, and on the other hand, rough construction techniques have led to serious damage to the environment and caused high resource consumption and wastage. Based on the maturity of BIM technology construction, AutoCAD, and other technologies [27–30], the combination of BIM technology and 3D printing technology should be considered for setting up an experimental study of indoor models. Additionally, it is necessary to conduct multi-factor orthogonal tests to visualize and digitally analyze the development of the freezing process.

Author Contributions: Conceptualization, J.H. and K.L.; Methodology, Y.W.; Software, K.L.; Validation, J.H., K.L. and Z.W.; Formal analysis, D.Z.; Investigation, D.Z.; Resources, Z.W.; Data curation, J.H.; Writing—original draft preparation, Y.W.; Writing—review and editing, Y.W.; Visualization, K.L.; Supervision, D.Z.; Project administration, K.L.; Funding acquisition, J.H. All authors have read and agreed to the published version of the manuscript.

Funding: This research was funded by the Innovative Research Team Project of the Natural Science Foundation of Hainan Province, P. R. China (522CXTD511), the High Technology Direction Project of the Key Research and Development Science and Technology of Hainan Province, P. R. China (ZDYF2021GXJS020), and the Characteristic Innovation (Natural Science) Projects of Scientific Research Platforms and Scientific Research Projects of Guangdong Universities in 2021 (2021KTSCX139). The authors also acknowledge support from the China Scholarship Council.

Informed Consent Statement: Not applicable.

Data Availability Statement: Not applicable.

Conflicts of Interest: The authors declare no conflict of interest.

References

1. Harris, J.S. *Ground Freezing in Practice;* Thomas Telford: London, UK, 1995.
2. Poetsch, F.H. *Das Gefrierverfahren: Methode für Schnelles, Sicheres und Lothrechtes Abteufen von Schächten im Schwimmsande und Überhaupt im Wasserreichen Gebirge; für Herstellung Tiefgehender Brückenpfeiler und für Tunnel-Bauten im Rolligen und Schwimmenden Gebirge;* Craz & Gerlach: Albaried, Germany, 1886.
3. Lackner, R.; Pichler, C.; Kloiber, A. Artificial Ground Freezing of Fully Saturated Soil: Viscoelastic Behavior. *J. Eng. Mech.* **2008**, *134*, 1–11. [CrossRef]
4. Hofmann, B.A.; Sego, D.C.; Robertson, P.K. In Situ Ground Freezing to Obtain Undisturbed Samples of Loose Sand. *J. Geotech. Geoenviron. Eng.* **2000**, *126*, 979–989. [CrossRef]
5. Alzoubi, M.A.; Madiseh, A.; Hassani, F.P.; Sasmito, A.P. Heat transfer analysis in artificial ground freezing under high seepage: Validation and heatlines visualization. *Int. J. Therm. Sci.* **2019**, *139*, 232–245. [CrossRef]
6. Yun, X.; Tang, B.; Greg, M.; Brian, M.; Brian, M. Radon bearing water protection in underground uranium mining—A case study. *Int. J. Min. Sci. Technol.* **2017**, *27*, 599–603. [CrossRef]
7. Zhou, J.; Tang, Y. Experimental inference on dual-porosity aggravation of soft clay after freeze-thaw by fractal and probability analysis. *Cold Reg. Sci. Technol.* **2018**, *153*, 181–196. [CrossRef]
8. Shen, Y.J.; Wang, Y.Z.; Zhao, X.D.; Yang, G.S.; Jia, H.L.; Rong, T.L. The influence of temperature and moisture content on sandstone thermal conductivity from a case using the artificial ground freezing (AGF) method. *Cold Reg. Sci. Technol.* **2018**, *155*, 149–160. [CrossRef]
9. Gallardo, A.H.; Marui, A. The aftermath of the Fukushima nuclear accident: Measures to contain groundwater contamination. *Sci. Total Environ.* **2016**, *547*, 261–268. [CrossRef]
10. Jardine, C.G.; Banfield, L.; Driedger, S.M.; Furgal, C.M. Risk communication and trust in decision-maker action: A case study of the Giant Mine Remediation Plan. *Int. J. Circumpolar Health* **2013**, *72*, 21184. [CrossRef]
11. Alzoubi, M.A.; Sasmito, A.P.; Madiseh, A.; Hassani, F.P. Freezing on Demand (FoD): An Energy Saving Technique for Artificial Ground Freezing. *Energy Procedia* **2019**, *158*, 4992–4997. [CrossRef]
12. Alzoubi, M.A.; Zueter, A.; Nie-Rouquette, A.; Sasmito, A.P. Freezing on demand: A new concept for mine safety and energy savings in wet underground mines. *Int. J. Min. Sci. Technol.* **2019**, *29*, 621–627. [CrossRef]
13. Alzoubi, M.A.; Sasmito, A.P. Thermal performance optimization of a bayonet tube heat exchanger. *Appl. Therm. Eng.* **2017**, *111*, 232–247. [CrossRef]
14. Alzoubi, M.A.; Sasmito, A.P. Development and Validation of Enthalpy-Porosity Method for Artificial Ground Freezing Under Seepage Conditions. In Proceedings of the ASME Joint US-European Fluids Engineering Summer Conference (FEDSM2018), Montreal, QC, Canada, 15–19 July 2018. [CrossRef]
15. Zhang, Y.; Ruan, L.; We, X.; Zhao, Y. Characteristics of Stratum Variations for Connecting Passage of Shield Tunnel with Artificial Freezing Method in Water Rich Sand Stratum. *Tunn. Constr.* **2021**, *41*, 106–114.
16. Qin, W.; Yang, P.; Jin, M.; Zhang, T.; Wang, H. Study on the application and measurement of artificial freezing method in the ultra-long contact channel of underground. *J. Undergr. Space Eng.* **2010**, *6*, 1065–1071.

17. Sun, L.; Zhang, D.; Qiao, X.; Luo, B. Research on freezing reinforcement and melt-sinking grouting for underground contac channels in soft and weak strata. *Mod. Tunn. Technol.* **2020**, *57*, 1002–1006.
18. Fu, Y.; Hu, J.; Wu, Y. Finite element study on temperature field of subway connection aisle construction via artificial grounc freezing method. *Cold Reg. Sci. Technol.* **2021**, *189*, 103327. [CrossRef]
19. Hu, J.; Liu, Y.; Wei, H.; Yao, K.; Wang, W. Finite-element analysis of the heat transfer of the horizontal ground freezing method ir shield-driven tunneling. *Int. J. Geomech. ASCE* **2017**, *17*, 04017080. [CrossRef]
20. Hu, J.; Liu, Y.; Li, Y.; Yao, K. Artificial ground freezing in tunnelling through aquifer soil layers: A case study in Nanjing Metrc Line 2. *KSCE J. Civ. Eng.* **2018**, *22*, 4136–4142. [CrossRef]
21. Fu, Y.; Hu, J.; Liu, J.; Hu, S.; Yuan, Y.; Zeng, H. Finite Element Analysis of Natural Thawing Heat Transfer of Artificial Frozen Soi in Shield-Driven Tunnelling. *Adv. Civ. Eng.* **2020**, *2020*, 2769064. [CrossRef]
22. Hu, J.; Liu, W.; Pan, Y.; Zeng, H. Site Measurement and Study of Vertical Freezing Wall Temperatures of a Large-Diameter Shielc Tunnel. *Adv. Civ. Eng.* **2019**, *2019*, 8231458. [CrossRef]
23. Qiu, P.; Li, P.; Hu, J.; Liu, Y. Modeling Seepage Flow and Spatial Variability of Soil Thermal Conductivity during Artificial Grounc Freezing for Tunnel Excavation. *Appl. Sci.* **2021**, *11*, 6275. [CrossRef]
24. Wu, Y.; Zeng, H.; Hu, J.; Ren, X.; Xue, X. Finite Element Study on Temperature Field of Underwater Dredging Devices via the Artificial Ground Freezing Method. *Geofluids* **2022**, *2022*, 7502693. [CrossRef]
25. Chen, Y.; Zhang, A.; Yang, T. Numerical simulation on the damage of buried thermal-pipeline under seismic loading based or thermal-mechanical coupling. *Syst. Eng. Procedia* **2011**, *2*, 428–433. [CrossRef]
26. Nisar, A.; Schmidt, M.J.J.; Sheikh, M.A.; Li, L. Three-dimensional transient finite element analysis of the laser enamelling process and moving heat source and phase change considerations. *Proc. Inst. Mech. Eng. Part B J. Eng. Manuf.* **2003**, *217*, 753–764 [CrossRef]
27. Borra, N.D.; Neigapula, V.S.N. Parametric optimization for dimensional correctness of 3D printed part using masked stereolithog raphy: Taguchi method. *Rapid Prototyp. J.* 2022; *ahead-of-print*. [CrossRef]
28. Wang, X.; Cao, J.; Cao, Y. A new multiobjective optimization adaptive layering algorithm for 3D printing based on demand oriented. *Rapid Prototyp. J.* 2022; *ahead-of-print*. [CrossRef]
29. Panahizadeh, V.; Ghasemi, A.H.; Asl, Y.D.; Davoudi, M. Optimization of LB-PBF process parameters to achieve best relative density and surface roughness for Ti6Al4V samples: Using NSGA-II algorithm. *Rapid Prototyp. J.* 2022; *ahead-of-print*. [CrossRef]
30. Linares, J.M.; Chaves-Jacob, J.; Lopez, Q.; Sprauel, J.M. Fatigue life optimization for 17-4Ph steel produced by selective laser melting. *Rapid Prototyp. J.* **2022**, *28*, 1182–1192. [CrossRef]

applied sciences

MDPI

Article

Orthogonal Test on the True Triaxial Mechanical Properties of Frozen Calcareous Clay and Analysis of Influencing Factors

Chuan-Xin Rong [1], Zhi Wang [1,2,*], Yi Cao [1], Qing Yang [1] and Wei Long [1]

[1] School of Civil Engineering and Architecture, Anhui University of Science and Technology, Huainan 232001, China
[2] Anhui Provincial Key Laboratory of Building Structure and Underground, Anhui Jianzhu University, Hefei 230601, China
* Correspondence: wangz519714795@163.com

Abstract: In the Huainan and Huaibei mining areas, a layer of calcareous clay is buried deep in the surface soil layer (at approximately 400 m). This layer is in a high-stress state and is prone to freezing pipe fractures in the freezing method. To obtain the true triaxial mechanical properties of this clay in its frozen state, this study conducted a cross test ($L_{16}(4^5)$) to explore the change law of the strength of frozen calcareous clay under the influence of multiple factors. The results showed that the true triaxial stress–strain curve of frozen calcareous clay was divided into three stages: the strain within 0.5% showed linear elasticity. Under compressive stress, ice crystals and their cements were damaged or melted and shrank. At approximately 5%, they showed plastic hardening. The soil particles and ice crystals in the frozen soil recombined and became denser, resulting in irreversible deformation. As the compression progressed, cracks bred and swelled. The failure stage was manifested as strain hardening due to the test loading conditions. As the deformation increased, the stress also slightly increased. The consistent strength-influencing factors could be obtained through range and hierarchy analyses. The primary and secondary order of influence of σ_1 was the confining pressure σ_3, water content ω, temperature T, Bishop parameter b, and salt content ψ. The influence weight of each factor was quantitatively calculated. In the significance analysis, when the interaction was not considered, the effects of the pressure and moisture content on the strength were always significant. The effect of temperature was significant only when the significance level $\Omega > 0.05$. The salt content and b value had no significant influence on the strength, and the significance of each factor followed the order of the results of the range analysis method and analytic hierarchy process; when considering the interaction, the interaction factors had different effects on the strength. When $\Omega > 0.01$, the influence of factor A (temperature T) $\times$ B (water content ω) on the strength showed significance, even exceeding that of temperature. This demonstrated that when studying the strength characteristics of frozen soil, it is necessary to comprehensively consider the various factors and their interaction to more accurately characterize the mechanical behavior of frozen solids.

Keywords: frozen calcareous clay; true triaxial orthogonal test; analytic hierarchical process; significance test; interaction

1. Introduction

The geotechnical strength index is an important survey parameter for the development and utilization of underground space. Under normal circumstances, conventional soil is considered a two- (saturated) or three-phase system, and laboratory testing and research have been conducted on such soils. For shallow rock and soil, the unconfined uniaxial compressive strength under different factors have been considered [1–3]. Such soils meet the engineering requirements; however, when the deep rock and soil are in a state of high stress, the contribution of the confining pressure to the strength cannot be ignored. To simulate a real stress environment, it was necessary to conduct false or true triaxial indoor tests [4–6]. The factors influencing the strength in such tests include the water content (or saturation), soil properties, loading rate, confining pressure, and dry density. The strength of conventional soil is generally low, and the influence of various factors on the strength was not significant. Frozen soil is a four-phase system. Artificial ground freezing curtains are used as temporary support measures in major underground projects such as coal mine shaft excavation, subway tunnels, and special underground spaces. The influence of various factors on the properties of frozen structures, such as thermal conductivity [7,8] and strength, is greatly enhanced and more complex. Cai [9] conducted an unconfined compressive strength test on frozen clay under different water and Na_2SO_4 contents. The test results showed that the deformation of frozen soil was significantly affected by the change in the water and salt contents. Its strength increased to the maximum value first and then decreased with the increase of water content and salt content. Through a series of triaxial tests, Niu [10] found that the deformation and strength of frozen soil were significantly affected by temperature, moisture, and pressure. When the moisture content was low, and the confining pressure increased, a silty clay transitioned from strain softening to strain hardening. With the increase in the water content, the initial tangent modulus gradually changed from a linear distribution to a parabolic one with the confining pressure. Lai [11] and Du [12] analyzed the change characteristics of sand strength with the moisture content at different temperatures and the dependence of sand strength on the temperature under the same moisture content; the influence of moisture content on the strength of frozen soil did not purely depend on the moisture content. For a certain type of soil, there is a fixed moisture content (optimal moisture content). When it is greater than this value, the strength of frozen soil decreases with the increase in the moisture content [9,13–17]. The temperature, water content, salt content, and confining pressure significantly affect the strength of frozen soil. In terms of true triaxiality, the influence of Bishop parameter b on the strength of frozen soil, particularly the second principal stress, cannot be ignored. Hence, scholars have conducted true triaxial tests that are more representative of the real stratum stress state. For example, Dai [18] discussed the effects of the water content, consolidated confining pressure, and intermediate principal stress ratio on the strength of branched clay and reached a reliable conclusion. As reported by Chen [19], when the consolidated confining pressure was less than the structural strength of the structural loess, the stress–strain curve was a softening type; otherwise, it was a hardening type. Ma [20] and Zhang [21] conducted a true triaxial test on frozen sand under the influence of Bishop parameter b and accordingly performed an analysis.

Many factors affect the strength of frozen soil, not just the ones mentioned in the above studies. Most scholars considered only two or three of the many factors, and research on the interaction effects of the various factors is lacking. Hence, in this study, we considered a layer of calcareous clay in the deep surface of the Huainan and Huaibei mining area (buried at approximately 400 m) as the research object. This layer is in a high-stress state and highly prone to freezing pipe fractures in the freezing method. To obtain the true triaxial mechanical properties of this clay in its frozen state, we conducted true triaxial orthogonal tests on calcareous clay under the influence of five factors: temperature, water content, salt content, confining pressure, and Bishop parameter b. The primary and secondary order of influence of the various factors on the strength of the frozen soil was analyzed, and the weight of the influence of each factor on the peak strength was quantitatively calculated through the analytic hierarchy process. Finally, the interaction and non-interaction between the factors were considered, and statistical theory was applied to elaborate the influence of each factor. The degree of significance provided a certain reference value for the design and construction of underground structures in frozen soil.

2. Test Design and Test Results

2.1. Orthogonal Experimental Design Scheme

Frozen soil is used as a curtain for coal mine shaft excavation, and its strength is an important design index. Studies have shown that many factors affect the strength of frozen soil, including water and salt content, confining pressure, temperature, and indoor test loading rate. In this study, we used a calcareous clay exhibiting strong plasticity as the test object. Table 1 presents the basic physical properties of this clay. Considering the temperature, water and salt content, confining pressure, and intermediate principal stress coefficient b, a true triaxial test was conducted on the frozen calcareous clay. Because of the many factors considered and the large number of tests required, an orthogonal test method was used to explore the influence of the various factors on the strength of this frozen soil. The orthogonal test design is presented in Table 2. A five-factor four-level ($L_{16}(4^5)$) orthogonal test was conducted. The corresponding levels of each factor were: (Factor A) Temperature $T = -5$–-20 °C; (Factor B) water content $\omega = 15$–22.5%; (Factor C) salt content $\psi = 0$–3%; (factor D) confining pressure $\sigma_3 = 1$–4 MPa; (Factor E) principal stress coefficient $b = 0, 0.33, 0.67$, and 1.

Table 1. Geotechnical test results of the studied frozen soil [22].

Soil Type	Moisture Content ω (%)	Wet Density ρ (g/cm^3)	Dry Density ρ_d (g/cm^3)	Specific Gravity G_s	Void Ratio	Liquid Limit ω_L (%)	Plastic Limit ω_P (%)	Plasticity Index I_P	Liquid Limit Index I_L
Calcium Clay	22.62	2.04	1.66	2.736	0.64	54	35	19	−0.65

Table 2. True triaxial test factors of calcareous clay and their corresponding levels. (Note: The salt content in this table was the salt content in 100 mL of pure water, that is, the mass percentage of anhydrous $CaCl_2$ [23]).

Five Factors \ Four Levels	L1	L2	L3	L4
Temperature T/°C (Factor A)	−5	−10	−15	−20
Moisture content ω/% (Factor B)	15	17.5	20	22.5
Salt content ψ/% (Factor C)	0.0	1.0	2.0	3.0
Confining pressure σ_3 /MPa (Factor D)	1	2	3	4
Bishop parameter b (Factor E)	0	0.33	0.67	1

2.2. Test Loading Mode

Sample preparation: The dimensions of the test block sample were 100 mm × 100 mm × 100 mm. After the undisturbed soil sample was dried, ground, and sieved, it was sprayed with different concentrations of $CaCl_2$ solution in layers, mixed evenly, placed in a Ziplock bag, and allowed to stand for 24 h. The sample was then controlled using the dry density and volume. The mixed soil samples of corresponding quality were packed in a grinding tool in layers and pressed to prepare standard samples for use, as shown in Figure 1. Before the test, the standard sample was frozen at the required low temperature for more than 48 h, and a true triaxial test was conducted.

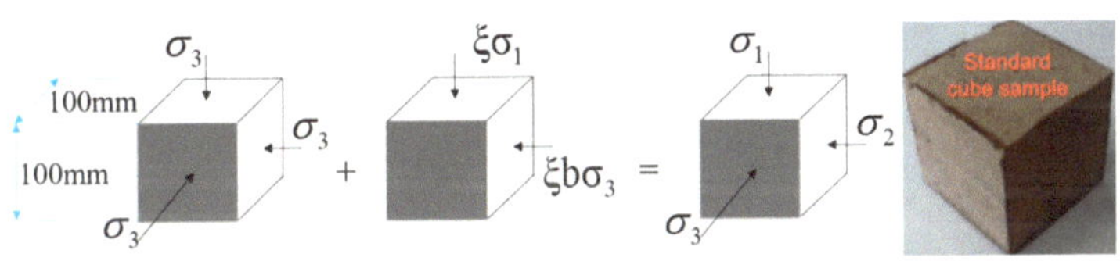

Figure 1. Test loading method.

Loading mode: For the test, we adopted a ZSZ-2000 microcomputer-controlled true triaxial frozen soil test platform. Figure 2 shows its loading device. The frozen soil triaxial test platform could realize independent sample loading in the three principal stress directions. The sample could be tested along different stress paths in the 3D stress space. The sample was loaded along the three principal stress directions using rigid plates. As shown in the three views in Figure 3, the F6 direction was the fixed end, which gave the reaction force in the F2 direction. Loading in the five directions, F1 to F5 was done independently by the servo cylinder through the force transmission rod. F3 and F5 were the loading directions of the first principal stress; F1 and F4 were the loading directions of the second principal stress; F2 and F6 were the third principal stress loading directions. Six steel plates were placed in a staggered overlap manner to realize the compression process of the sample. This test adopted a loading method with a constant principal stress ratio b in the control, where $b = (\sigma_2 - \sigma_3)/(\sigma_1 - \sigma_3)$. The parameter σ_3 was a constant. By shifting the term and taking the derivative of this formula, we find that $\xi(\sigma_2) = b\xi(\sigma_1)$, and ξ were the loading rate in the direction of the first principal stress, and $\xi = 25\mathrm{N/s}$. The specific loading process of the test is shown in Figure 1. After the sample reached the required temperature for the test and stabilized for 30 min, a three-way synchronous pressure σ_3 was applied to place the specimen in a state of equal stress in the three directions. The specimen was then stabilized and consolidated until its deformation was unchanged. Different loading rates were subsequently set in the σ_1 and σ_2 directions. The ratio of the two loading rates was equal to b to realize the true triaxial test process while keeping b unchanged. The test failure threshold was when the deformation in the direction of the major principal strain reached 15 mm, that is $\varepsilon_1 = 15\%$, or when the test monitoring curve shows that the sample was evidently damaged, that is, when the strain in the first principal strain–first principal stress curve grew almost vertically, the corresponding strength was the failure strength.

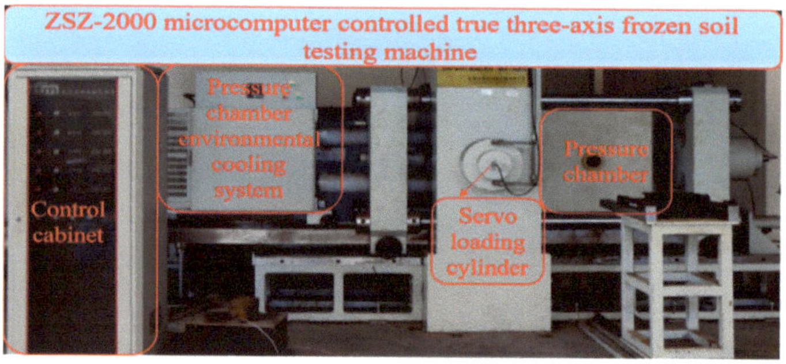

Figure 2. True triaxial test platform for frozen soil.

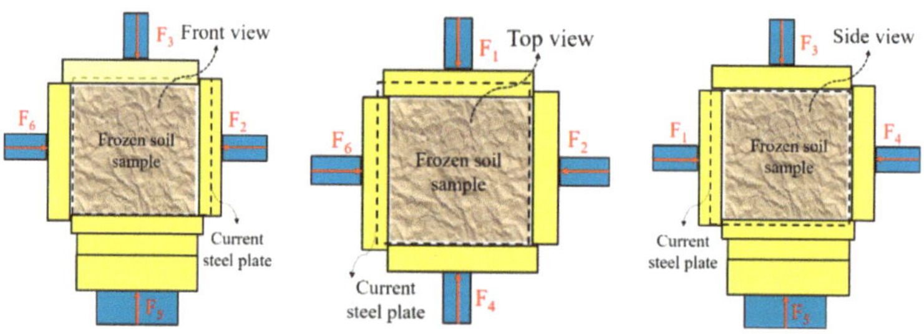

Figure 3. Three views of lapped rigid plates for frozen soil samples.

2.3. Test Results and Discussion

Considering that the test factors and loading environment of each sample of this true triaxial orthogonal test were different, there were many test curves; therefore, only some test equivalent stress–strain curves were selected for the analysis, as shown in Figure 4, where q was the von Mises stress, see Equation (1). The first principal strain ε_1 was the sum of the strains in the F3 and F5 directions; the volume strain ε_v was the sum of the strains in directions F1 to F6, and the sign indicated that the pressure was positive.

$$q = \sqrt{\frac{1}{2}\left[(\sigma_1 - \sigma_2)^2 + (\sigma_2 - \sigma_3)^2 + (\sigma_3 - \sigma_1)^2\right]} \tag{1}$$

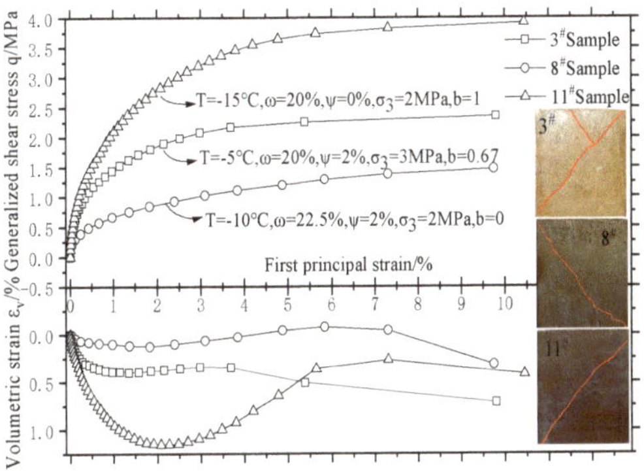

Figure 4. True triaxial stress–strain and volume deformation curves of frozen calcareous clay.

Figure 4 shows that the equivalent stress–strain relationship of the frozen calcareous clay was divided into three stages. The first was the linear elastic stage. Due to the high water content in the three selected samples, the samples were consolidated and frozen for a long duration. When the first principal strain was less than 0.5%, there was a certain instantaneous strength between the soil particles and cementation between the soil particles and ice crystals. Frozen soil samples exhibited good integrity and resistance to small deformations and rapid stress growth. The second was the plastic hardening stage. During this period, due to the increase in stress, the phenomenon of ice crystal pressure melting occurs. The soil and ice crystal particles, and pores reappeared in the frozen soil sample. The sample became denser, the bearing capacity was further improved, and plastic deformation from the first principal strain occurred, reaching approximately 5%. The third was the strain hardening failure stage. The occurrence of this stage depended on the test loading conditions. As mentioned above, a six-direction independent rigid loading mode was applied to control the load applied to the frozen soil samples. Frozen soil typically has certain residual stress after failure; if the frozen soil sample does not completely lose its bearing capacity or the test process was terminated, the force acting on the sample will continue to increase. As shown in Figure 4, the reason for the gradual growth in the deviator stress in the later stage was the rapid growth and expansion of the cracks in the sample under the action of the high stress applied (close to the ultimate bearing capacity). The samples expanded first and then shrank during the entire loading process. The expansion increased the gap between the rigid plates. When the specimen was placed between the rigid plates became incompressible, it overflowed from the gap between the plates, resulting in large deformation and unloading of the frozen soil specimens. This was also the reason for the evident shrinkage of the body.

Table 3 presents the results of the true triaxial orthogonal test conducted on the saline calcareous clay. The range analysis of the data listed in Table 3 was performed. Tables 4 and 5 list the analysis results. From a comparison of the R-values in Table 5, the primary and secondary order of influence affecting the peak strength σ_1 of frozen calcareous clay was as follows: pressure σ_3, water content ω, temperature T, Bishop parameter b, and salt content ψ. The primary and secondary order affecting σ_2 is shown in Table 6: the Bishop parameter b, confining pressure σ_3, water content ω, temperature T, and salt content ψ. Comparing the value of L, factor C (that is, the salt content ψ) had a certain effect on the peak strength and second principal stress of frozen calcareous clay compared with the other four factors, but the degree of influence was weaker, and the four levels show a trend of first increasing and then decreasing. The Bishop parameter reflected the magnitude of the second principal stress. Therefore, the influence of this factor on the second principal stress was the most significant. This was self-evident; the sample was more compact, the existence of an external force increased the friction on the sliding surface, the cohesive force increased, and the peak strength increased. Existing research results show that [24,25] temperature was the main factor affecting the strength of frozen soil; this was because the moisture in the sample freezes in a low-temperature environment. The lower the temperature, the greater the volume percentage of the ice crystal cement. The lower the unfrozen water content, which increased the strength of the cementation between the soil particles, the higher the strength. However, the presence of salt changes the concentration of the aqueous solution in the sample, thereby changing its freezing temperature, and the water was difficult to freeze. Therefore, the greater the concentration of the salt solution, the greater the decrease in the peak intensity.

Table 3. Design and results of true triaxial orthogonal test conducted on saline calcareous clay [23].

Five-Factor and Four-Level ($L_{16}(4^5)$) Orthogonal Test on Saline Calcareous Clay							
Factor / Test Number	Temperature T/°C	Moisture Content ω/%	Salt Content ψ/%	Confining Pressure σ_3/MPa	Bishop Parameter b	Test Results	
						σ_{1max}/MPa	σ_{2max}/MPa
1	−5	15	0	1	0	3.789	1
2	−5	17.5	1	2	0.33	5.694	3.22
3	−5	20	2	3	0.67	5.35	4.58
4	−5	22.5	3	4	1	6.11	6.11
5	−10	15	1	3	1	10.209	10.209
6	−10	17.5	0	4	0.67	9.568	7.734
7	−10	20	3	1	0.33	3.064	1.683
8	−10	22.5	2	2	0	3.474	2
9	−15	15	2	4	0.33	10.396	6.102
10	−15	17.5	3	3	0	6.509	3
11	−15	20	0	2	1	5.914	5.914
12	−15	22.5	1	1	0.67	3.071	2.387
13	−20	15	3	2	0.67	7.94	5.981
14	−20	17.5	2	1	1	5.197	5.208
15	−20	20	1	4	0	8.411	4
16	−20	22.5	0	3	0.33	6.941	4.302

Table 4. Analysis of the degree of influence of various factors on the major principal stress σ_1 [23].

Factor Level		A	B	C	D	E	Optimal Level	Factor Priority
First principal stress σ_{1max}/MPa	L1	5.236	8.083	6.553	3.780	5.546	$A_4B_1C_2D_4E_4$	D > B > A > E > C
	L2	6.579	6.742	6.846	5.755	6.524		
	L3	6.473	5.685	6.104	7.252	6.482		
	L4	7.122	4.899	5.906	8.621	6.857		
	R	1.886	3.184	0.940	4.841	1.311		

Table 5. Analysis of the degree of influence of various factors on the intermediate principal stress σ_2 [23].

Factor Level		A	B	C	D	E	Optimal Level	Factor Priority
Second principal stress σ_{2max}/MPa	L1	3.728	5.823	4.737	2.570	2.500	$A_2B_1C_2D_4E_4$	E > D > B > A > C
	L2	5.406	4.790	4.954	4.279	3.827		
	L3	4.351	4.044	4.473	5.523	5.171		
	L4	4.873	3.700	4.194	5.987	6.860		
	R	1.678	2.123	0.760	3.417	4.360		

Table 6. Critical values of $F_{\Omega}(f_1,f_2)$ for F verification [26].

Significance Level Ω	0.1	0.05	0.01	0.005
$f_1 = 3, f_2 = 12$	2.606	3.490	5.953	7.230
$f_1 = 3, f_2 = 19$	2.397	3.13	5.010	\
$f_1 = 9, f_2 = 19$	1.984	2.42	3.523	\

To systematically study the relationship between the peak stress σ_1 and the intermediate principal stress σ_2 of frozen saline calcareous clay and the confining pressure σ_3, water content ω, temperature T, Bishop parameter b, and salt content ψ, according to the dimension analysis theory, an exponential function was used to compare the test data in Table 4. The calculation formula could be obtained by fitting in Origin as follows:

$$\sigma_1 = 86.6208T^{0.1975} \cdot \omega^{-1.23085} \cdot \psi^{-0.00776} \cdot \sigma_3^{0.59654} \cdot b^{0.02066}, R^2 = 0.966 \quad (2)$$

To more intuitively reflect the significant effects of the confining pressure and water content on the peak stress and the second principal stress in the range analysis, the confining pressure σ_3 and water content ω were drawn for a temperature $T = -10$ °C and salt content $\psi = 1\%$. Figures 5 and 6 show the 3D trend charts of the peak stress σ_1 and second principal stress σ_2 under different b values, respectively.

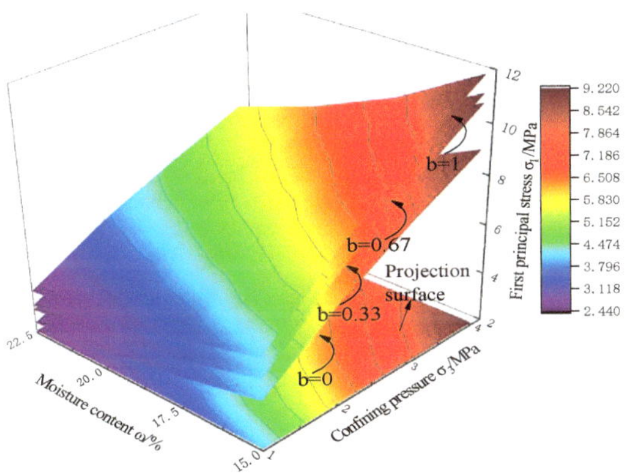

Figure 5. Three-dimensional trend chart of water content, confining pressure, and high principal stress under different b values [23].

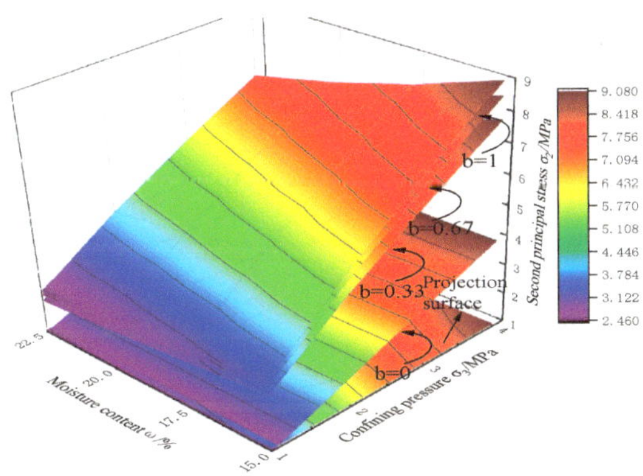

Figure 6. Three-dimensional trend chart of water content, confining pressure, and intermediate principal stress under different b values [23].

Figure 5 shows that as the value of b increases, the slope of the curved surface becomes steeper, indicating that an increase in b directly increased the stress in the σ_2 direction, the average principal stress increased, and the frozen soil was compacted. Although the second principal stress also increased, it accelerated the melting effect of ice crystals in the frozen soil and increased the restraint and occlusion of soil particles. Therefore, the strength significantly increased when $b = 0$ or $b \neq 0$, while the peak stress between b = 0.33 and 1 gradually decreased. This was because although the stress in the σ_2 direction increased, the occlusion between the soil particles was increased to a certain extent, and the strength was increased; however, the σ_3 direction constraint remained unchanged. The principal stress σ_1 significantly increased, and the strength of the sample decreased as the moisture content increased; the change in the moisture content cannot significantly affect the first principal stress σ_1. For example, when the confining pressure was 1 MPa, the first principal stress was 4.895 MPa for 15% water content, and the first principal stress was 2.972 MPa for 22.5% water content, which was reduced by 1.923 MPa, and the rate of change was 39.3%. The confining pressure had a significant influence on the first principal stress σ_1. For example, when the water content was 15%, the first principal stress with a confining pressure of 1 MPa was 4.895 MPa, and the first principal stress with a confining pressure of 4 MPa was as high as 11.193 MPa, which was an increase of 6.298 MPa, and the rate of change was 56.3%.

The 3D trend chart of σ_2 under different b values, shown in Figure 6, was the same as that of the peak stress σ_1 because σ_2 and σ_1 were dependent on each other. When the stress in the σ_1 direction reached the peak, the sample was broken, and there was no room for growth in σ_2. As the confining pressure increased, the sample was sufficiently dense, and it was difficult for dilatancy to occur. The restraint and compaction effect of σ_2 became insignificant; therefore, the distance between the curved surfaces became smaller as the value of b increased; the water content increased under freezing conditions. The most intuitive effect of the enhancement in the peak stress of the soil was the volume content of the ice crystals in the frozen soil and the cementation of the ice crystals on the soil particles. However, the increase in the confining pressure also caused compression and melting and weakened the contribution of the increase in the water content to the increase in the peak stress. Therefore, the confining pressure was the main significant factor for the increase in the peak stress and the second principal stress, whereas, for the stress in the σ_2 direction, the b value was the most significant factor. For example, when the confining pressure was 4 MPa, the Bishop parameter b increased from 0 to 1, and the corresponding second principal stress increased from 4 to 9.128 MPa, which was an evident increase.

3. Levels of Analysis

The range analysis method was conducted to obtain the order of priority among the influencing factors without considering the interaction of the various factors. However, it was not possible to obtain the influence of the factors and their levels on the test results' influence weight. Therefore, the analytic hierarchy process was introduced to explore the weights of the above five factors on the strength of the frozen calcareous clay.

3.1. Hierarchical Analysis Theory

Similar to Table 2, considering its more general situation, we supposed that there were k influencing factors, namely $A^{(1)}, A^{(2)}, \ldots, A^{(k)}$, and the number of levels of each factor was $n_1, n_2, \ldots, n_k$. As shown in Figure 7, the AHP model of the orthogonal experiment had a three-layer structure comprising the index, factor, and level layers.

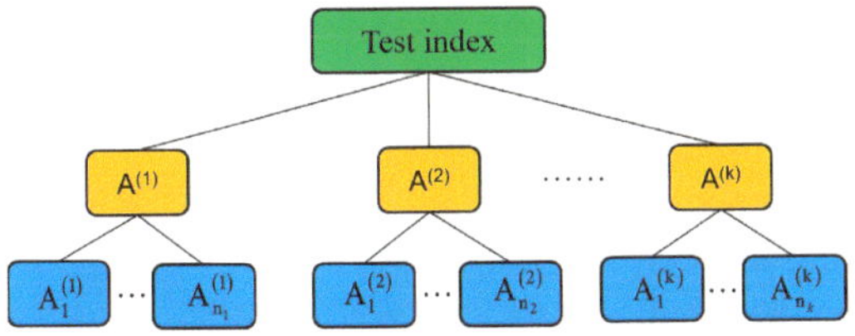

Figure 7. Orthogonal test AHP model.

Therefore, the sum of the test data under the j-th level of the defined factors were called the influence effect of the j-th level of the factor on the test, where $i = 1, 2 \ldots, k$; $j = 1, 2 \ldots, n_k$. If the value of the inspection test index was high, let $M_{ij} = K_{ij}$; otherwise, let $M_{ij} = 1/K_{ij}$. Among them, the effect matrix of the horizontal layer on the experiment is shown in Equation (3), and S is expressed in Equation (4).

$$A = \begin{pmatrix} M_{11} & 0 & \cdots & 0 \\ M_{21} & 0 & \cdots & 0 \\ \cdots & \cdots & \cdots & \cdots \\ M_{n_1 1} & 0 & \cdots & 0 \\ 0 & M_{12} & \cdots & 0 \\ 0 & M_{22} & \cdots & 0 \\ \cdots & \cdots & \cdots & \cdots \\ 0 & M_{n_2 2} & \cdots & 0 \\ \cdots & \cdots & \cdots & \cdots \\ 0 & 0 & \cdots & M_{1k} \\ 0 & 0 & \cdots & M_{2k} \\ \cdots & \cdots & \cdots & \cdots \\ 0 & 0 & \cdots & M_{n_k k} \end{pmatrix} \quad (3)$$

$$S = \begin{pmatrix} 1/t_1 & 0 & \cdots & 0 \\ 0 & 1/t_2 & \cdots & 0 \\ \cdots & \cdots & \cdots & \cdots \\ 0 & 0 & \cdots & 1/t_k \end{pmatrix} \quad (4)$$

where $t_j = \sum_{i=1}^{n_j} M_{ij} \quad (j = 1, 2, \cdots, k)$.

Therefore, matrix A right multiplied by matrix S was the normalization process of each column of matrix A, and matrix AS was called the level standard influence effect matrix. The range of the factors was R_i ($i = 1, 2, \ldots, m$), which was called the influence of factors on the experiment. The weight matrix of the influence of factors on the experiment is expressed in Equation (5):

$$C = \begin{pmatrix} \frac{R_1}{\sum_{i=1}^{m} R_i} & \frac{R_2}{\sum_{i=1}^{m} R_i} & \cdots & \frac{R_m}{\sum_{i=1}^{m} R_i} \end{pmatrix} \quad (5)$$

Finally, the influence weight of each factor level on the index $\varpi = ASC^T$ was obtained by the analytic hierarchy process. ϖ was ($n^{1+} n^{2+} \ldots n^{k+}$) $\times$ 1 vector, which indicated the influence of each level of the factor on the experiment.

3.2. Discussion on the AHP Result

The above method yields the following:

$$C = \begin{pmatrix} 0.155 & 0.262 & 0.077 & 0.398 & 0.108 \end{pmatrix} \quad (6)$$

$$S = \begin{pmatrix} 0.039 & 0 & \cdots & 0 \\ 0 & 0.039 & \cdots & 0 \\ \cdots & \cdots & \cdots & \cdots \\ 0 & 0 & \cdots & 0.039 \end{pmatrix}_{5\times5} \quad (7)$$

$$A = \begin{pmatrix} 5.236 & 0 & 0 & 0 & 0 \\ 6.579 & 0 & 0 & 0 & 0 \\ 6.473 & 0 & 0 & 0 & 0 \\ 7.122 & 0 & 0 & 0 & 0 \\ 0 & 8.084 & 0 & 0 & 0 \\ 0 & 6.742 & 0 & 0 & 0 \\ 0 & 5.685 & 0 & 0 & 0 \\ 0 & 4.899 & 0 & 0 & 0 \\ 0 & 0 & 6.533 & 0 & 0 \\ 0 & 0 & 6.846 & 0 & 0 \\ 0 & 00 & 6.104 & 0 & 0 \\ 0 & 0 & 5.906 & 0 & 0 \\ 0 & 0 & 0 & 3.780 & 0 \\ 0 & 0 & 0 & 5.756 & 0 \\ 0 & 0 & 0 & 7.252 & 0 \\ 0 & 0 & 0 & 8.621 & 0 \\ 0 & 0 & 0 & 0 & 5.546 \\ 0 & 0 & 0 & 0 & 6.524 \\ 0 & 0 & 0 & 0 & 6.482 \\ 0 & 0 & 0 & 0 & 6.856 \end{pmatrix} \quad (8)$$

Therefore, the degree of influence of the various factors on the peak intensity σ_1 was as follows:

$$\varpi_1 = ASC^T = \begin{pmatrix} 5.236 & 0 & 0 & 0 & 0 \\ 6.579 & 0 & 0 & 0 & 0 \\ 6.473 & 0 & 0 & 0 & 0 \\ 7.122 & 0 & 0 & 0 & 0 \\ 0 & 8.084 & 0 & 0 & 0 \\ 0 & 6.742 & 0 & 0 & 0 \\ 0 & 5.685 & 0 & 0 & 0 \\ 0 & 4.899 & 0 & 0 & 0 \\ 0 & 0 & 6.533 & 0 & 0 \\ 0 & 0 & 6.846 & 0 & 0 \\ 0 & 00 & 6.104 & 0 & 0 \\ 0 & 0 & 5.906 & 0 & 0 \\ 0 & 0 & 0 & 3.780 & 0 \\ 0 & 0 & 0 & 5.756 & 0 \\ 0 & 0 & 0 & 7.252 & 0 \\ 0 & 0 & 0 & 8.621 & 0 \\ 0 & 0 & 0 & 0 & 5.546 \\ 0 & 0 & 0 & 0 & 6.524 \\ 0 & 0 & 0 & 0 & 6.482 \\ 0 & 0 & 0 & 0 & 6.856 \end{pmatrix} \begin{pmatrix} 0.039 & 0 & \cdots & 0 \\ 0 & 0.039 & \cdots & 0 \\ \cdots & \cdots & \cdots & \cdots \\ 0 & 0 & \cdots & 0.039 \end{pmatrix}_{5\times5} \begin{pmatrix} 0.155 \\ 0.262 \\ 0.077 \\ 0.389 \\ 0.108 \end{pmatrix} = \begin{pmatrix} 0.0317 \\ 0.0389 \\ 0.0391 \\ 0.0431 \\ 0.0826 \\ 0.0689 \\ 0.0581 \\ 0.0501 \\ 0.0197 \\ 0.0206 \\ 0.0183 \\ 0.0177 \\ 0.0587 \\ 0.0893 \\ 0.1126 \\ 0.1338 \\ 0.0234 \\ 0.0275 \\ 0.0289 \\ 0.0273 \end{pmatrix} \quad (9)$$

$$\varpi_1 = ASC^T = \begin{pmatrix} 3.728 & 0 & 0 & 0 & 0 \\ 5.406 & 0 & 0 & 0 & 0 \\ 4.351 & 0 & 0 & 0 & 0 \\ 4.873 & 0 & 0 & 0 & 0 \\ 0 & 5.823 & 0 & 0 & 0 \\ 0 & 4.790 & 0 & 0 & 0 \\ 0 & 4.044 & 0 & 0 & 0 \\ 0 & 3.700 & 0 & 0 & 0 \\ 0 & 0 & 4.737 & 0 & 0 \\ 0 & 0 & 4.954 & 0 & 0 \\ 0 & 00 & 4.473 & 0 & 0 \\ 0 & 0 & 4.194 & 0 & 0 \\ 0 & 0 & 0 & 2.570 & 0 \\ 0 & 0 & 0 & 4.279 & 0 \\ 0 & 0 & 0 & 5.523 & 0 \\ 0 & 0 & 0 & 4.194 & 0 \\ 0 & 0 & 0 & 0 & 2.500 \\ 0 & 0 & 0 & 0 & 3.827 \\ 0 & 0 & 0 & 0 & 5.171 \\ 0 & 0 & 0 & 0 & 6.860 \end{pmatrix} \begin{pmatrix} 0.054 & 0 & \cdots & 0 \\ 0 & 0.054 & \cdots & 0 \\ \cdots & \cdots & \cdots & \cdots \\ 0 & 0 & \cdots & 0.054 \end{pmatrix}_{5\times 5} \begin{pmatrix} 0.136 \\ 0.172 \\ 0.062 \\ 0.277 \\ 0.354 \end{pmatrix} = \begin{pmatrix} 0.0274 \\ 0.0397 \\ 0.0320 \\ 0.0358 \\ 0.0541 \\ 0.0445 \\ 0.0376 \\ 0.0344 \\ 0.0159 \\ 0.0166 \\ 0.0150 \\ 0.0140 \\ 0.0384 \\ 0.0640 \\ 0.0826 \\ 0.0896 \\ 0.4460 \\ 0.6828 \\ 0.9226 \\ 1.2239 \end{pmatrix} \tag{10}$$

From the calculation result obtained using Equation (9) without considering the interaction, the four levels of each factor have the highest influence weight on the peak stress σ_1: factor A was A_4 = 0.0431; factor B was B_1 = 0.0826; factor C was C_2 = 0.0206; factor D was D_4 = 0.1338; and factor E was E_4 = 0.0289. Thus, the optimal level of the experiment was $A_4B_1C_2D_4E_4$, and the primary and secondary order of influence of each factor was determined by the sum of the weights at each level of the five factors. The order of the primary and secondary influence of each factor was D > B > A > E > C. Similarly, the degree of influence of each factor on the intermediate principal stress σ_2 was obtained. As expressed in Equation (10), the weights with the highest influence were A_2 = 0.397; B_1 = 0.0541; C_2 = 0.0166; D_4 = 0.0896; and E_4 = 1.2239. The optimal level of σ_2 was E > D > B > A > C. Both these results were consistent with the range analysis results. Moreover, the influence weight of each level of the factor on the peak intensity was obtained through the analytic hierarchy process.

4. Significance and Interaction Analysis

4.1. Significance Analysis Method

The magnitude of the variation was characterized by the sum of the squares of the variation, which was defined as:

$$S_i = \sum_{i=1}^{n} (x_i - \tilde{x})^2 \tag{11}$$

where, $\tilde{x} = \frac{1}{n}\sum\limits_{i=1}^{n} x_i$.

The estimated value of the deviation SS_i was expressed as:

$$SS_i = Q_i \Big/ f_i \tag{12}$$

where f was the degree of freedom, and $f_T = N - 1$, N was the total number of trials, $f_i = k - 1$, k was the level of factor i, $f_e = f_T - f_i$.

Based on the above deviation definition method, the calculation process of the significance level of the five-factor four-level orthogonal test reported in this paper was as follows:

The sum of the squared deviations S_T was:

$$S_T = \sum_{i=1}^{16} x_i^2 - \frac{1}{16}\left(\sum_{i=1}^{16} x_i\right)^2 \quad (13)$$

Let $\eta = \frac{1}{16}\left(\sum_{i=1}^{16} x_i\right)^2$. From the data listed in Table 3, we obtained η= 645.63; then, the sum of the squared deviations of each factor $S_A, S_B, \ldots, S_E$ was expressed as:

$$\begin{aligned} S_A &= \tfrac{1}{4}\sum_{i=1}^{4} T_{Ai}^2 - \tfrac{1}{16}\left(\sum_{i=1}^{16} x_i\right)^2 \\ S_B &= \tfrac{1}{4}\sum_{i=1}^{4} T_{Bi}^2 - \tfrac{1}{16}\left(\sum_{i=1}^{16} x_i\right)^2 \\ &\cdots\cdots \\ S_E &= \tfrac{1}{4}\sum_{i=1}^{4} T_{Ei}^2 - \tfrac{1}{16}\left(\sum_{i=1}^{16} x_i\right)^2 \end{aligned} \quad (14)$$

In the formula: $T_{Ai}, T_{Bi}, \cdots, T_{Ei}$ were the sum of the test values of each factor at a certain level, e.g., T_{Ai} was the sum of the four groups of test values at the level of factor 1.

The sum of squares of the error effect deviation S_e was:

$$S_e = S_T - (S_A + S_B + \cdots + S_E) \quad (15)$$

The sum of the squared deviations of the error effect reflected the fluctuation value of the test error in this set of data. From Equations (13)–(15), the sum of the squared deviations corresponding to each factor was obtained. For example, the calculation process of S_A was:

$$\begin{aligned} T_{A1} &= (3.789 + 5.694 + 5.35 + 6.11) = 20.943 \\ S_A &= (20.9432 + 26.3152 + 25.892 + 28.4892)/4 - 645.63 = 7.621 \end{aligned} \quad (16)$$

Table 7 presents the rest of the calculation results. Referring to the corresponding degrees of freedom of the factors according to Equation (12), the estimated value of the deviation SS_i of each factor was determined.

Table 7. Significance test on influencing factors of peak intensity σ_1.

Factor	Sum of Squares	Degrees of Freedom	F Ratio	Significance
A	7.621	3	3.494	III
B	22.826	3	10.466	II
C	2.181	3	1.000	V
D	51.719	3	23.713	I
E	3.808	3	1.746	IV
Error	2.18	12	0.01	\
Sum	88.152	15	\	\

Considering the ratio of deviation estimates:

$$F_i = \frac{SS_i}{SS_e} = \frac{S_i/f_i}{S_e/f_e} \quad (17)$$

For a given significance level Ω, if

$$F_i \geq F_\Omega(f_i, \ \ f_e) \quad (18)$$

The influence of factor i on the test index was significant; otherwise, it was not significant.

4.2. Significance Analysis without Considering Interaction

The test data listed in Table 3 were calculated and tested using the above calculation method, and the critical values of the F test listed in Table 6 were compared. Table 7 lists the test results.

The results of the significance study of the peak stress σ_1 showed that under complex stress conditions (under true triaxial conditions), the confining pressure and water content had significant effects on the strength, and the temperature effect was significant when $\Omega > 0.05$. The influences of the salt content and b value on the intensity were not very significant. The order of significance of each factor was consistent with the results of the range analysis method and analytic hierarchy process.

4.3. Significance Analysis Considering Interaction

In the real state, the influence of frozen soil strength was not the result of single factors that were independently affected and then simply superimposed. The interactive influence of various factors could not be ignored. For example, the water content was the factor determining the volumetric ice content of frozen soil. However, an increase in the salt content reduced the freezing temperature of the water in the soil, making it difficult to freeze, and the volumetric ice content naturally decreased, thereby reducing the strength. At the same time, the application of a confining pressure caused the compression–thaw effect in the frozen soil, and the ice crystal cementation and melting affected the intensity. There were complex interactions between the various factors affecting the intensity. The above analysis results that did not consider the interaction between the factors could only be qualitatively analyzed to a certain extent. Hence, it was necessary to make the following significance analysis considering these interactions.

According to the basic idea of the orthogonal experiment design, the $L_{32}(4^9)$ orthogonal experiment table was used to design the experiment. The orthogonal experiment table head design was implemented in strict accordance with the experimental procedures. Due to many factors, the table head was too large, and the table head design is not presented herein. This study only considered the interaction between factors A, B, and C. Some of the 32 groups of effective tests were consistent with the test conditions listed in Table 3, and the test results could be used. Most of the remaining test results were calculated from Equation (2). The data obtained were analyzed for significance using the method described in Section 4.1. Table 8 presents the results.

Table 8. Significance test of factors affecting the peak intensity σ_1 when the interaction was considered.

Factor	Sum of Squares	Degrees of Freedom	F Ratio	Significance
A	5.286	3	2.259	V
B	46.950	3	20.064	II
A × B	8.259	9	3.529	IV
C	2.340	3	1.000	VIII
A × C	4.606	9	1.968	VI
B × C	2.547	9	1.088	VII
D	81.744	3	34.933	I
E	18.564	3	7.933	III
Error	2.34	19	0.01	\
Sum	173.296	31	\	\

The test results showed that for the significance levels listed in Table 6, the effects of the confining pressure, water content, and intermediate principal stress coefficient b on the peak stress were all significant; the interactions A × B, A × C, and B × C had different degrees of influence on σ_1, and the significance of the three interactions exceeded the degree of influence of the salt content. When $\Omega > 0.01$, the influence of A × B on the equivalent stress was significant, and its influence even exceeded that of the temperature. From the results listed in Table 8, we concluded that the order of significance of the influence of various factors on the peak stress was as follows: confining pressure D, water content B, intermediate principal stress coefficient *b*, interaction A × B, temperature A, interaction A × C, interaction B × C, and salt content C. Hence, when studying the strength of frozen soil, it was necessary to comprehensively consider the various factors and their interactions to more accurately characterize the mechanical behavior of frozen soil.

5. Conclusions

In this study, a true triaxial orthogonal test was conducted on frozen calcareous clay to explore its strength change characteristics under the influence of multiple factors, including the degree of influence of each factor. The following conclusions were drawn from the results:

(1) The characteristics of the stress–strain curve of frozen calcareous clay could be divided into three stages: the strain was within 0.5%, showing linear elasticity, and under the effect of pressure, the ice crystals, and their cements were damaged or compressed, and they shrink; at approximately 5%, they showed plastic hardening. The soil particles and ice crystals in the frozen soil recombined and became denser, resulting in irreversible deformation. With the compression process, cracks bred and expanded. The failure stage was manifested as strain hardening due to the test loading conditions. As the deformation increased, the stress also slightly increased. In this stage, the soil sample was squeezed out, and the volume deformation could not represent its true change.

(2) Through analytic hierarchy process, the peak stress σ_1 of each factor under four levels was quantitatively calculated, and the maximum value of influence weight of σ_1 was A_4 = 0.0431; B_1 = 0.0826; C_2 = 0.0206; and D_4 = 0.1338; E_4 = 0.0289, the maximum and minimum weight was 0.1338 and 0.0197, respectively, and the optimal level was $A_4B_1C_2D_4E_4$. By comparing the sum of the weights of the four levels of each factor, the order of primary and secondary influence consistent with the range method was obtained, which was D > B > A > E > C. Similarly, the maximum and minimum weight of σ_2 was 1.2239 and 0.014, respectively, and the order of σ_2 was E > D >B > A > C.

(3) When the interaction was not considered under true triaxial conditions, the effects of the confining pressure and water content on the strength were always significant. The temperature effect was significant only when the significance level $\Omega > 0.05$. The salinity and *b* value had a significant effect on the strength. The impact was not significant, and the ranking of the significance of each factor was consistent with the results of the range analysis method and the analytic hierarchy process. When considering the interaction, the interaction factors had different effects on the intensity. When $\Omega > 0.01$, A × B was equivalent, and the effect of stress was significant, even exceeding that of the temperature. When studying the strength of frozen soil, it was necessary to comprehensively consider the various factors and the interaction between them, to more accurately characterize the mechanical behavior of frozen soil.

Author Contributions: Conceptualization, Z.W. and C.-X.R.; methodology, Z.W. and C.-X.R.; software, Z.W. and Y.C.; validation, C.-X.R., Z.W. and Y.C.; formal analysis, Z.W.; investigation, Q.Y.; resources, Z.W.; data curation, W.L.; writing—original draft preparation, Z.W.; writing—review and editing, Z.W. and W.L.; visualization, Q.Y.; supervision, C.-X.R.; project administration, Z.W.; funding acquisition, C.-X.R. All authors have read and agreed to the published version of the manuscript.

Funding: This research was funded by the National Natural Science Foundation of China (grant numbers 51878005 and 51374010); and the Engineering Research Center of the Ministry of Education of Anhui University of Science and Technology in 2020 (grant number: JYBGCZX2020209).

Institutional Review Board Statement: Not applicable.

Informed Consent Statement: Not applicable.

Data Availability Statement: Not applicable.

Acknowledgments: This work was supported by Grants-in-Aid for scientific research from the National Natural Science Foundation of China (51878005, 51374010), Engineering Research Center of the Ministry of Education of Anhui University of Science and Technology in 2020 (JYBGCZX2020209).

Conflicts of Interest: The authors declare no conflict of interest.

References

1. Lu, L.N.; Fan, H.H.; Chen, H.; Ma, D.L.; Wang, Z.N.; He, Z.Q. Experimental study on influencing factors of uniaxial tensile strength of dispersive soil. *Chin. J. Geotech. Eng.* **2014**, *36*, 1160–1166.
2. Lv, H.B.; Zeng, Z.T.; Ge, R.L.; Zhao, Y.L. Experimental study on tensile strength of swelling and shrinking soil. *Rock Soil Mech.* **2013**, *34*, 615–620, 638.
3. Liu, B.; Sun, Y.D.; Yuan, Y.F.; Liu, X.C.; Bai, X.T.; Fang, T.J. Strength characteristics and strengthening mechanism of frozen sandstone with different water content. *J. China Univ. Min. Technol.* **2020**, *49*, 1085–1093, 1127.
4. Chen, L.H. Correlation and non-correlation regression statistics of probability characteristics of shear strength. *J. Geotech. Eng.* **2013**, *35*, 1397–1402.
5. Chen, C.L.; Zhang, H.M. Some Problems in Triaxial Test of Saturated Sand. *Chin. J. Geotech. Eng.* **2000**, *22*, 659–663.
6. Gen, H.H.; Wu, J.Y.; Shan, R.L.; Huang, B. Study on the Suction Contribution Form of Unsaturated Soil Strength. *Chin. J. Geotech. Eng.* **2012**, *34*, 1598–1603.
7. Li, K.Q.; Li, D.Q.; Liu, Y. Meso-scale investigations on the effective thermal conductivity of multi-phase materials using the finite element method. *Int. J. Heat Mass Transf.* **2020**, *151*, 119383. [CrossRef]
8. Li, K.Q.; Miao, Z.; Li, D.Q.; Liu, Y. Effect of mesoscale internal structure on effective thermal conductivity of anisotropic geomaterials. *Acta Geotech.* **2022**, *17*, 3553–3566. [CrossRef]
9. Cai, Z.Y.; Wu, Z.Q.; Huang, Y.H.; Cao, Y.Y.; Wei, Y.B. Geotechnical Engineering Department, Influence of water and salt contents on strength of frozen soils. *Chin. J. Geotech. Eng.* **2014**, *36*, 1580–1586.
10. Niu, Y.Q.; Lai, Y.M.; Wang, X.; Liao, M.K.; Gao, J. Study on the influence of initial moisture content on the deformation and strength of frozen silty clay. *Rock Soil Mech.* **2016**, *37*, 499–506.
11. Lai, Y.M.; Zhang, Y.; Zhang, S.J.; Jin, L.; Chang, X.X. The influence of supersaturated water content and temperature on the strength of frozen sand. *Rock Soil Mech.* **2009**, *30*, 3665–3670.
12. Du, H.M.; Ma, W.; Zhang, S.; Zhou, Z.W. Study on the influence characteristics of confining pressure and water content on the failure strain energy density of frozen sand. *Rock Soil Mech.* **2017**, *38*, 1943–1950.
13. Xiao, H. The relationship between the uniaxial compressive strength of artificial frozen soil and temperature and water content. *Geotech. Eng.* **2008**, *11*, 62–63, 76.
14. Yin, Z.Z.; Chen, Y.L.; Wang, P. Uniaxial unconfined compressive strength test on artificially frozen clay in Shanghai. *Rock Soil Mech.* **2012**, *33*, 788–792.
15. Sun, L.Q.; Lu, J.X.; Li, H.; Yan, S.W.; Jia, X.; Han, S.Z. Experimental study on the influence of water content and salt content on the strength characteristics of artificial frozen soil. *Chin. J. Geotech. Eng.* **2015**, *37*, 27–31.
16. Ge, X.X.; Yang, Z.H.; Still, B. Mechanical properties of naturally frozen silty soil for seismic design of pile foundations.10th International Symposium on Cold Regions Development. *Reston: Am. Soc. Civ. Eng.* **2013**, *32*, 215–227.
17. Ma, X.J.; Zhang, J.M.; Chang, X.X.; Zheng, B.; Zhang, M.Y. Experimental research on strength of warm and ice-rich frozen clays. *Rock Soil Mech.* **2008**, *229*, 2498–2502.
18. Dai, J.Q.; Su, Z.J.; Zhao, M.C.; Xiang, Y.H. Study on true triaxial test and strength characteristics of silty clay. *Rock Soil Mech.* **2016**, *37*, 2534–2540, 2546.
19. Chen, C.L.; Shao, S.J.; Zhang, Z. Study on true triaxial test of artificially prepared structural loess. *Rock Soil Mech.* **2013**, *34*, 2231–2237.
20. Ma, S.L.; Yao, Z.M.; Liu, S.; Pan, X. Damage Constitutive Model of Frozen Sand under the Influence of Intermediate Principal Stress Coefficient. *Coal Field Geol. Explor.* **2020**, *48*, 130–136.
21. Zhang, M.; Xu, C.S.; Du, X.L.; Wang, G.S., Lu, D.C. Study on true triaxial test of the influence of intermediate principal stress coefficient and stress path on the shear characteristics of sand. *J. Hydraul. Eng.* **2015**, *46*, 1072–1079.
22. Wang, Z.; Rong, C.; Du, M.; Ma, M.; Liu, X. The development law of freezing temperature field of calcareous clay layer. *Math. Probl. Eng.* **2020**, *2022*, 3902054. [CrossRef]
23. Cao, Y.; Rong, C.; Wang, Z.; Shi, X.; Wang, B.; Long, W. Experimental Study on True Triaxial Mechanical Properties of Frozen Calcareous Clay under the Influence of Multiple Factors. *Crystals* **2022**, *12*, 328. [CrossRef]
24. Bai, R.; Xu, X.; Hua, S.; Wang, J. Significant analysis of the influence factors on strength of frozen soil base on multivariable linear regression model. *J. Glaciol. Geocryol.* **2019**, *41*, 416–423.
25. Chou, Y.L.; Jiang, X.G.; He, B.B.; Sheng, Y. Experimental study on mechanical properties of frozen loess based on structure. *J. Glaciol. Geocryol.* **2014**, *36*, 913–921.
26. Teng, S.Z.; Feng, J.H. *Mathematical Statistics*; Dalian University of Technology Press: Dalian, China, 2005.

applied sciences

MDPI

Article

Analysis of Freeze–Thaw Response and Pore Characteristics of Artificially Frozen Soft Soil under Combined Formation Seepage

Jie Zhou [1,2,*], Zhongqiu Guo [1], Chuanhe Wang [1], Zeyao Li [1], Huade Zhou [1] and Wansheng Pei [3]

[1] Department of Geotechnical Engineering, College of Civil Engineering, Tongji University, 1239 Siping Road, Shanghai 200092, China

[2] Key Laboratory of Geotechnical and Underground Engineering, Tongji University, Ministry of Education, 1239 Siping Road, Shanghai 200092, China

[3] State Key Laboratory of Frozen Soil Engineering, Northwest Institute of Eco-Environment and Resources, CAS, Lanzhou 730000, China

* Correspondence: zhoujie1001@tongji.edu.cn

Abstract: Artificial ground freezing (AGF) is a widely used method in coastal tunnel construction and reinforcement. With more and more underground construction in coastal areas, clay–sand combined formation, which is common in coastal areas, brings more challenges to AGF. In this paper, the frost–thaw characteristics of soft clay during AFG under the construction of combined formation seepage were studied by model test. It was found that the shape of the freezing curtain changed under the condition of seepage, and the water content of the upper soft soil layer decreased markedly after settlement. Subsequently the micro characteristics of melted soil by CT were also conducted for further mechanism analysis, and it was indicated that the distribution of CT number had obvious segmentation characteristics along the height. Finally, the 3D structure of melted clay was reconstructed, and a method was proposed to calculate freeze–thaw settlement through CT numbers.

Keywords: AGF; combined formation seepage; freeze–thaw settlement; CT numbers

1. Introduction

Artificial ground freezing (AGF) is widely used in urban rail transit construction. However, there are many problems in the use of AGF, such as uneven ground settlement, tunnel segment leakage, and so on. Sand clay combination strata are common in China's coastal areas, and the seepage in the sand layer aggravates the uncertainty of the problems above.

Sudisman [1] found that the existence of seepage delays the closure of freezing curtain. Xiao et al. [2] obtained that the symmetry of frozen walls decreased, and the formation time of the frozen wall increased under the condition of seepage from model test. Wang et al. [3] also reached a similar conclusion through model test. Jian et al. [4] used COMSOL to simulate the formation process of freezing wall under different seepage velocities and found that the freezing temperature field inhomogeneity increases as the initial stratum seepage velocity increases. The maximum thickness of the downstream frozen wall initially increases and then decreases. However, the average thickness of the frozen wall linearly decreases, and when the seepage velocity is greater than 7 m/d, the frozen wall can hardly be formed. Some scholars put forward the concept of critical seepage velocity; that is, when the seepage velocity is greater than this threshold, the frozen curtain does not close. M. Vitel et al. [5] proposed that when the seepage velocity is greater than 2 m/d, the stratum is difficult to be frozen according to their coupled thermo-hydraulic model. Su et al. [6] draw the conclusion that the frozen curtain does not close with seepage velocity up to 2.5 m/d by model test and Bakholdin's analytic solution of temperature field. Seepage also affects the formed frozen curtain. In sudden seepage, Song et al. [7] found that when the seepage velocity was 0–30 m/d, it caused obvious damage to the upper reaches of the frozen curtain, while the wall was strengthened in the middle and lower reaches to a small extent, but when the flow velocity was more than 30 m/d, the frozen curtain was completely damaged. Reasonable arrangement of freezing pipes increases the critical seepage velocity to adapt to more complex situations. After encrypting the freezing pipes in the range of 120 degrees upstream of the outer ring of the double-loop pipes, Dong et al. [8] shortened the freezing time and increased the thickness of frozen curtain. Liu et al. [9] found that when the freezing pipe layout coincides with the seepage direction, reducing the freezing pipe spacing cannot shorten the freezing time. Coolant properties are also important influencing factors. When the cooling medium is brine, the critical seepage velocity is 1–2 m/d [10]. AGF using liquid nitrogen as refrigeration circulating liquid can be applied when the seepage velocity is 20 m/d and 50 m/d [11].

As a non-destructive testing technology, computed tomography (CT) has been widely used in geotechnical field in recent years [12]. The change of micro-structure in different parts of soil can be measured by the statistical characteristics of CT number [13]. Konrad studied the ice thickness and spacing of different soil in different freezing time through X-ray test [14]. The ice thickness and spacing in frozen soil increase with the decrease of temperature, and the ice thickness and spacing of different types of soil are different. Zhang Shijian [15] reflected the damage of rock through CT image and CT number, studied the relationship between strength change and damage expansion of mudstone and water supply and loading and unloading process, and put forward the concept of damage rate. CT scanning can reduce the extraction of pore data by artificial intervention. Based on genetic algorithm, Liu et al. [16] determined the gray-level threshold of CT image segmentation and achieved the recognition of water, ice, and rock, which assured the digital expression of water and ice content. Lang et al. [17] used a median filtering algorithm to remove the noise of CT image and made it better reflect the complex pore structure in the rocks. More accurate and intuitive data can be obtained by improving the spatial and temporal resolution of CT images, improving image processing software and developing new image reconstruction algorithms [18]. Sun [19], using Avizo, an advanced three-dimensional visualization image processing software application, reconstructed the core by scanning method and made it digitalized and visualized. The reconstructed core mode can show the pore size, pore throat size and connection mode more intuitively, and at the same time, it can get the quantitative distribution state of porosity ratio, matrix skeleton structure ratio and other data. Luo et al. [20] quantified the three-dimensional pore network in soil column by using the improved method through Avizo version 5 software, including the continuous change of pore network along depth, macropore size distribution, network density, surface area, length density, length distribution, average hydraulic radius, curvature, inclination (angle), and connectivity (path number and node density). On this basis, Luo et al. [21] found that large porosity, path number, hydraulic radius, and large pore angle are the best predictors of hydraulic parameters of silty sand.

When the construction process of AGF is applied in the clay layer, the boundary conditions (such as temperature boundary and water replenishment condition) with seepage are more complex than the case where there is no seepage. In order to explore the response characteristics of soft clay in AGF construction under the presence of seepage, a model test was designed to obtains the micro characteristics of frozen and thawed soil by CT, and puts forward a settlement calculation method based on CT in this paper.

2. The Model Test Program

2.1. Engineering Background

The AGF project chosen was applied in the cross passage in the Satellite Hall of Pudong Airport. (Figure 1. Left) Freezing pipe crossed the sandy silt layer and mucky clay layer. The top row of freezing pipes was selected as the research object (Figure 1. Right). Table 1 shows the site construction parameters after generalization.

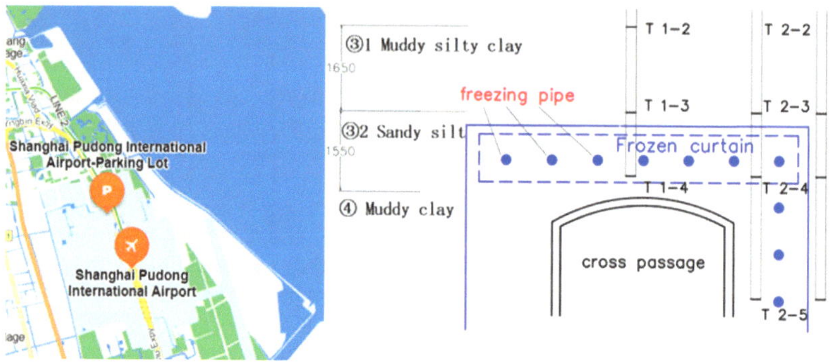

Figure 1. Prototype location and working condition.

Table 1. Simplified site construction parameters.

Prototype Parameters	Value
Buried depth of freezing pipe	8.8 m
Frozen curtain thickness	1.7 m
Freezing pipe length	7.3 m
Freezing pipe spacing	0.9 m
Number of freezing pipe	36
Total flow of freezing pipe	200 m^3/h
Single pipe flow of freezing pipe	5.56 m^3/h
Outer diameter of freezing pipe	89 mm
Wall thickness of freezing pipe	8 mm
Average brine temperature	−29 °C
Average temperature of frozen soil	−10 °C

The layer of mucky clay is widely distributed in Shanghai, which is crossed by many underground projects. Mucky clay has the characteristics of high-water content, low strength, and sensitivity to frost heave and thaw settlement, so it is an unfavorable layer which should be emphasized. When designing the model test, considering the most unfavorable factors, the freezing pipes were placed in the mucky clay. The basic physical and mechanical properties of mucky clay are shown in Table 2.

Table 2. The basic physical and mechanical properties of mucky clay.

Engineering Index	Mucky Clay
Water content, W(%)	48.1
Natural bulk density, $\gamma(kN/m^3)$	16.9
Specific gravity, G_s	2.75
Void ratio, e	1.363
Saturation, Sr(%)	97
Liquid limit, $W_L(\%)$	43.9
Plastic limit, $W_p(\%)$	23.1
Vertical permeability, K(cm/s)	2.19×10^{-7}

2.2. Model Test Design

The selection of model similarity ratio is very important in model test. Considering the actual conditions in the laboratory and the accuracy of the model test, the geometrical ratio of this model test chosen was C_L = 10. According to the above generalized prototype construction model and the selected geometric similarity constants, the similarity ratio of some basic physical quantities can be obtained (Table 3).

Table 3. Similarity ratio.

Physical Parameters	Dimension	Similarity Constant
Geometric length, l	L	$C_L = 10$
Density, ρ	ML^{-3}	$C_\rho = 1$
Displacement, δ	L	$C_\delta = C_\varepsilon C_L = 10$
Temperature, T	Θ	$C_T = 1$
Permeability coefficient, k	LT^{-1}	$C_k = \frac{C_K C_\rho C_g}{C_\mu} = 1$
Time, t	T	$C_t = C_L^2 = 100$

The artificial freezing model experimental system we used consists of three parts: a refrigeration system, model test box, and acquisition system. The schematic diagram of the experimental system is shown in Figure 2. The model test box is independently developed, and it can control the velocity of water in the underlying layer of clay (sand layer) at 1.2 m/d, the widespread seepage velocity in the engineering stratum of Shanghai.

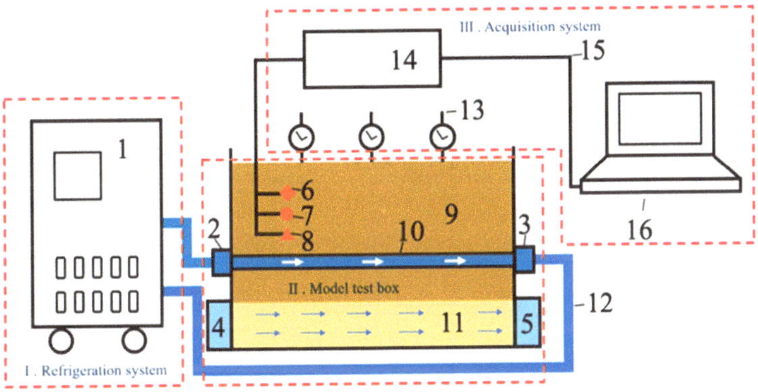

Figure 2. The schematic diagram of the whole experimental system. (1. Refrigerated circulation pump; 2. Refrigerant inlet; 3. Refrigerated outlet; 4. Upstream flume; 5. Downstream flume; 6. Temperature sensor; 7. Earth pressure sensor; 8. Pore pressure sensor; 9. The clay; 10. Freezing pipe; 11. Sand; 12. Refrigerant; 13. Dial indicator; 14. Acquisition instrument; 15. USB cable; 16. Information processing terminal).

2.3. Model Test Process

In the test process, we used remolded soil samples which was similar to the actual construction situation. The mucky clay used was taken from the foundation pit at the southeast corner of the intersection of Tiantong road and Fujian north road in Jing'an District, Shanghai, with a depth of 12.5 m. After laying the sand layer at a thickness of 20 cm at the bottom, the freezing pipes were set up on the upper part and fill the test box with clay. The freezing pipe is set 10 cm above the junction of sand and clay. The water content and degree of consolidation of soil samples are similar to those of undisturbed soil samples by remodeling and compaction.

After the preloading consolidation settlement was stable, we removed the load and started the freeze–thaw cycle. The seepage velocity of the lower sand layer was controlled through the inflow and outflow of water, and collected the data including settlement moisture content and temperature change through the sensor. In order to obtain the physical and mechanical parameters after freeze-thaw, after the test, samples were taken in layers in the model box to measure the moisture content and dry density at different depths. We also explored a method of taking large original soil sample in model box and using CT to analyze the structural characteristics of undisturbed samples.

After the model test, the undisturbed soil was taken out from the model box with a sampler and an iron cylinder, as shown in Figure 3a. The diameter of the iron bucket used for soil borrowing was designed 200 mm high and 127 mm wide. The bottom surface of the undisturbed soil taken out is shown in Figure 3b. The structure of the sample bottom surface is clear, which means this method can maintain the structure of the soil well. Four groups of soil samples were taken, and the sampling locations are shown in Figure 4 and Table 4. After that, the internal structures of soil samples were obtained through CT in the Cold and Arid Regions Environmental and Engineering Research Institute, which is shown in Figure 5.

(a) (b)

Figure 3. Sampling in model box. (**a**) Sampling and (**b**) sample bottom.

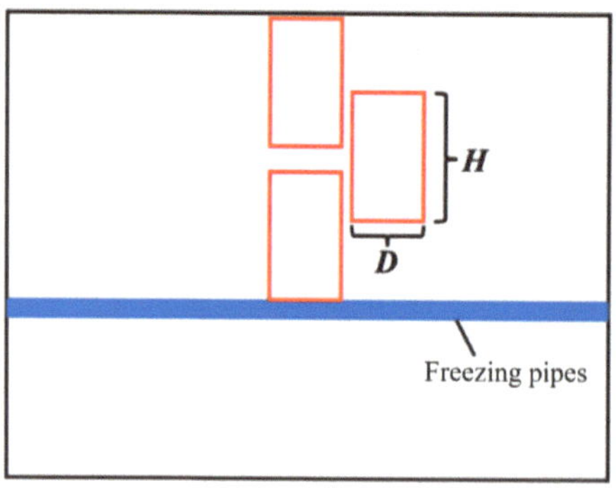

Figure 4. CT sampling diagram.

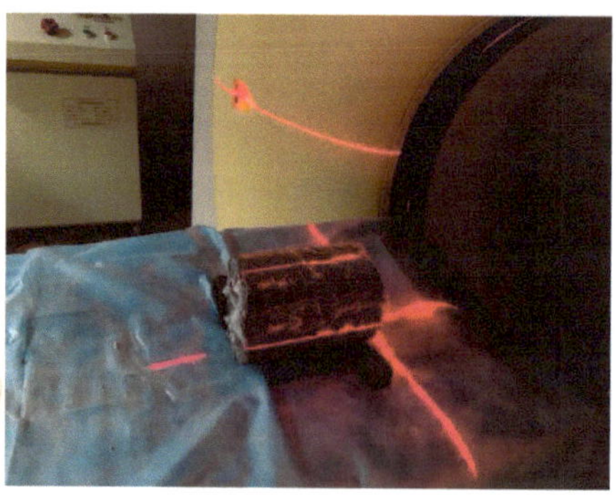

Figure 5. CT scan sample.

Table 4. Sampling location.

Number	Y1	Y2	Y3	Y4
Height (mm)	5–200	5–200	110–310	290–490

3. Result and Discussion

3.1. Change of the Temperature Field in Model Test

The original data of temperature field of model test are shown in Figure 6. According to the temperature data in the box, the temperature at each point decreases rapidly after freezing, and the temperature near the freezing pipe decreases most rapidly. After stopping freezing, the temperature of frozen soil increases rapidly in a short time, and there is an interval near 0 °C. When the temperature rises to this interval, the temperature rise rate slows down, which is roughly [−1.4, −0.6]. This is because in this interval, a large amount of ice changes into water and absorbs latent heat. This interval is the main phase change interval of the muddy clay in Shanghai.

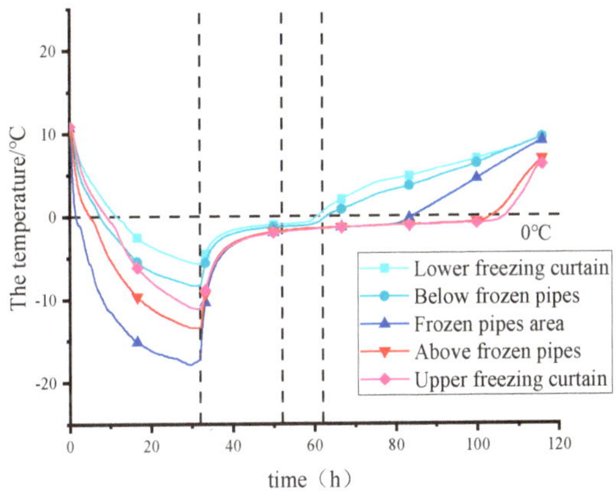

Figure 6. Temperature distribution during the model test.

In order to more intuitively show the distribution of the temperature field, the cloudy map of the freezing process is shown in Figure 7 and that of the melting process is shown in Figure 8.

According to the temperature cloudy map in the freezing process box, it can be seen that with the passage of freezing time, the expansion of the freezing front above the freezing pipe becomes slower and slower.

Due to the influence of seepage, the temperature field distribution in the freezing process is not symmetrical about the freezing pipe. At 32 h, the −10 °C isotherm above the freezing pipe is 8 cm away from the freezing pipe, while the −10 °C isotherm below the freezing pipe is only 5 cm away from the freezing pipe. There is strong convection under the freezing pipe, and the freezing curtain is thinner than that above the freezing pipe.

In the freezing process of combined formation with seepage, the heat transfer can be divided into two parts, including heat conduction in clay layer and heat convection at the junction of the clay layer and sand layer. During the freezing process, the temperature at the liquid inlet of the freezing pipe is 5 °C lower than that at the liquid outlet, which causes the temperature of the clay layer near the liquid inlet to be lower than that at the liquid outlet. However, the existence of seepage brings the cold energy of the upstream to the downstream, resulting in the downstream temperature lower than that of the upstream. It can be seen from the test data that under the seepage velocity of 1.2 m/d, the temperature cloudy map in the box shows a "bell mouth" opening into the upstream. The upstream temperature is lower than the downstream, which is particularly prominent at the beginning of freezing. This shows that when the seepage velocity is 1.2 m/d, the influence of heat conduction in clay layer on soil temperature is greater than that of heat convection.

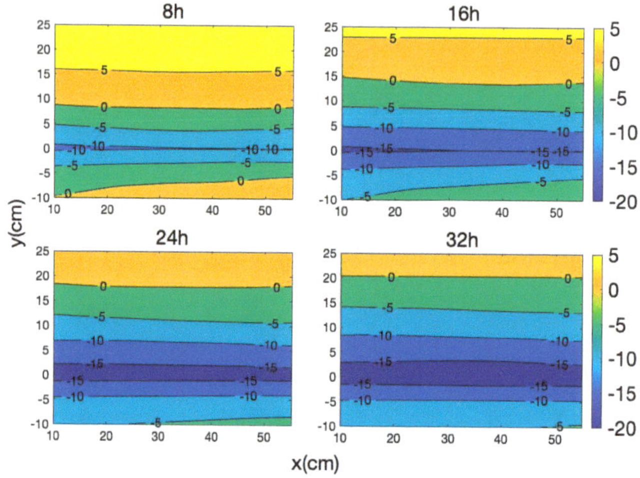

Figure 7. Temperature distribution nephogram during the freezing stage (**x** stand for the horizontal position and **y** stand for the distance from the freezing center).

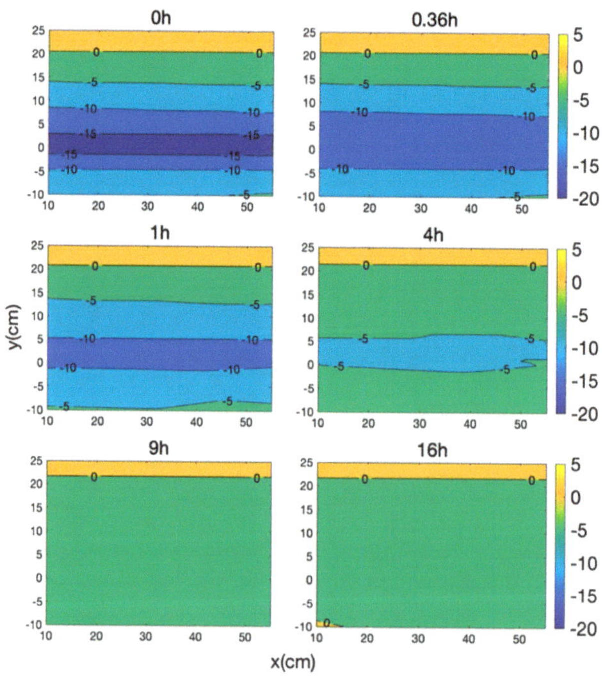

Figure 8. *Cont.*

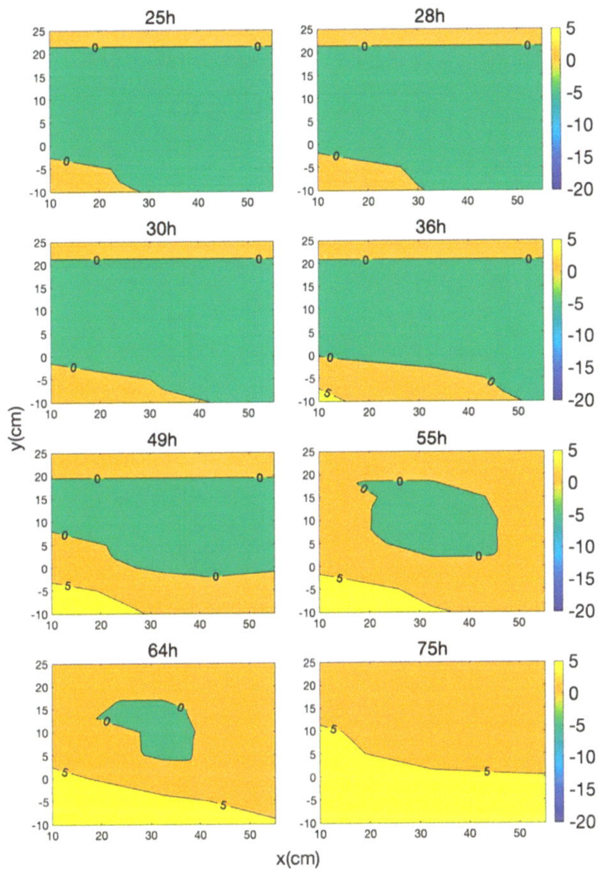

Figure 8. Temperature distribution nephogram during the melting stage (**x** stand for the horizontal position and **y** stand for the distance from the freezing center).

Within 16 h after stopping freezing, the temperature in the area around the freezing pipe increases, and the temperature in the freezing area is gradually uniform, but the area with a temperature less than 0 °C does not shrink. During the melting process, the influence of seepage is more obvious. As can be seen from Figure 8, seepage accelerates the melting of the upstream freezing curtain and brings the cold energy to the downstream, delaying the melting of the downstream.

3.2. Frost Heave, Thawing Settlement, and Change of Moisture Content in Model Test

During the test, the surface displacement is monitored, and the frost heave and thaw settlement at different distances from the upstream box wall are shown in Figure 9.

The colored solid lines in the figure above show the surface displacement recorded by dial indicators at different distances from the upstream box wall. The dotted line on the left is the time when the pump is turned off, and the dotted line on the right is the time when the soil in the box is completely melted (the temperature of any part of the model inward is >0 °C).

Frost heave develops with the beginning of freezing, but after freezing stopped, the area of the frozen area did not decrease, and water in the unfrozen area still migrated to the frozen area and changed into ice. After the pump is shut down for about 15 h, the surface heave reaches max, at 8 mm. Then the melting settlement increased to 40 mm, which was much larger than the frost heave. Combined with the cloud diagram analysis, it can be concluded that when there is an area with temperature >0 °C in the frozen clay, the settlement begins to develop rapidly. When the soil in the box is completely melted (the temperature at each point in the box is >0 °C), the settlement tends to be stable in a short time.

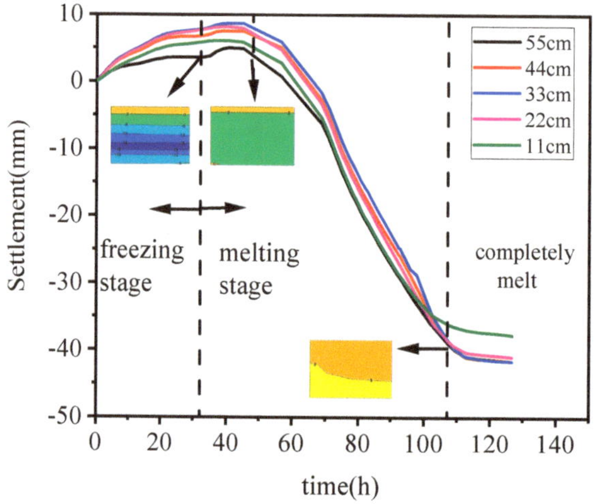

Figure 9. Frost heaving and thawing settlement.

The thaw settlement develops first at 11 cm and the settlement reaches stability first, indicating that the area near the upstream melts first and completes the consolidation first under the action of seepage. That also confirms the temperature development trend in the temperature cloud map, the upstream melts first and gradually develops to the downstream area.

Before and after the test, some soil samples were extracted to determine the moisture content. The test results are shown in Figure 10.

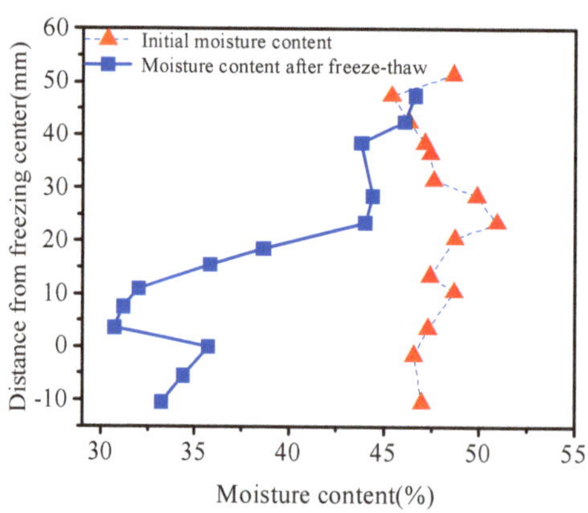

Figure 10. Change of water content.

From the water content distribution before and after freezing and thawing, the water content of each soil layer during filling is near the set value of 48%, mainly distributed in the range of 45–50%, which proves that the soil quality of remolded soil is relatively uniform. After the freeze–thaw cycle, the moisture content of the soil layer far from the freezing pipe changes little compared with that of the filling soil. The water content in the frozen area is low, and the lowest water content is only 30%, indicating that a large amount of liquid water in the frozen area is discharged in the melting stage, which is also the cause of melting settlement. Compared with other areas within the freezing range, the water content in the freezing pipe is higher, at 39%. The same trend was also observed in the dry density distribution shown in Figure 11. The dry density increases in the frozen area but decreases in the freezing pipe area.

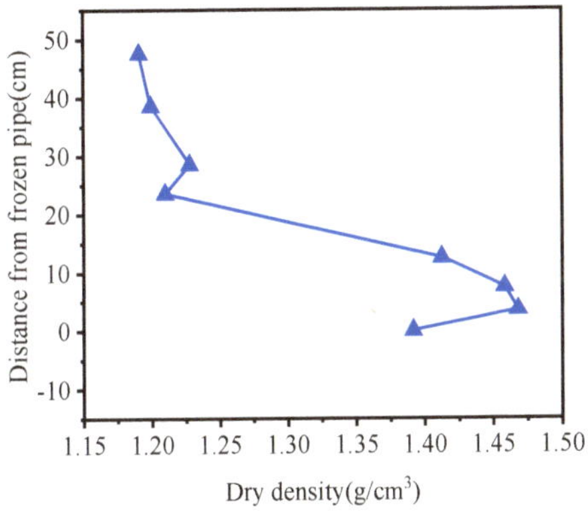

Figure 11. Dry density obtained by cutting ring method.

3.3. Relationship between CT Number and Soil Settlement

In order to obtain the 3D distribution characteristics of pores in the clay, the software "Aviso" was used to realize the 3D reconstruction of CT image. In order to reduce the error, we selected the region in the red ring as the basis of soil structure remodeling in every CT image, as in Figure 12. The 3D reconstruction results are shown in Figure 13.

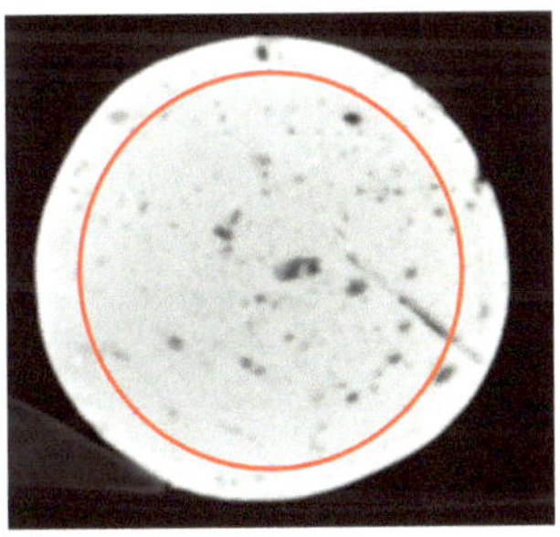

Figure 12. Image cutting.

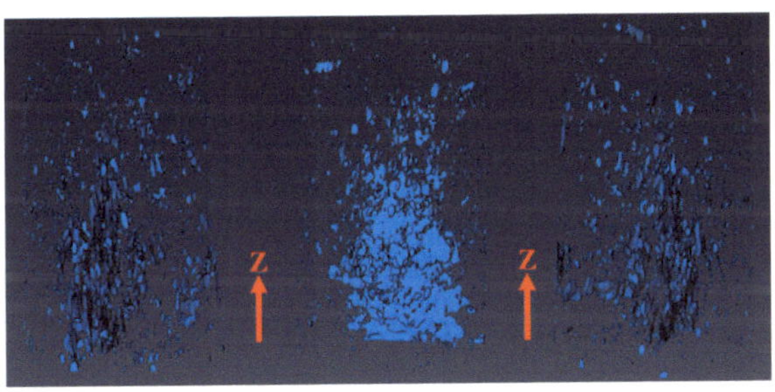

Figure 13. Pore 3D reconstruction results.

As can be seen from Figure 14, the distribution of CT numbers along the height can be divided into three areas:

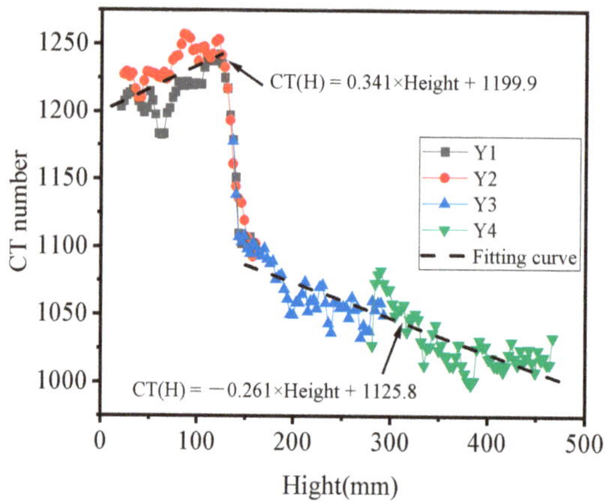

Figure 14. Distribution of CT number along height.

① The average value of CT number in 0–125 mm area is higher than other areas, the CT number near the freezing front is the largest, and the soil is the densest. There is a linear relationship between CT number CT (H) and height *h*:

$$\mathrm{CT(H)} = 0.341 \times h + 1199.9 \tag{1}$$

② In the 125–150 mm section, the CT number decreases rapidly with the increase of height. After melting, the final position of the freezing front is 150 mm from the centerline of the freezing pipe. The water in the unfrozen area above the freezing front migrates to the freezing area below the freezing front. The water in the freezing area is continuously enriched during freezing, and the clay drainage consolidation in the melting stage. The relationship between CT number CT (H) and height h in this section is as follows:

$$\mathrm{CT(H)} = -6.1318 \times h + 2010.6 \tag{2}$$

③ In the 150–485 mm section, the CT number in this section decreases slowly with the increase of height. During the freezing process, the moisture in the unfrozen area near the front decreases continuously, the soil loses water and compresses, and the soil layer becomes dense. The CT number in this interval is negatively correlated with the distance from the freezing front:

$$\mathrm{CT(H)} = -0.261 \times h + 1125.8 \tag{3}$$

Actually, the CT number reflects the density of the object. The CT numbers collected from each layer were connect with the dry density and obtained the fitting curve shown in Figure 15.

It can be seen from Figure 15 that there is a good linear relationship between CT number and dry density:

$$\rho_{\mathrm{dry}} = 0.0012\mathrm{CT(H)} - 0.0341 \tag{4}$$

In the above formula, the correlation coefficient $R^2 = 0.9483$, so the dry density value can be calculated according to the CT number. It should be noted that in Figure 15, the experimental data at the point near CT (H) = 1150 are missing because the CT number decreases rapidly in the 125–150 mm section away from the freezing pipe, and the interval range is only 25 mm, while the height of the ring knife sample for measuring the dry density is 20 mm, and the dry density is not measured in this interval.

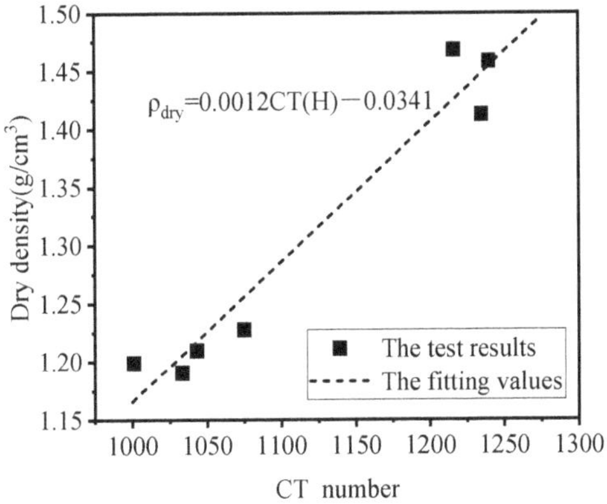

Figure 15. The relationship between CT number and dry density.

The void ratio $e(h)$ of soil can be calculated according to the relationship between dry density and height h:

$$e(h) = \frac{\rho_w G_s}{\rho_{dry}(h)} - 1 \tag{5}$$

The calculated void ratio distribution along the height is shown in Figure 16.

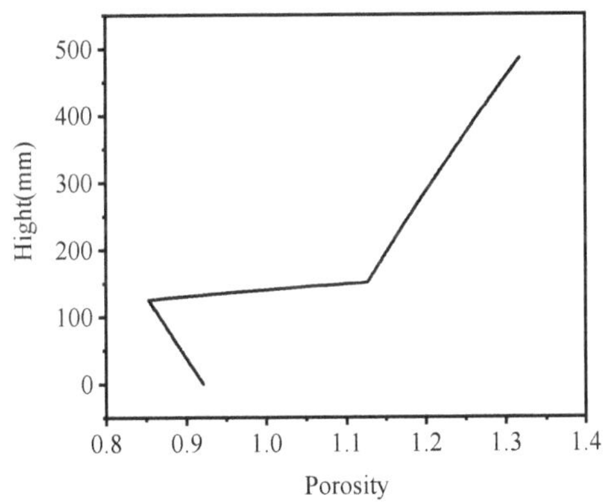

Figure 16. Void ratio along height.

Without considering the lateral deformation, the thawing settlement of soil belongs to a one-dimensional compression problem:

$$S = \frac{e_1 - e_2}{1 + e_1} H \tag{6}$$

where e_1 and e_2 is the void ratio of soil before and after freezing and thawing, and H is the thickness of soil layer.

According to the CT test results, the spatial distribution $e(h)$ of void ratio along the vertical direction after melting has been obtained, and the melting settlement can be obtained by integration:

$$S = \int_{h_1}^{h_2} \frac{e_1 - e(h)}{1 + e_1} h \, dh \tag{7}$$

where h_1 and h_2 are the upper and lower limit height of the frozen area.

The void ratio at the height of h = 485 mm (at the surface) is taken as the initial value e_1 before freezing and thawing. According to the settlement calculated by the void ratio, the settlement of the soil layer (0–485 mm) from the freezing pipe to the surface is 41.2 mm.

It should be pointed out that there is also 100 mm of thick, remolded mucky clay under the freezing pipe in the model test. Theoretically, this part of soil is also greatly compressed during freezing and thawing. However, due to the limitation of sampling conditions, CT samples of 100 mm thick clay under the freezing pipe were not taken. In order to obtain the settlement of this layer, layered settlement markers were buried on the horizontal plane where the center line of the freezing pipe is located during filling. After the test, the height of layered settlement markers was recorded. The compression value of 100 mm thick soil layer under the freezing pipe in the freezing and thawing stage was 5 mm.

The total settlement is 46.24 mm by adding the soil layer compression value above the freezing pipe calculated from the CT test results to the measured compression value below the freezing pipe, and the measured surface settlement is 42 mm. The settlement calculated from CT data is very close to the measured value, indicating that the settlement calculation method proposed in this paper is reliable.

4. Conclusions

In this test, sensors were used to determine the temperature, water content, and dry density at different depths and explored the method of obtaining undisturbed large-scale soil column in the model box, which can well maintain the undisturbed structure of soil sample. The CT tomography tests were also carried out on the soil sample in the model test and obtained the full section CT test data above the freezing pipe. By analyzing all the data, we found that:

1. Seepage has a very significant impact on the thickness of the freezing curtain. In the freezing process, due to the existence of seepage, the freezing curtain on the side near the sand layer below the freezing pipe is 40% thinner than that above. When melting, the seepage brings the upstream cooling capacity to the downstream. The process of drainage consolidation at the upstream is faster than the downstream.
2. The freezing–thawing cycle under the influence of seepage leads to the change of soil moisture content, especially in the area close to the frozen area, the moisture content of soft clay has decreased by 9% to 18%. The decrease of water content before and after freezing–thawing may be due to the fact that pores caused by freezing–thawing cycle promote the settlement of soil layer and then promote the drainage of soil water.
3. The distribution of the CT number along the height has obvious segmentation characteristics, and the CT number in the frozen area is significantly higher than that in the unfrozen area. We combined the CT number with the dry density obtained from the test and proposed a method for calculating the melt settlement according to the CT number and the real settlement of the model test. The calculated value is very close to the measured value, which means that the fitting method is accurate and effective.

Author Contributions: Conceptualization, J.Z. and C.W.; methodology, C.W. and Z.G.; software, Z.L. and H.Z.; validation, Z.G., C.W. and H.Z.; formal analysis, Z.G., C.W. and H.Z.; investigation, Z.G. and C.W.; resources, W.P.; data curation, Z.L. and H.Z.; writing—original draft preparation, C.W. and Z.G.; writing—review and editing, J.Z. and Z.G.; visualization, Z.L.; supervision, J.Z.; project administration, J.Z. and W.P.; funding acquisition, W.P. All authors have read and agreed to the published version of the manuscript.

Funding: The research work herein was supported by the National Natural Science Foundation of China (No. 41702299) and the Foundation of State Key Laboratory of Frozen Soil Engineering (No. SKLFSE201916).

Institutional Review Board Statement: Not applicable.

Informed Consent Statement: Not applicable.

Data Availability Statement: Not applicable.

Conflicts of Interest: The authors declare no conflict of interest.

References

1. Sudisman, R.A.; Osada, M.; Yamabe, T. Heat Transfer Visualization of the Application of a Cooling Pipe in Sand with Flowing Pore Water. *J. Cold Reg. Eng.* **2016**, *31*, 04016007. [CrossRef]
2. Yang, X.; Ji, Z.; Zhang, P.; Qi, J. Model test and numerical simulation on the development of artificially freezing wall in sandy layers considering water seepage. *Transp. Geotech.* **2019**, *21*, 100293. [CrossRef]
3. Wang, B.; Rong, C.; Cheng, H.; Cai, H. Experimental investigation on heat transfer law of multiple freezing pipes in permeable stratum with high seepage velocity. *Int. J. Heat Mass Transf.* **2022**, *182*, 121868. [CrossRef]
4. Lin, J.; Cheng, H.; Cai, H.B.; Tang, B.; Cao, G.Y. Effect of seepage velocity on formation of shaft frozen wall in loose aquifer. *Adv. Mater. Sci. Eng.* **2018**, *2018*, 2307157. [CrossRef]
5. Vitel, M.; Rouabhi, A.; Tijani, M.; Guérin, F. Thermo-hydraulic modeling of artificial ground freezing: Application to an underground mine in fractured sandstone. *Comput. Geotech.* **2016**, *75*, 80–92. [CrossRef]
6. Su, Y.L.; Yue, Z.R.; Li, X.K. Model Test Study on Freezing of Sand Layer under Seepage Condition. *Railw. Stand. Des.* **2020**, *64*, 94–100+108.
7. Zhang, S.; Yue, Z.; Sun, T.; Han, Y.; Gao, W.; Hu, T.; Han, Y. Evolution laws for frozen wall formation under conditions of sudden seepage. *Math. Probl. Eng.* **2020**, *2020*, 8836149. [CrossRef]
8. Dong, Y.B.; Rong, C.X.; Wang, B.; Yang, F. Study on Optimal Arrangement Method of Multi—Circle Frozen Hole Under Action of Large Velocity Groundwater. *Saf. Coal Mines* **2020**, *51*, 18–25.
9. Liu, Y.; Li, K.Q.; Li, D.Q.; Tang, X.S.; Gu, S.X. Coupled thermal–hydraulic modeling of artificial ground freezing with uncertainties in pipe inclination and thermal conductivity. *Acta Geotech.* **2022**, *17*, 257–274. [CrossRef]
10. Andersland, O.B.; Ladanyi, B. *Frozen Ground Engineering*; John Wiley & Sons: Hoboken, NJ, USA, 2003.
11. Harris, J.S. Ground Freezing in Practice. *Int. J. Rock Mech. Min. Sci. Geomech. Abstr.* **1995**, *11*, 173A.
12. Chen, S.J.; Zhao, S.P.; Ma, W. Studying frozen soil with CT technology: Present studies and prospects. *J. Glaciol. Geocryol.* **2013**, *35*, 193–200.
13. Zhao, S.; Wei, M.A.; Zheng, J.F. CT real-time monitoring on frozen Lanzhou loess at different temperatures and under uniaxial loading. *Rock Soil Mech.* **2010**, *31*, 92–97.
14. Konrad, J.-M. Frost heave in soils: Concepts and engineering. *Can. Geotech. J.* **1994**, *31*, 223–245. [CrossRef]
15. Zhang, S.; Lai, Y.; Zhang, X. Study on the damage propagation of surrounding rock from a cold-region tunnel under freeze—Thaw cycle condition. *Tunn. Undergr. Space Technol.* **2004**, *19*, 295–302. [CrossRef]
16. Liu, H.; Yang, G.; Ye, W.; Shen, Y.; Zhang, H.; Tang, L. Analysis of water and ice content and damage characteristics of the frozen rock during freezing based on the three-valued segmentation of CT images. *J. Min. Saf. Eng.* **2016**, *33*, 1130–1137.
17. Lang, Y.; Liang, Z.; Duan, D.; Cao, Z. Three-dimensional parallel numerical simulation of porous rocks based on CT technology and digital image processing. *Rock Soil Mech.* **2019**, *40*, 1204–1212.
18. Wen, H.; Fan, S.; Ma, L.; Guo, J.; Cheng, X.; Tong, X. CT Scanning Technology on coal-rock damage: A comprehensive review. *Coal Sci. Technol.* **2019**, *47*, 44–51.
19. Sun, Z. A Method of Extracting Pore Data of Tight Sandstone Based on 3D CT Scanning Image. *J. North China Inst. Sci. Technol.* **2020**, *17*, 6–13.
20. Luo, L.; Lin, H.; Li, S. Quantification of 3-D soil macropore networks in different soil types and land uses using computed tomography. *J. Hydrol.* **2010**, *393*, 53–64. [CrossRef]
21. Luo, L.; Lin, H.; Schmidt, J. Quantitative Relationships between Soil Macropore Characteristics and Preferential Flow and Transport. *Soil Sci. Soc. Am. J.* **2010**, *74*, 1929–1937. [CrossRef]

applied sciences

MDPI

Article

The Quantification and Evolution of Particle Characteristics of Saturated Silt under Freeze–Thaw Cycles

Jie Zhou [1,*], Zeyao Li [1] and Wansheng Pei [2]

[1] Department of Geotechnical Engineering, Tongji University, Shanghai 200092, China
[2] State Key Laboratory of Frozen Soil Engineering, Northwest Institute of Eco-Environment and Resources, Chinese Academy of Sciences, Lanzhou 730099, China
* Correspondence: zhoujie1001@tongji.edu.cn; Tel.: +86-13636651710; Fax: +86-021-65987079

Abstract: Freeze–thaw action is a complicated process. How it affects particle characteristics of saturated silt may provide a much clearer understanding of its internal mechanism. A series of specific apparatus were developed for sample reconstitution, including sand pluviation device, freeze–thaw device, and special sampling device. After reconstituting samples by sand pluviation method and a specific parameter-controlled freeze–thaw testing, scanning electron microscope (SEM) and laser scattering and transmissometry (LST) tests were conducted to explore the particle characteristics of silt under freeze–thaw cycles. The test results show that freeze–thaw action could probably induce the particles' (60–200 μm) breakage, also affecting the clay particles' (less than 5 μm) aggregation. With the increase of freeze–thaw times, freeze–thaw action on the particle impact decreases. The larger the effective confining pressure, the lower the freezing temperature, greater the compaction degree, and higher the fine content, which can all aggravate the effects of freeze–thaw action on silt particles. Finally, two characteristic evolution modes of particle structure under freeze–thaw cycles have been inferred based on particle interaction during the freeze–thaw process, which could provide a reference for long-term durability evaluation of pavements in cold regions.

Keywords: freeze–thaw cycle; scanning electron microscope (SEM) and laser scattering and transmissometry (LST); saturated silt; particle crushing and aggregating; particle characteristic index

1. Introduction

Cohesionless soil is widely distributed in the middle and lower reaches of rivers, such as silt. They are typical frost-susceptible soils, which have a significant silt/fine content allowing the additional water rise of capillary during freezing. This water can turn to ice lenses in winter and further water is drawn up from frozen front to balance the capillary forces. This cycle ultimately leads to heaving at the surface which causes pavement cracking and uplifting (Figure 1).

It is further complicated as the upper ice thaws in summer. The volume of water held as ice is many times greater than water held by the soil under saturated conditions (due to moisture migration). The water cannot drain down through the soil as it is still frozen. The result is a further weakening of the highway as the subgrade's ground-bearing capacity is diminished. Even though freeze–thaw action is a complicated multi-field multi-scale process, the inherent cause of this bearing weakening can be traced to the particle variation characteristics of saturated cohesionless soil deposited in water environment, such as particle breakage, rearrangement, and rounding, by freeze–thaw cycles.

So far, the effects of freeze–thaw action on saturated cohesionless soil particles have been studied over several decades. Arturo [1,2] conducted horizontal and vertical freezing experiments on gravel soil. The test results showed that the freeze–thaw action could cause a certain sorting phenomenon. Freeze–thaw action would cause large particles to move against freezing direction, and small particles to move in freezing direction. Edwin and

Anthony [3] studied the freeze–thaw characteristics of cohesionless soils. These results showed that freezing and thawing can change the pore structure of the soil, thereby increasing the permeability of the soil. Konrad [4] conducted the freeze–thaw cycle test on saturated clayey silt, and qualitatively analyzed the structural characteristics of the freeze–thaw soil, concluding that freeze–thaw cycles could destroy the bonding force between soil particles and induce particle rearrangement. Qi et al. [5] used SEM to quantitatively study the microscopic pore characteristics of silt soil in upper reaches of the Yellow River under freeze–thaw cycle. The results showed that the freeze–thaw cycle can make small pores smaller and large pores larger. Mu et al. [6] scanned the loess with open system freeze–thaw by electron microscope, and quantitatively analyzed its pore characteristics, obtaining the process and mechanism of the effects of freeze–thaw cycles on pore structures. Tan et al. [7] analyzed the effects of freeze–thaw on saturated silt pore microstructure using nitrogen adsorption and mercury injection. These results showed that freeze–thaw cycles have little effect on pores with diameters less than 10^{-8} m, whilst having great influence on 10^{-8}–10^{-4} m pores. Tang and Li [8] studied the pore structure characteristics of saturated freeze–thaw silt and silt through scanning electron microscope and mercury intrusion test. The results showed the accumulated pore volume and pore area of frozen–thawed silty sand increased along with decreasing freezing temperature; the most probable pore size and probability decrease along with increasing freezing temperature. Jin et al. [9] analyzed the dynamic structure characteristics of salty silt under freeze–thaw cycles. Zheng et al. [10] qualitatively investigated the particle gradation and pore characteristics of the Yellow River flooded area under freeze–thaw cycle using nuclear magnetic resonance. Ren and Vanapalli [11] measured the SFCC (including both freezing and thawing branches) of two kinds of cohesionless soil for different freeze–thaw cycles and analyzed the effect of freeze–thaw cycles on soil texture. The results show that the effect of freeze–thaw on soil texture is mainly the production of lens. Leuther and Schlüter [12] analyzed the impact of freeze–thaw cycles on soil structure and soil hydraulic properties, concluding that freeze–thaw has a greater impact on soil pores and can increase soil permeability. The effects of initial freeze–thaw are most obvious. Zhou et al. [13] conducted mercury intrusion experiments on saturated artificial frozen–thawed silty sand deposited in middle and lower reaches of the Yangtze River and obtained the influence of freeze–thaw action on the pore characteristics of Nanjing saturated silt fine sand. The results showed that freezing and thawing could loosen the silty fine sand structure and increase the soil average.

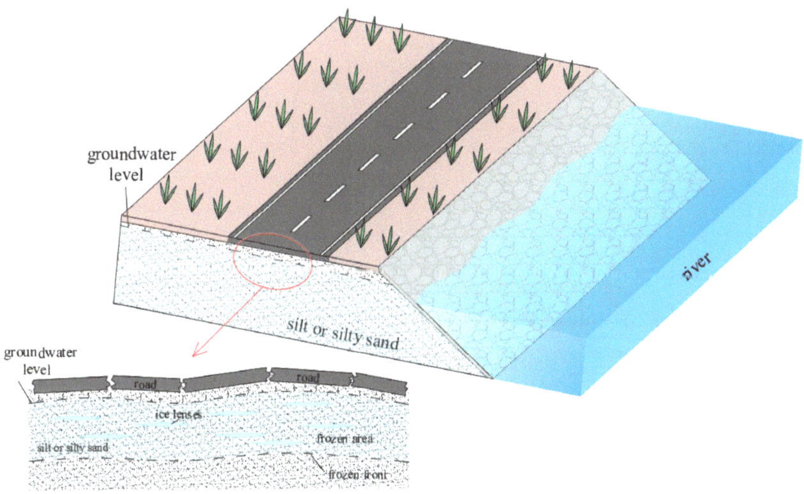

Figure 1. Pavement damage during freeze–thaw cycles in saturated cohesionless soil.

It can be seen that research about the effects of freeze–thaw action on cohesionless soil has mainly concentrated on macrostructure and pore characteristics. However, there are relatively few studies to quantitatively evaluate the effects of freeze–thaw action on the particle characteristics of cohesionless soil with fine particles, especially combined with other impact factors, such as confining pressure, compaction degree, etc.

The particle characteristics directly affect the macro-physical and mechanical properties of soil. Studying the effects of freeze–thaw action on the particle characteristics of saturated cohesionless soil deposited in water environment can therefore help us to understand the freeze–thaw mechanism on the physical and mechanical properties of soil, especially when multiple impact factors are involved, so as to assist traffic engineering construction along rivers in cold regions or artificial ground freezing to ensure engineering safety. Hence, after reconstituting samples by sand pluviation method and a specific parameter-controlled freeze–thaw testing, scanning electron microscope (SEM) test and the laser scattering and transmissometry (LST) were conducted for silt before and after freeze–thaw cycles, to explore the effects of freeze–thaw action on the particle characteristics of saturated silt deposited in water environments.

2. Test Materials

Saturated cohesionless soil is widely distributed in the Yangtze River coastal area. The cohesionless soil layers deposited in river facies such as silt and silty sand (less fine) are important soil layers for urban engineering construction. This paper focuses on the third layer (mezzanine) of silt near the Shangyuanmen subway station of Nanjing Metro Line $3^{\#}$ along the Yangtze River. In this research area, groundwater level is high, and the soil layers are saturated. The main component of soil particles is silicate minerals, and the soils are anisotropic in orientation. Through geotechnical investigation of Shangyuanmen subway station of Nanjing Metro Line $3^{\#}$, the particle shape characteristics of Nanjing saturated silt are shown in Figure 2; the particle grading curves are shown in Figure 3. The basic physical parameters are shown in Table 1:

Figure 2. The shape of soil particles of saturated silt along Yangtze River.

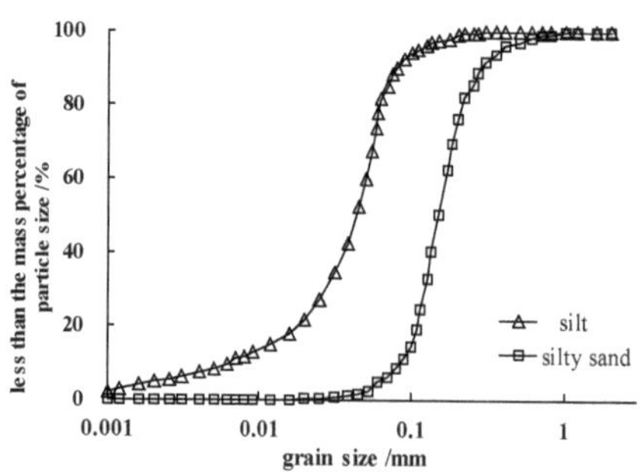

Figure 3. Soil particle grading curves (a. silt; b. silty sand).

Table 1. Physical and mechanical properties of soils.

	Water Content/%	Unit Weight/ (kN/m³)	Natural Void Ratio	Particles Specific Gravity	Maximum Void Ratio	Minimum Void Ratio	Cohesion/ kPa	Internal Friction Angle/°	Compression Modulus/ kPa	Uniformity Coefficient	Fine Particle Content/%
silt	27.3	18.9	0.797	2.69	1.31	0.69	7.2 *	28.1 *	10.27	6.09	88.3
silty sand	24.8	19.4	0.734	2.70	1.18	0.65	5.3 *	30.6 *	10.82	1.89	7.2

Note: * is obtained by direct shear test.

3. Experimental Program

To ensure the orientation of sample particles, the sand pluviation method is used for sample reconstitution by taking compaction degrees, fine contents, etc., into account; then, the reconstituted samples were frozen and thawed under different confining pressures and freezing temperatures; finally, scanning electron microscope (SEM) and laser scattering and transmissometry (LST) tests were both conducted to comprehensively explore the particle characteristics of saturated silt under long-term freeze–thaw cycles. The testing procedures were as follows:

① Firstly, dry and smash the undisturbed soil into soil particles, and pass it through 2.0 mm fine sieve for particle separation;

② Use the sample preparation device of sand pluviation method (as shown in Figure 4) for sample preparation:

The sample preparation device consists of a sand-spreading device, a water tank, a movable trolley platform, and some double valves. The water level of water tank is 1 m higher than the double valves [13]. To avoid sorting due to different particle diameters during the initial spreading period, a movable trolley platform is specially set up. After sand spreading for a period of time (30 s) move the trolley platform to the sand accumulation area to accumulate sand. The compactness degree of sand samples could be controlled initially by falling distance and falling amount, and then accurately controlled by sample compactor. The length and diameter of sand samples are 80.0 mm and 39.1 mm, respectively.

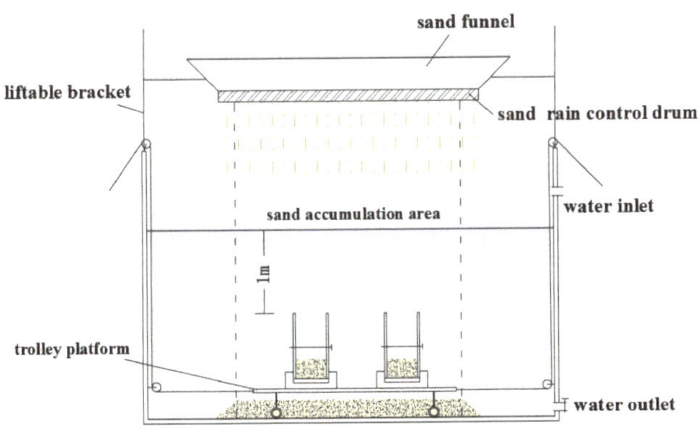

Figure 4. Sample preparation device of sand pluviation method for soil reconstitution.

③ Store the reconstituted samples in a vacuum saturator to saturate for 12 h;

④ Place the saturated samples on the freeze–thaw device (as shown in Figure 5) and set the preloading:

First, place the saturated samples on the freeze–thaw device. Open the drainage channel of freeze–thaw device and put an appropriate amount of weight on loading frame according to the effective confining pressure of the experimental program. Close the drainage channel after loading for 2 h. Then add another appropriate amount of weight to make the loading reach the designed total overburden stress value.

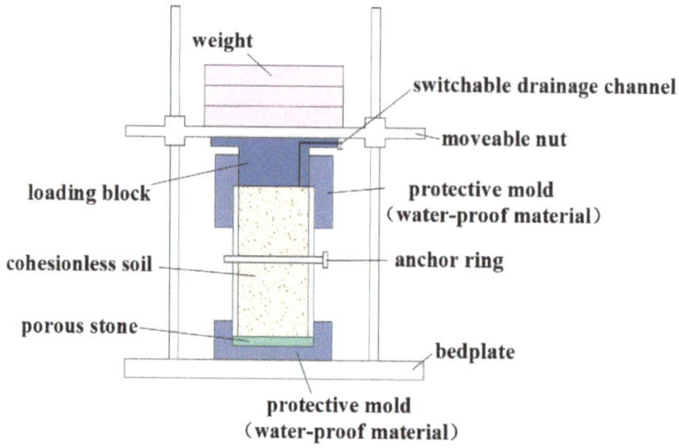

Figure 5. Freeze–thaw device diagram.

⑤ Put the freeze–thaw device with sample under a certain confining pressure of weights into a DW-40 type low temperature test chamber and freeze for 12 h. The freezing temperature matches the experimental program. After freezing, take it out to room temperature (constant temperature 20 °C by air conditioner) and thaw for 12 h. In this way, cycle the freeze–thaw operation until the design freeze–thaw cycles are reached.

⑥ Make the SEM sample and observe its microstructure:

The soil structure of silt can hardly be preserved without special device and techniques. In this paper a special SEM sampling device and a kind of curing agent, menthol, are used for SEM sample preparation. Evenly press the SEM sampling device by using a pallet into the soil sample. The SEM sampling device is 10 mm diameter, 5 mm height, and 0.1 mm thickness by No. 400 filter mesh (0.038 mm), as shown in Figure 6. Heat menthol to melt, then sprinkle the liquid menthol evenly on the surface of soil sample through the filer mesh with a small sprinkling can. After the menthol solidifies, it preserves the soil structure of silt at best. Set the soil sample into the sample chamber of SEM, start vacuum pump to sublimate the solid menthol, then run the SEM and take clear SEM images.

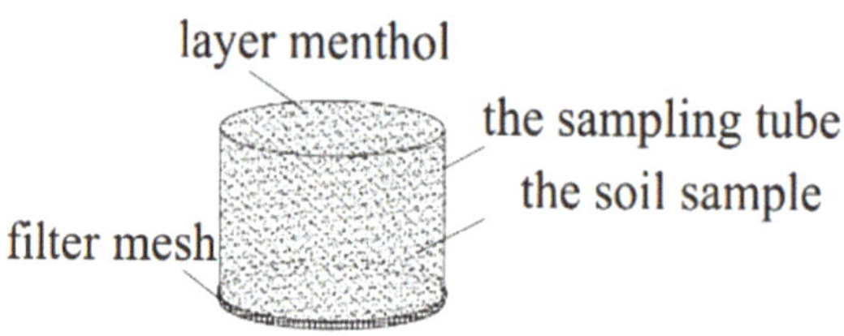

Figure 6. The sampling device of SEM.

⑦ Use LST to further determine the particle characteristics of soil:

Weigh 2 g of the dried frozen–thawed soil sample and put it into a 500 mL beaker with 200 mL distilled water. After soaking overnight, add 5 mL of 4% sodium hexametaphosphate, then add distilled water to 500 mL. Turn on the magnetic stirrer until the particles in suspension are evenly distributed. Finally, use the LST to determine the particle characteristics of the suspension.

The depth of natural frozen soil in Nanjing area is generally less than 1 m, and the depth of artificial frozen soil is about 7–15 m when using artificial ground freezing to construct the subway cross passages [13]. There are two test groups: the silt test group (A0–A16) and the prepared soil with less fine (from silty sand to silt) test group (C1–C4). The silt test group sets vertical confining pressure of 150 kPa, effective confining pressure of 75 kPa (the buried depth is about 7.8 m), relative bulk density of 0.827, freeze–thaw cycles of 20 times, fine particle content of 88.3% as the basic test conditions for comparison according to field conditions. The whole experimental program is designed to explore the effects of freeze–thaw cycles (*N*), effective confining pressure (σ), freezing temperature (*T*), compaction degree (D_r), and fine particle content (*L*) on the particle characteristics of cohesionless soil deposited in water environments (shown as Table 2).

Table 2. Experimental program.

Sample Label	Soil Type	Freeze–Thaw Cycles	Simulation Depth/m	Vertical Confining Pressure & Effective Confining Pressure/kPa	Freezing Temperature/°C	Compaction Degree	Fine Particle Content/%	Remark
A0	silt	0	7.8	150&75	/	0.827	88.3	
A1	silt	10	7.8	150&75	−20	0.827	88.3	
A2	silt	20	7.8	150&75	−20	0.827	88.3	Basic test
A3	silt	30	7.8	150&75	−20	0.827	88.3	
A4	silt	50	7.8	150&75	−20	0.827	88.3	
A5	silt	100	7.8	150&75	−20	0.827	88.3	
A6	silt	20	0.0	0&0.0	−20	0.827	88.3	
A7	silt	20	3.8	75&37.5	−20	0.827	88.3	
A8	silt	20	11.6	225&112.5	−20	0.827	88.3	
A9	silt	20	15.4	250&125	−20	0.827	88.3	
A10	silt	20	7.8	150&75	−20	0.300	88.3	
A11	silt	20	7.8	150&75	−20	0.500	88.3	
A12	silt	20	7.8	150&75	−20	0.700	88.3	
A13	silt	20	7.8	150&75	−20	0.900	88.3	
A14	silt	50	7.8	150&75	−5	0.827	88.3	
A15	silt	50	7.8	150&75	−10	0.827	88.3	
A16	silt	100	7.8	150&75	−30	0.827	88.3	
C1	prepared soil	20	7.8	150&75	−20	0.830	23.4	Silty sand
C2	prepared soil	20	7.8	150&75	−20	0.833	39.6	Silty sand
C3	prepared soil	20	7.8	150&75	−20	0.836	55.9	Silt
C4	prepared soil	20	7.8	150&75	−20	0.839	72.1	Silt

4. Test Results

4.1. Qualitative Analysis of Particle Structure of Silt under Freeze–Thaw Cycles by SEM

The microscopic particle structures of saturated silt deposited in water environments (scanning magnification of electron microscope 500 times) with different freeze–thaw cycles are shown in Figure 7.

It can be seen from Figure 7 that the unfrozen silt is mainly composed of irregular sand particles and silt particles, and a small amount of clay particles is randomly distributed. The soil particles have a certain orientation; the flaky soil particles are arranged in the horizontal direction and form relatively stable staggered structure. Compared with unfrozen silt, the fine content of silt after freeze–thaw is significantly increased, irregular flocs are formed, and the orientation characteristic of soil particles is weakened. The more freeze–thaw cycles the silt experienced, the more fine flakes (aggregation) the silt contains, more irregular flocculent content and less obvious particles orientation the soil has. This shows that freeze–thaw action can cause main particles' breakage and some fine aggregation, generate irregular flocs, and weaken the orientation of particles on silt. It should be noted that long-term freeze–thaw cycles can not only induce particle crushing and maybe some fine aggregate, but also affect the particle shapes, resulting in less regular orientations.

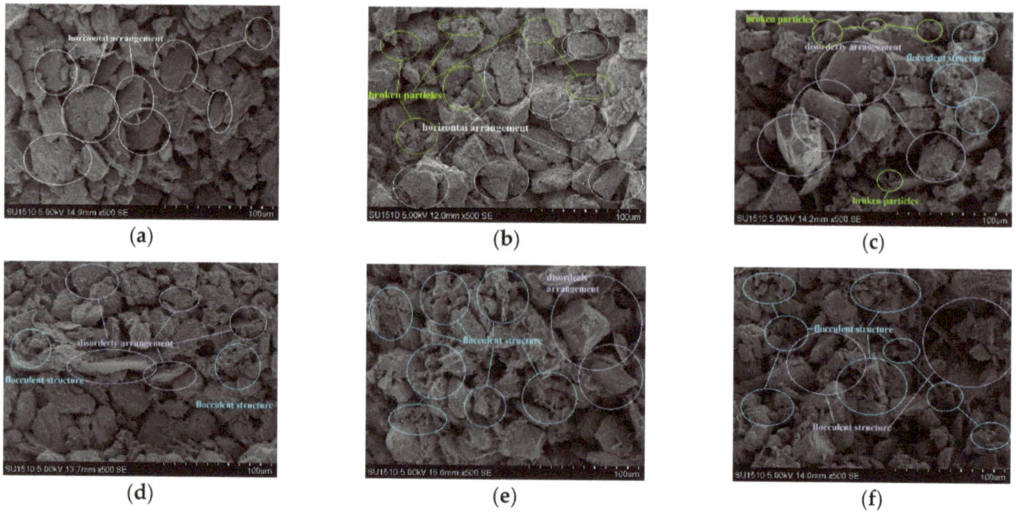

Figure 7. The SEM images of silt under different freeze–thaw cycles. (**a**) Unfrozen; (**b**) freeze–thaw 10 times; (**c**) freeze–thaw 20 times; (**d**) freeze–thaw 30 times; (**e**) freeze–thaw 50 times; (**f**) freeze–thaw 100 times.

4.2. *Quantitative Analysis of Particle Structure of Silt under Freeze–Thaw Cycles by LST*

4.2.1. Particle Characteristic Evaluation Parameters Establishment

The particle characteristics of soil could be further measured by LST. Under the condition of vertical confining pressure 150 kPa, effective confining pressure 75 kPa, freezing temperature −20 °C, and relative bulk density 0.827, the particle bulk density curves of silt with different freeze–thaw cycles are shown in Figure 8. It can be seen that as the silt experienced more freeze–thaw cycles, the content of clay particles smaller than 5 μm is reduced; the content of 5–10 μm particles changes in a stable fluctuating state; the content of 5–50 μm particles increases significantly; the content of 50–60 μm particles remains basically unchanged; the content of 60–200 μm particles decreases; the content of particles larger than 200 μm does not change significantly. Combined with the SEM images of silt under long-term freeze–thaw cycles (Figure 7), it can see that the freeze–thaw action probably induces the clay particles (less than 5 μm) to aggregate and form flocculent structure; probably inducing the part silt and sand particles (60–200 μm) in breakage. The diameters of formed flocculent structures (aggregated) and broken particles are distributed from 5–60 μm.

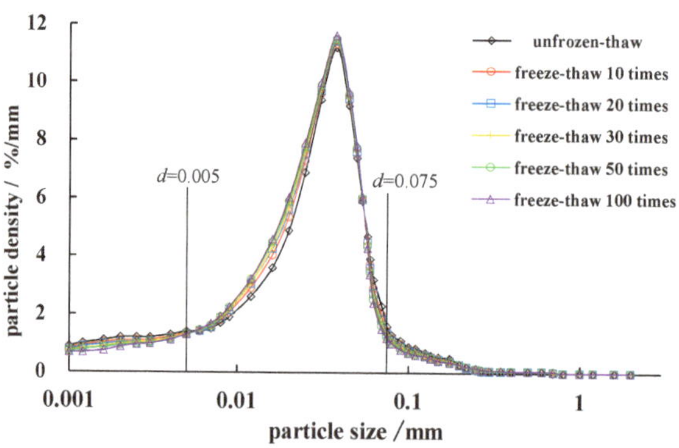

Figure 8. The particle bulk density curves of silt.

As early as 1985, Hardin [14] defined a parameter crushing index B_r to evaluate the particle size variation in soil shearing. It took the area enclosed by d = 0.074 mm, P = 100% (P is the percentage content of particles less than a certain size (%), d is the soil particle size (mm)), and the gradation curve before particle breakage in P-lg$d$ coordinate system as the crushing potential S_p, and the reduced area of gradation curve after testing was taken as the particle crushing amount S, and the two were divided to obtain the crushing index B_r.

$$B_r = S/S_p \quad (1)$$

The model is mainly aimed at particle breakage size greater than 0.074 mm of sand particles [15,16] (he regarded that the shear failure of soil could not affect particles below 0.074 mm [10,17]). However, in our research, the fine particles of soil account for 88.3%. As shown in the SEM images (Figure 7) and preliminary analysis of Figure 8, for silt under long-term freeze–thaw cycles, the soil particles were main influenced at a range of 1–2000 μm and soil particles both could break and aggregate in 5–60 μm, so a particle size index B_f of silt is defined in this paper to describe the total particle size variation characteristics of silt under long-term freeze–thaw cycles:

$$B_f = (|S1| + |S2|)/S_p \quad (2)$$

where S is the amount of particle size variation (%•mm), which is characterized by the area between the particle accumulation curve of silt before and after freeze–thaw cycles in P-lg$d$ coordinate system (note: this area is the sum of the variation amount including particles aggregation $|S1|$ and particles breakage $|S2|$. Shown in Figure 9); S_p is the variation potential (%•mm), which is characterized by the area along with d = 0.001 mm, P = 100%, and particle accumulation curve in P-lg$d$ coordinate system.

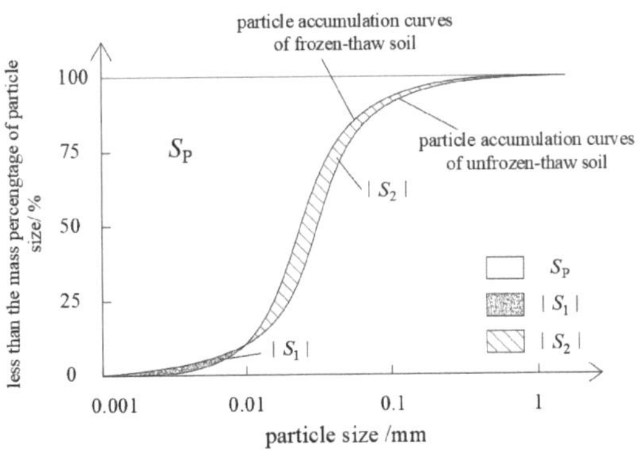

Figure 9. Schematic diagram of total particle size variation index.

From SEM image observing (Figure 7) and preliminary analysis of LST (Figure 8), freeze–thaw cycles not only change particle size, but also affect particle shape. Analogously, the concept of sphericity potential is proposed in this paper. The variation degree of total particle shape is characterized by the ratio of sphericity variation amount to sphericity potential.

The total particle shape index B_s of silt could be defined to describe the particle shape variation characteristics under long-term freeze–thaw cycles:

$$B_s = Q/Q_p \quad (3)$$

where Q is the amount of particle sphericity variation (mm), which is characterized by the area between the sphericity curve of silt before and after freeze–thaw cycles in S_s-lg$d$ coordinate system, shown in Figure 10; Q_p is the particle sphericity potential (mm), which is characterized by the area along with d = 0.001 mm, S_s = 1, and sphericity curve of silt before freeze–thaw in S_s-lg$d$ coordinate system; S_s is the sphericity of soil particles [4]:

$$S_s = \frac{P_{epqc}}{P_{real}} = 2\sqrt{\pi \cdot A}/p_{real} \tag{4}$$

where S_s is between 0 and 1. The smaller the S_s, the more irregular the soil particle shape; P_{epqc} is the perimeter of equivalent projection circle measured by LST (m); P_{real} is the actual boundary perimeter of particle projected image (m); A is the projected area of particle (m).

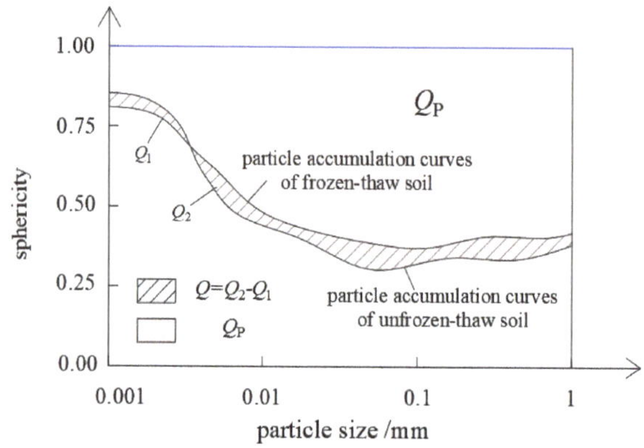

Figure 10. Schematic diagram of total particle shape variation rate.

Under the condition of vertical confining pressure 150 kPa, effective confining pressure 75 kPa, freezing temperature −20 °C, and relative bulk density 0.827, the sphericity curves of silt with different freeze–thaw cycles are shown in Figure 11. It shows that the sphericity of silt ranges from 0.46–0.90, and the shapes of small particles are characterized by roundness, while the shapes of large particles are relatively poor. The sphericity of particles smaller than 40 μm with different freeze–thaw cycles hardly change; that means freeze–thaw action has no obvious effect on the shape of particles smaller than 40 μm in silt. For particles larger than 40 μm, freeze–thaw action could cause a certain statistical increase of sphericity, especially the sphericity of particles larger than 100 μm, which show more statistically significant increases.

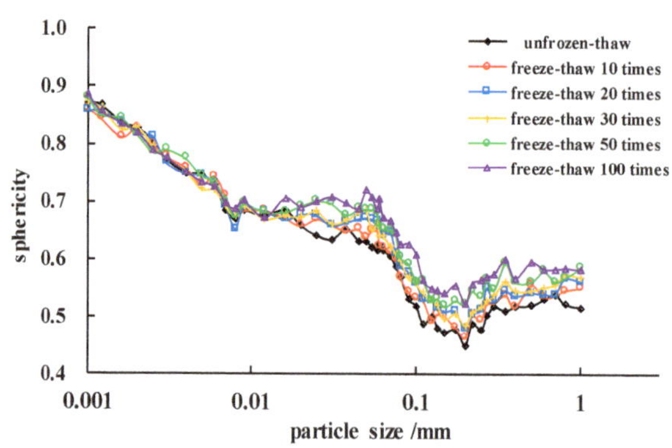

Figure 11. The sphericity curves of silt.

4.2.2. The Particle Characteristics of Silt under Freeze–Thaw Cycles

Under the condition of vertical confining pressure 150 kPa, effective confining pressure 75 kPa, freezing temperature −20 °C, and relative bulk density 0.827, the values of B_f of silt with different freeze–thaw cycles are shown in Figure 12. It shows that total particle size index increases with freeze–thaw cycles. The initial freeze–thaw cycles have a greater degree of influence on total particle size index. With the more freeze–thaw cycles the silt has experienced, the influence degree of single freeze–thaw cycle on the particle size becomes smaller.

To quantitatively describe the effects of freeze–thaw action on the total particle size index, a hyperbolic type total particle size variation model is established under freeze–thaw cycle number:

$$B_f = \frac{N^a}{b + cN^a} \tag{5}$$

where a, b, c are model parameters, which respectively relate to accumulative total particle size index, total particle size index in the first freeze–thaw cycle, and the final total particle size index.

A hyperbolic type total particle size variation model was used to fit the relationship between total particle size index of silt and freeze–thaw cycles (show as Equation (6)), and the data fits well with the correlation coefficient R^2 value of 0.98.

$$B_f = \frac{N^{0.973}}{1.113 + 0.056N^{0.973}} \quad R^2 = 0.98 \tag{6}$$

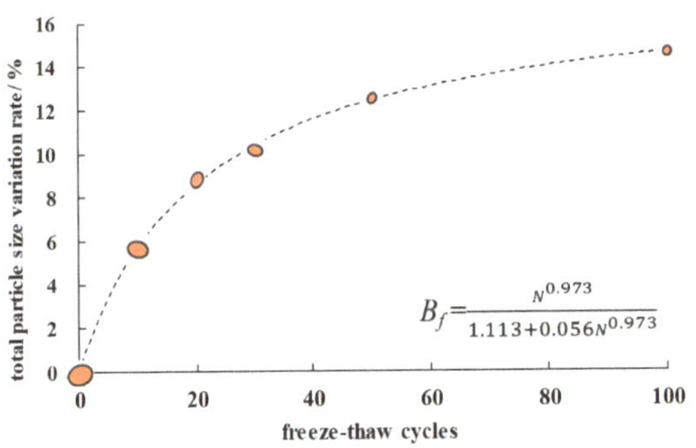

Figure 12. The effect of freeze–thaw cycles on the total particle size of silt.

Under the condition of vertical confining pressure 150 kPa, effective confining pressure 75 kPa, freezing temperature −20 °C, and relative bulk density 0.827, the total particle shape index values of silt with different freeze–thaw cycles are shown in Figure 13. It shows that the total particle shape index increases with freeze–thaw cycles. This means that freeze–thaw action causes the total particle shape of silt to become more spherical. The initial freeze–thaw cycles have a greater degree of influence on total particle shape index. With more freeze–thaw cycles, the influence degree of a single freeze–thaw cycle on the particle shape of silt becomes smaller and smaller.

Similarly, a hyperbolic type model is established to describe the effects of freeze–thaw cycles on total particle shape index of silt:

$$B_s = \frac{N^d}{e + fN^d} \tag{7}$$

where d, e, f are model parameters, which are respectively related to accumulative growth rate of total particle shape variation, total particle shape index in the first freeze–thaw cycle, and the final total particle shape index.

A hyperbolic type total particle shape variation model was used to fit the relationship between total particle shape index of silt and freeze–thaw cycles (shown as Equation (8)), and the data fit relatively well with the correlation coefficient R^2 value of 0.97.

$$B_s = \frac{N^{1.249}}{10.45 + 0.091N^{1.249}} R^2 = 0.97 \quad (8)$$

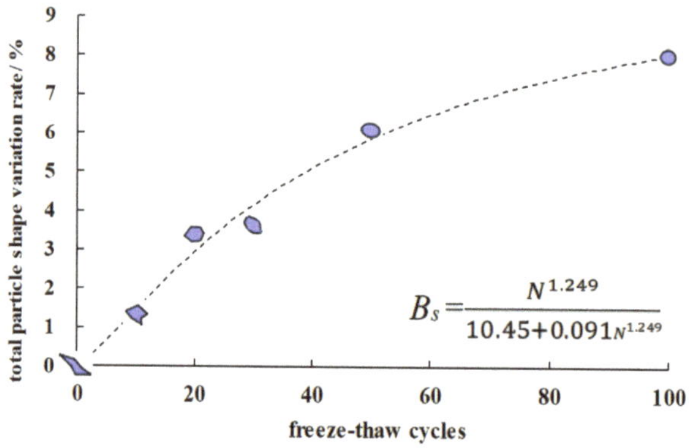

Figure 13. The effect of freeze–thaw cycles on the total particle shape of silt.

4.3. Particle Characteristics of Freeze–Thaw Effect under Different Impact Factors

4.3.1. Confining Pressure

Under the conditions of freeze–thaw cycles 20, freezing temperature −20 °C, and relative bulk density 0.827, the total particle size index values of silt with different freeze–thaw cycles are shown in Figure 14. It shows that within the depth range of natural frozen soil and artificial frozen soil, the greater the effective confining pressure, the higher the total particle size index of silt. With the higher effective confining pressure, the influence degree of effective confining pressure on total particles size is smaller and smaller. The quadratic function model is used to fit the relationship between total particle size index of silt and effective confining pressure (shown as Equation (8)), and the data fitting effect is relatively good.

$$B_f = -3.1 \times 10^{-4}\sigma^2 + 0.0896\sigma + 3.813 R^2 = 0.99 \quad (9)$$

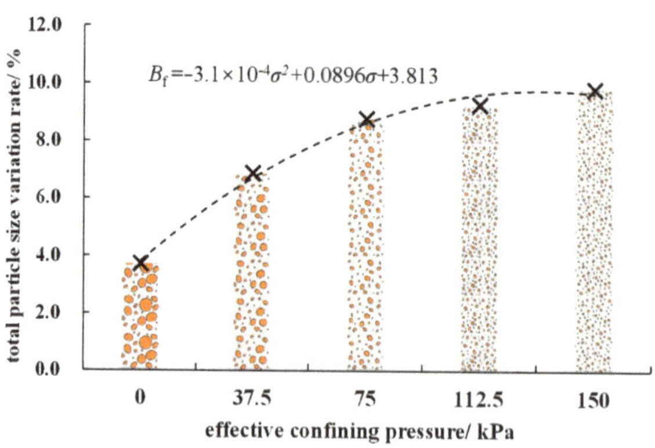

Figure 14. The effect of effective confining pressure on the total particle size of silt.

Under the conditions of freeze–thaw cycles 20, freezing temperature −20 °C, and relative bulk density 0.827, the total particle shape index values of silt with different freeze–thaw cycles are shown in Figure 15. It shows that within the depth range of natural frozen soil and artificial frozen soil, the greater the effective confining pressure, the higher the total particle shape index of silt. With higher effective confining pressure, the variation of effective confining pressure has less and less influence degree on total particle shape. The quadratic function model is used to fit the relationship between total particle shape index of silt and effective confining pressure (shown as Equation (9)), and the data fitting effect is relatively good.

$$B_s = -6.1 \times 10^{-5}\sigma^2 + 0.0198\sigma + 2.248R^2 = 0.99 \quad (10)$$

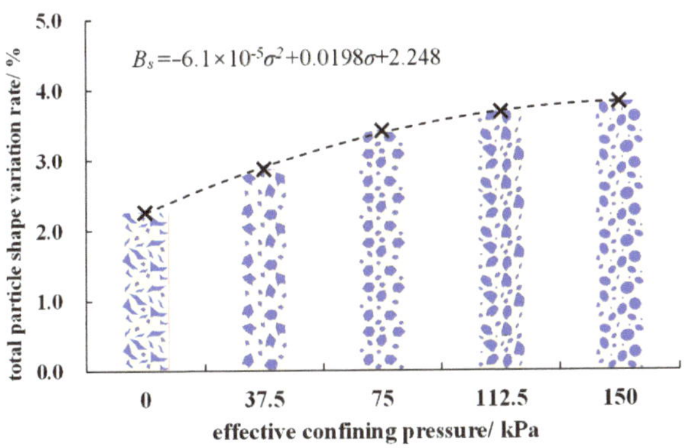

Figure 15. The effect of effective confining pressure on the total particle shape of silt.

4.3.2. Freezing Temperature

Under the condition of vertical confining pressure 150 kPa, effective confining pressure 75 kPa, and relative bulk density 0.827, the total particle size index values of silt with 20 freeze–thaw cycles at different temperatures are shown in Figure 16. It shows that within the temperature range of natural frozen soil and artificial frozen soil, the freezing temperature and the total particle size index of silt are roughly linearly correlated. The total particle size index of silt linearly increases with the decrease of freezing temperature. Linear fitting is performed on total particle size index of silt and freezing temperature (shown as Equation (11)), and the data fitting effect is relatively good.

$$B_f = -0.150T + 5.768 \quad R^2 = 0.98 \quad (11)$$

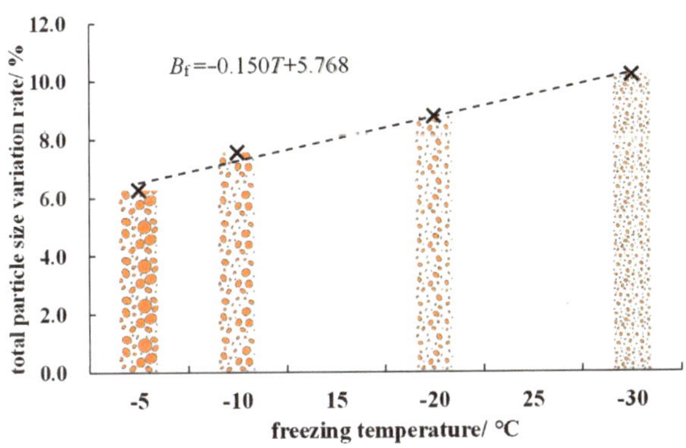

Figure 16. The effect of freezing temperature on the total particle size of silt.

Under the condition of vertical confining pressure 150 kPa, effective confining pressure 75 kPa, and relative bulk density 0.827, the total particle shape index values of silt with 20 freeze–thaw cycles at different temperatures are shown in Figure 17. The freezing temperature and total particle shape index are roughly linearly correlated. The lower the freezing temperature, the higher the total particle shape index. Linear fitting is performed on total particle shape index of silt and freezing temperature (shown as Equation (12)), and the data fitting effect is relatively good.

$$B_s = -3.3 \times 10^{-2}T + 2.713 \quad R^2 = 0.98 \tag{12}$$

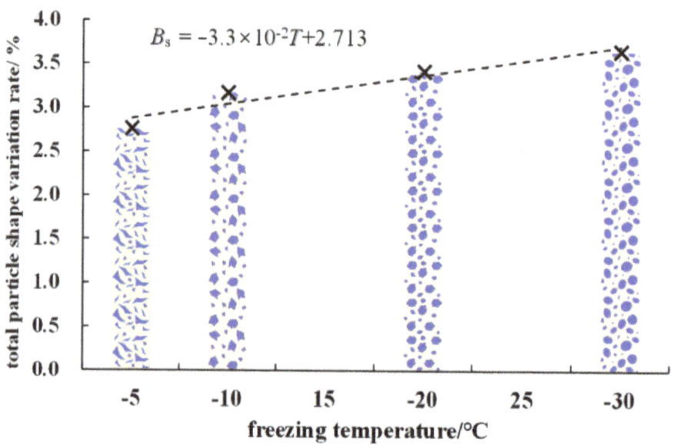

Figure 17. The effect of freezing temperature on the total particle shape of silt.

4.3.3. Compaction Degree

Figure 18 shows the total particle size index values of silt with different densities under the conditions of vertical confining pressure 125 kPa, effective confining pressure 75 kPa, and freezing temperature −20 °C after 20 freeze–thaw cycles. It can be seen from Figure 17 that the silt with higher relative bulk density would have higher total particle size index. The higher the relative bulk density, the smaller the influence degree of relative bulk density changing on total particles size. The quadratic function model is used to fit the relationship between total particle size index of silt and relative bulk density (shown as Equation (13)), and the data fitting effect is relatively good.

$$B_f = -9.25{D_r}^2 + 16.96D_r + 1.160 \quad R^2 = 0.98 \tag{13}$$

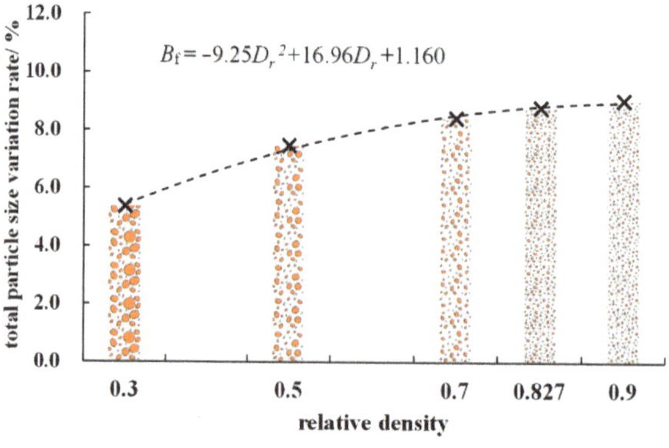

Figure 18. The effect of soil bulk density on the total particle size of silt.

Figure 19 shows the total particle shape index values of freeze–thaw silt with different densities under the conditions of vertical confining pressure 150 kPa, effective confining pressure 75 kPa, and freezing temperature −20 °C after 20 freeze–thaw cycles. It shows that the higher the relative bulk density, the higher the total particle shape index of silt. With the increase of relative bulk density, the variation of relative bulk density has less and less influence on total particle shape. The quadratic function model is used to fit the relationship between total particle shape index of silt and relative bulk density (shown as Equation (14)), and the data fitting effect is relatively good.

$$B_s = -1.66{D_r}^2 + 3.04D_r + 2.029 \quad R^2 = 0.98 \tag{14}$$

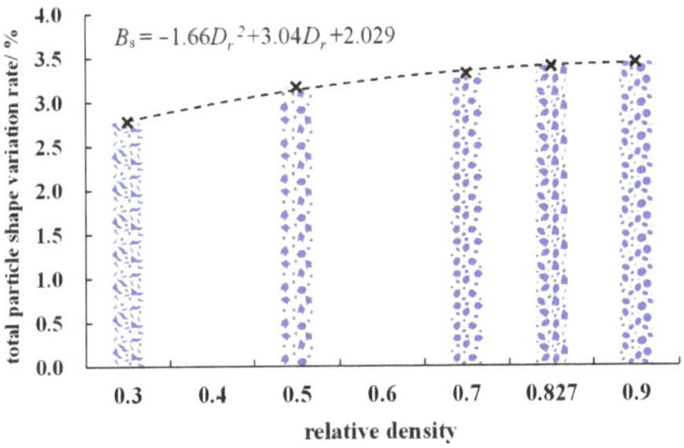

Figure 19. The effect of soil bulk density on the total particle shape of silt.

4.3.4. Fine Content

By mixing the silt and silty sand for a certain proportion, the cohesionless soil with different fine particle contents would be obtained. Under the condition of −20 °C freezing temperature for 20 freeze–thaw cycles, the total particle size index values of cohesionless soil with different fine particles content are shown in Figure 20. It can be seen that the total particle size index of cohesionless soil with higher fine particles content is greater. As the fine particles content increases, the variation of fine particles content has less and less influence degree on total particle size. The quadratic function model is used to fit the relationship between total particle size index of silty sand and fine particles content (shown as Equation (15)), and the data fitting effect is relatively good.

$$B_f = -1.1 \times 10^{-3}L^2 + 0.195L + 0.1368 \quad R^2 = 0.97 \tag{15}$$

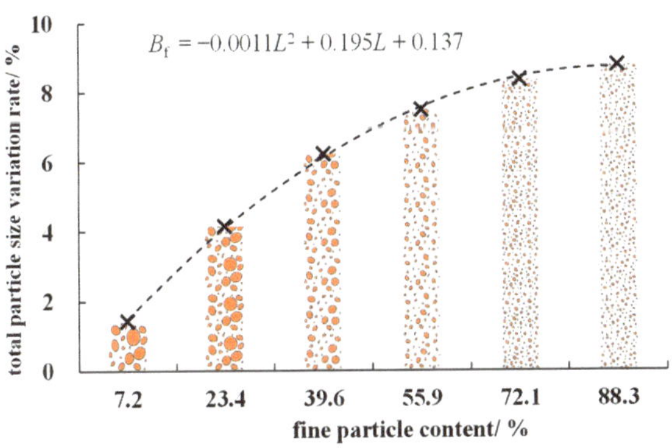

Figure 20. The effect of fine content on the total particle size.

Under the condition of −20 °C freezing temperature for 20 freeze–thaw cycles, the total particle size index values of cohesionless soil with different fine particles content are shown in Figure 21. It shows that the total particle shape index of the freeze–thaw cohesionless soil with higher fine particles content is greater. As the fine particles content increases, the variation of fine particles content has less influence degree on total particle shape. The quadratic function model is used to fit the relationship between total particle shape index and fine particles content (shown as Equation (16)), and the data fitting effect is relatively good.

$$B_s = -4.1 \times 10^{-4} L^2 + 0.078L + 0.182 \quad R^2 = 0.99 \tag{16}$$

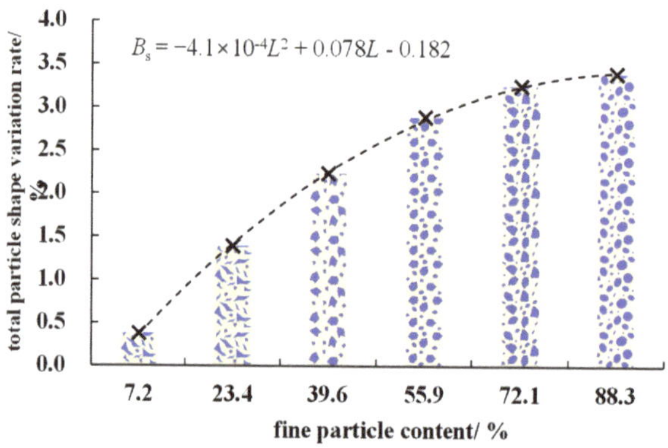

Figure 21. The effect of fine content on the total particle shape.

5. Mechanism and Discussion

When saturated soil is freezing, the phase variation of pore water would cause volume expansion which generates wedge force to increase the distance of soil particles [18]. Due to the capillary force and van der Waals force, the capillary water and film water in soil are less likely to be frozen than gravity water. The smaller the pores, the greater the proportion of capillary water and film water in these pores and the less likely they are to be frozen [19]. This induces the large pores near the freezing front to freeze first. With the continuous expansion of the freezing front, the medium pores and small pores nearby are frozen successively [20].

The soil of silt not only contains a large amount of silt particles (5–75 μm), but also contains some clay particles (<5 μm) and sand particles (>75 μm) (Figure 22a). Clay particles are smaller in size, larger in specific heat capacity, more spherical in shape, and the diameters of their nearby pores are small. Due to the small particle size, small diameter of nearby pores, and large specific heat capacity, pore water surrounding clay particles are less likely to be frozen under the same conditions than silt particles and sand particles. During the freezing process, the wedge force formed by the growth of ice crystals first acts on sand particles and silt particles, causing more frost heave displacements for large and medium particles, and increasing the diameter of nearby pores [21,22]. This exacerbates the possibility of mutual attraction and aggregation for clay particles (Figure 22b). Because of their small diameter and spherical shape of clay particles, the wedge force of ice crystal growth does not easily impact the clay particles themselves, so the shape variations of clay particles are not obvious. Due to the large particle size and irregular shape of silt particles and sand particles, the wedge force of ice crystal growth is more likely to make their structure looser and induces irregular boundary collapse of particles. Hence, it produces the phenomenon of silt particles and sand particles becoming broken, the total particle arrangement becoming messy, and the total particle shape becoming spherical (Figure 22c). Lowering the freezing temperature could cause the wedge force for ice crystal growth to become greater. Increasing the effective confining pressure has the same impact as lowering the freezing temperature. The higher the compaction degree, the greater the friction between soil particles, and soil particles are more likely to break rather than slip. Therefore, greater effective confining pressure, lower freezing temperature, and greater compactness could all exacerbate the impact of freeze–thaw action on total particle size and total particle shape to a certain extent. The result would be that freeze–thaw action has greater effects on total particles characteristics of silt with greater effective confining pressure, lower freezing temperature, and greater compaction degree.

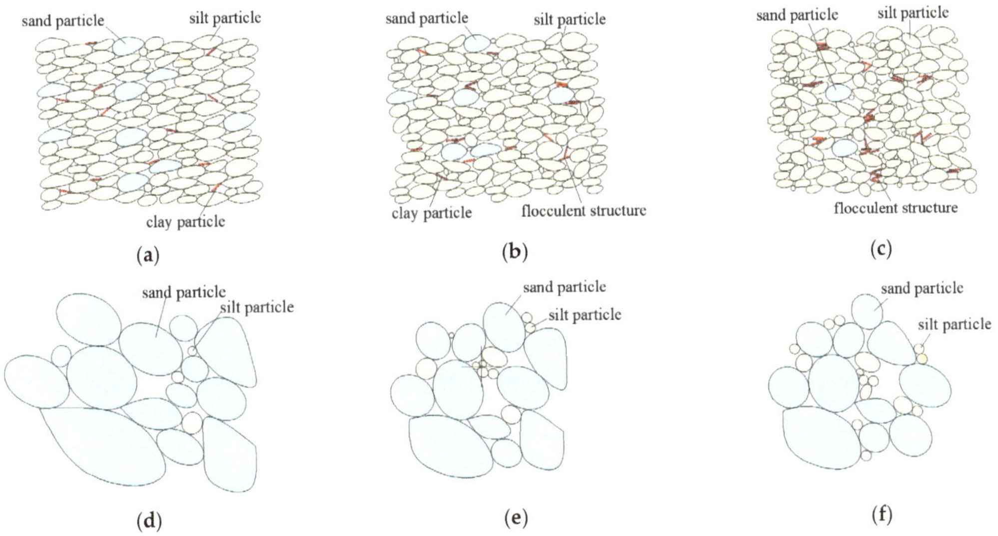

Figure 22. Schematic diagram of soil structure under freeze–thaw action. (**a**) Unfrozen silt; (**b**) silt after initial freeze–thaw cycle; (**c**) silt after long-term freeze–thaw cycles; (**d**) unfrozen sand; (**e**) sand after initial freeze–thaw cycle; (**f**) sand after long-term freeze–thaw cycles.

The content of clay particles in silty sand is very small, and total particle size is much larger than that of silt. Because of the small content of clay particles, SEM could not clearly observe the flocculent structure in silty sand after freeze–thaw, which only shows the characteristics of particle rearrangement and particle fragmentation (Figure 22e,f). The effects of freeze–thaw action on silt particles and sand particles in silty sand is similar to that in silt. Because the total particle size of silty sand is larger, the ability of individual particles to resist crushing deformation is stronger. Therefore, the effects of freeze–thaw action on silty sand is similar to that of silt, but the influence degree is smaller. The main distinction is the role of fine aggregate, especially for clay particles.

6. Conclusions

The SEM and LST tests were conducted to explore the effects of freeze–thaw action on the particle characteristics of saturated silt deposited in water environments. Several different impact factors were considered during sample reconstitution and freeze–thaw processes. The conclusions could be drawn as follows:

(1) Different for the only sand particle crushing in conventional soil shearing, freeze–thaw action could probably induce the particles' (60–200 μm) breakage, and also affect the clay particles' (less than 5 μm) aggregation and form flocculent structure. The diameters of flocculent structures and broken particles range from 5–60 μm. With continuous freeze–thaw cycles, this phenomenon could be more obvious. Freeze–thaw action could hardly influence the shape of particles smaller than 40 μm, and could

make the shape of particles larger than 40 μm more spherical (especially larger than 100 μm).

(2) Two parameters of particle size index and particle shape index are defined and specifically improved to quantitively evaluate the particle characteristics of silt under freeze–thaw action. They both hyperbolically increase with freeze–thaw cycles. Within 100 times, the influence degree of single freeze–thaw cycle on the particle size/shape of silt becomes smaller and gets more stable.

(3) Greater effective confining pressure, lower freezing temperature, greater compaction degree, and higher fine content could aggravate the influence of freeze–thaw cycles on particles' size and particles' shape for silt to a certain extent. From another aspect, the durability of pavements under freeze–thaw cycles could be comprehensively evaluated and controlled by these impact factors.

(4) From a perspective of freeze–thaw mechanism on different soil particles, two micro-structural evolution modes of particle characteristics under freeze–thaw action from initial freeze–thaw cycle to long-term freeze–thaw cycles have been inferred. The main distinction is the role of fine aggregate, especially for clay particles.

Author Contributions: Conceptualization, J.Z. and Z.L.; methodology, J.Z. and Z.L.; validation, J.Z. and W.P.; formal analysis, Z.L.; investigation, W.P.; resources, J.Z. and Z.L.; data curation, Z.L.; writing—original draft preparation, Z.L.; writing—review and editing, Z.L.; visualization, J.Z. and Z.L.; supervision, W.P.; project administration, J.Z.; funding acquisition, J.Z. All authors have read and agreed to the published version of the manuscript.

Funding: The research work herein was supported by National Natural Science Foundation of China (No. 41702299) and Foundation of State Key Laboratory of Frozen Soil Engineering (No. SKLFSE201916).

Conflicts of Interest: The authors declare no conflict of interest.

References

1. Arturo, E.C. The Frost Behavior of Soils. Part I: Vertical Sorting. In *Highway Research Board Bulletin*; U.S. Army Cold Regions Research and Engineering Laboratory: Hanover, NH, USA, 1962; pp. 9–34.
2. Arturo, E.C. The Frost Behavior of Soils: Laboratory and Field Data for a New Concept. Part II: Horizontal Sorting. In *Highway Research Board Bulletin*; U.S. Army Cold Regions Research and Engineering Laboratory: Hanover, NH, USA, 1962; pp. 46–64.
3. Edwin, J.C.; Anthony, J.G. Effect of freezing and thawing on the permeability and structure of soils. *Eng. Geol.* **1979**, *13*, 73–92.
4. Konrad, J.M. Physical processes during freeze-thaw cycles in clayey silts. *Cold Reg. Sci. Technol.* **1989**, *16*, 291–303. [CrossRef]
5. Qi, J.L.; Zhang, J.M.; Zhu, Y.L. The significance of soil mechanics of the effects of freezing and thawing on soil structure. *Chin. J. Rock Mech. Eng.* **2003**, *S2*, 2690–2694.
6. Mu, Y.H.; Ma, W.; Li, G.Y.; Mao, Y.C. Quantitative analysis of impacts of freeze-thaw cycles upon microstructure of compacted loess. *Chin. J. Geotech. Eng.* **2011**, *33*, 1919–1925.
7. Tan, Y.Z.; Wu, P.; Fu, W.; Wan Z, W.; Zhang, H.; Zhang, Z.H. Strength and micromechanism of improved silt under freeze-thaw cycle effect. *Rock Soil Mech.* **2013**, *34*, 2827–2834.
8. Tang, Y.Q.; Li, J.Z. Test method and application for microstructures of undisturbed silty sand and sandy silt. *Environ. Earth Sci.* **2018**, *77*, 657. [CrossRef]
9. Jin, Q.; Zheng, Y.J.; Cui, X.Z.; Cui, S.Q.; Qi, H.; Zhang, X.N.; Wang, S. Evaluation of dynamic characteristics of silt in Yellow River Flood Field after freeze-thaw cycles. *J. Cent. South Univ.* **2020**, *27*, 2113–2122. [CrossRef]
10. Zheng, Y.J.; Jin, Q.; Cui, X.Z.; Zhang, H.; Liu, Z.Q.; Zhang, J. Dynamic behavior and meso-damage evolution of saturated saline silt from yellow river flooded area under freeze-thaw cycle. *China J. Highw. Transp.* **2020**, *33*, 32–44.
11. Ren, J.; Vanapalli, S.K. Effect of freeze-thaw cycling on the soil-freezing characteristic curve of five Canadian soils. *Vadose Zone J.* **2020**, *19*, 1–8. [CrossRef]
12. Leuther, F.; Schlüter, S. Impact of freeze-thaw cycles on soil structure and soil hydraulic properties. *Soil* **2021**, *7*, 179–191. [CrossRef]
13. Zhou, J.; Li, Z.Y.; Tian, W.J.; Sun, J.W. Effects of Artificial Freezing on Liquefaction Characteristics of Nanjing Sand. *China Railw. Sci.* **2021**, *42*, 28–38.
14. Hardin, B.O. Crushing of soil particles. *J. Geotech. Eng.* **1985**, *111*, 1177–1192. [CrossRef]
15. Cavarretta, I.; Coop, M.R.; O'sullivan, C. The influence of particle characteristics on the behaviour of coarse grained soils. *Géotechnique* **2010**, *60*, 413–423. [CrossRef]
16. Marsal, R.J. Large-scale testing of rockfill materials. *ASCE J. Soil Mech. Found. Eng.* **1967**, *93*, 27–43. [CrossRef]

17. Zhang, S.; Tong, C.X.; Li, X.; Sheng, D. A new method for studying the evolution of particle breakage. *Géotechnique* **2015**, *65*, 911–922. [CrossRef]
18. Miller, R.D. Freezing and heaving of saturated and unsaturated soils. *Highw. Res. Rec.* **1972**, *399*, 1–11.
19. Christ, M.; Kim, Y.C. Experimental study on the physical-mechanical properties of frozen silt. *KSCE J. Civ. Eng.* **2009**, *13*, 317–324. [CrossRef]
20. Kim, S.Y.; Hong, W.T.; Lee, J.S. Silt fraction effects of frozen soils on frozen water content, strength, and stiffness. *Constr. Build. Mater.* **2018**, *183*, 565–577. [CrossRef]
21. Coop, M.R.; Sorensen, K.K.; Freitas, T.B.; Georgoutsos, G. Particle breakage during shearing of a carbonate sand. *Géotechnique* **2004**, *54*, 157–163. [CrossRef]
22. Fan, W.; Yang, P. Ground temperature characteristics during artificial freezing around a subway cross passage. *Transp. Geotech.* **2019**, *20*, 100250. [CrossRef]

applied sciences

MDPI

Article

An Analytical Solution to Steady-State Temperature Field in the FSPR Method Considering Different Soil Freezing Points

Yin Duan [1,2,*], Chuanxin Rong [1,2], Xianwen Huang [2,3] and Wei Long [1,2]

[1] State Key Laboratory of Mining Response and Disaster Prevention and Control in Deep Coal Mines, Huainan 232001, China
[2] School of Civil Engineering and Architecture, Anhui University of Science and Technology, Huainan 232001, China
[3] School of Civil Engineering, Suzhou University of Science and Technology, Suzhou 215009, China
* Correspondence: yinduan@aust.edu.cn

Abstract: Taking the freeze-sealing pipe roof method (FSPR) adopted in the Gongbei Tunnel project as the background, this study develops a simplified calculation model by considering different soil freezing points, tube layout, and site conditions. The analytical solution of the linear single row tubes is then used to formulate the analytical solution of the freezing temperature field of two kinds of linear single row tubes, with equal spacing in the image plane. This is achieved through conformal mapping and the variable separation method. Finally, the analytical solution to the steady-state temperature field of FSPR in the object plane is obtained. The numerical solutions of common freezing parameters in freezing engineering are analyzed to evaluate the accuracy of the analytical solution, and the influence of parameter differences on the freezing temperature field are also discussed, to provide a theoretical reference for popularization and application of similar construction methods.

Keywords: freeze-sealing pipe roof method; analytical solution; steady-state temperature field; freezing point

1. Introduction

The theoretical analysis of artificial freezing temperature fields has been extensively investigated in the ground freezing engineering field, particularly in heat conduction problems, including "phase transition", "hydrothermal coupling", "temporal and spatial effect" and other factors [1,2]. The currently applied research methods include analytical, experimental and numerical methods [3–7]. These analytical methods use mathematical and physical equations to establish accurate functional relationships for research problems. The relationships between variables in the function are clear and can be directly solved, and they can be applied by engineers and technicians during the design stage and to evaluate the effect of on-site freezing construction. Therefore, they can be utilized to study artificial freezing temperature fields [8–10]. However, considering the differences in the number and arrangement of freezing tubes, freezing front movement, water migration and other factors, the analytical method still has some limitations when applied in mathematical solutions, and can only analyze the temperature field of single tube freezing. In academic and engineering fields, many cases [11] have confirmed that the artificial freezing temperature field develops very slowly, at the final stage of the freezing process. Since the freezing tube will have reached the equilibrium state of cooling and heat absorption, the size of the frozen wall almost does not change, and would be very close to the steady-state heat transfer temperature field [12]. Therefore, the steady-state heat transfer model can be used for approximate calculation.

Extensive research and derivation have been performed on the analytical solutions of freezing steady-state temperature field. Currently, the commonly used analytical solutions include the single-tube freezing steady-state temperature field [13], the two-to-five-tube equidistant linear arrangement steady-state temperature field [14–16] presented by Bakholdin, the symmetric and asymmetric steady-state temperature fields of linear single-row tubes [17], linear double-row tubes and three-row tubes [18–20], and the annular single- circle-tube and double- circle-tube temperature field [21,22]. The aforementioned results are based on the classic Trupak single-tube freezing steady-state temperature field analytical solution formula, obtained through potential function superposition, separation variable method, conformal mapping and other processing methods, combined with practical engineering situations to simplify the corresponding model, and improve its application value in engineering. However, with the continuous and rapid development of urban construction in China, higher requirements have been proposed regarding the formation and functioning of underground engineering structures. For many engineering fields that involve difficult construction, it is challenging to establish strong support methods to handle complex geological conditions. Most analytical solutions also need to be constantly revised or optimized to meet higher requirements such as various geological conditions, accurate and fast calculation of temperature and frozen soil curtain thickness [23–25]. Therefore, continuous in-depth studies should be performed to establish a theoretical basis that is in line with engineering practices.

To overcome the challenges associated with constructing shallow buried and concealed excavation tunnels with large sections in the water-rich soft soil of coastal cities, Chinese experts and scholars have put forward a new tunnel construction method; the freeze-sealing pipe roof method (FSPR) that integrates the pipe roof method (PRM) and the artificial ground freezing method (AGF) [26,27]. In Figure 1, a plurality of closely arranged large-diameter steel pipes are jacked into the strata at both ends of the tunnel section to form a pipe roof. Subsequently, the surrounding water-containing soft soil layer is artificially frozen by installing a freezing tube in the inner wall of the jacking pipe. A closed freezing curtain is created within the scope of the tunnel excavation section, which finally constitutes a large-scale composite support structure of "frozen soil curtain and jacking pipe" [28].

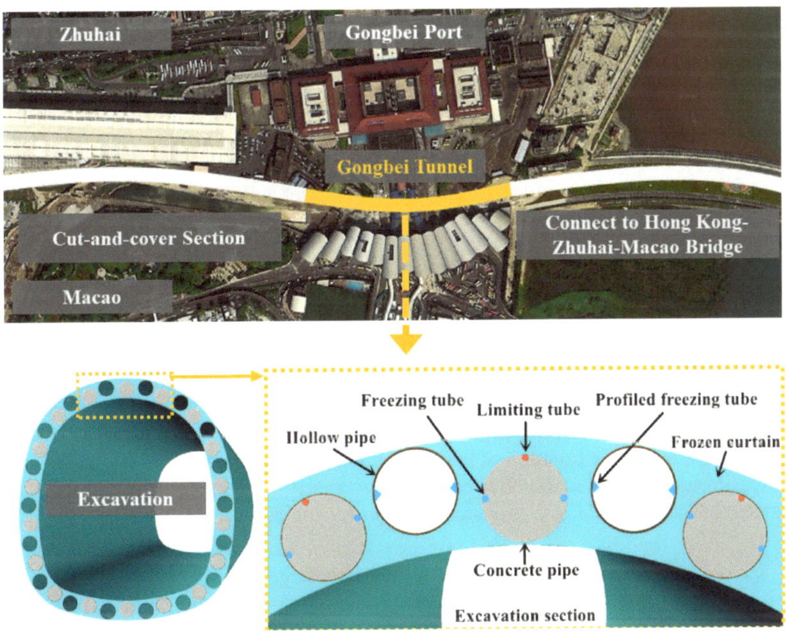

Figure 1. Schematic presentation of FSPR applied in the Gongbei Tunnel.

In Figure 1, the underground excavation section of the Gongbei Tunnel, the key project of the Zhuhai link of the Hong Kong-Zhuhai-Macao Bridge in China, achieved successful application of FSPR for the first time in the world, with good engineering results. In the construction scheme of this project, artificial ground freezing technology was used to freeze the ground to form a frozen soil curtain, thereby sealing the water between the pipes and improving the bearing capacity of the pipe roof. The freezing construction period takes 180 days, and considering the damage caused by excessive frost heaving, the thickness of the frozen soil curtain is designed in advance, and the cooling capacity is strictly controlled [29]. According to the field measured data, after 90 days of freezing, the thickness of the frozen soil curtain remains unchanged, because the cooling supply and heat absorption of the freezing tube reach a state of balance, so the temperature field after that can be regarded as a quasi-steady state. Model analysis and calculation of the freezing temperature field is the basis of theoretical research on FSPR, which can provide strong support for similar methods in freezing construction parameter designs, process monitoring, and target prediction [30]. During the whole freezing construction process, it is necessary to grasp the frozen soil curtain thickness and temperature distribution law based on the calculation and analysis of the temperature field, and to evaluate the reliability of water sealing between pipes [31].

In this article, inspired by the existing analytical results and to better adapt to the complex ground freezing conditions in the practical project, we have established a simplified model by considering different soil freezing points and tube placement forms of FSPR. Through the conformal mapping function and separation of variables solution, the analytical solution of the freezing temperature field of two kinds of freezing tubes with equal spacing in a straight line in the image plane is derived, after which the analytical solution of steady-state temperature field in the object plane is obtained. The accuracy of the analytical solution is verified by comparing the numerical solution of this project in the range of soil freezing point $0\sim-1.5$ °C, and the influence of the parameter differences on the freezing temperature field is also discussed to provide a theoretical reference for popularization and applications of similar construction methods.

2. Establishment of the Calculation Model for FSPR Steady State Temperature Field

2.1. Model Simplifications and Assumptions

In Figure 2, 18 concrete pipes and 18 hollow pipes were alternately arranged to form a super-large section pipe roof during the freezing construction of the pipe curtain of Gongbei Tunnel. The arrangement axis of the 18 hollow top pipes is a 5-segment circular arc with left and right symmetry, while the arc lengths and radii are Arc_1 = 5.45 m and R_1 = 9.86, Arc_2 = 4.87 m and R_2 = 6.96 m, Arc_3 = 11.75 m and R_3 = 20.96 m, Arc_4 = 3.86 m and R_4= 3.86 m, Arc_5 = 6.17 m and R_5 = 18.86 m, respectively. The axes of the 18 concrete pipes are offset by 30 cm inward, thus, there is a slight dislocation of the circular freezing tubes in concrete pipes, and of the profiled freezing tubes in hollow pipes. The soil between the pipes is frozen through the cryogenic refrigerant circulation method in these two types of freezing tubes. The limiting tube is arranged on the outside of the axis of the concrete pipe; excess cold can be removed by circulating hot brine in the tube, while the thickness of the outside of the frozen soil curtain can be controlled to reduce frost heave of the stratum. The structural form and function of the FSPR method are extremely complex, therefore, appropriate assumptions and model simplification are required to ensure the feasibility of the analytical solution of the steady-state temperature field, including:

(1) The entire length of the underground excavation section of the Gongbei Tunnel is 255 m long and is curved. The actual tube curtain freezing is a three-dimensional heat conduction problem. The temperature deviation of longitudinal freezing is ignored, and it can be simplified to a two-dimensional plane problem.

(2) Ignoring the irregular shape of the pipe curtain section and the slight offset between the hollow and concrete pipe axes, all 36 pipes are considered to be arranged on the same circumferential line, that is, the pipe curtain section is simplified to a circle.

(3) In the actual project, due to the arrangement of the pipes, the outline of the frozen soil curtain is irregularly wavy. Considering the steady-state temperature field, the end of the freezing process is studied. For mathematical derivation convenience, it is assumed that the contour line of the frozen soil curtain is approximately a circle, and its rationality can be evaluated by verifying the analytical solution.

(4) The profiled freezing tube in the hollow pipe contains a non-circular section, and its size is smaller compared with that of the jacking pipe. It is estimated to have the same section and size as the circular freezing tube in the concrete pipe. Flow and temperature differences of the low-temperature refrigerant in the two types of freezing pipes in the freezing process are ignored, and only the two types of freezing tubes with the same tube wall temperature are considered during derivation of the analytical solution. The effects of hollow and concrete pipes on the freezing temperature field are also ignored, and only the effects of freezing tubes are considered.

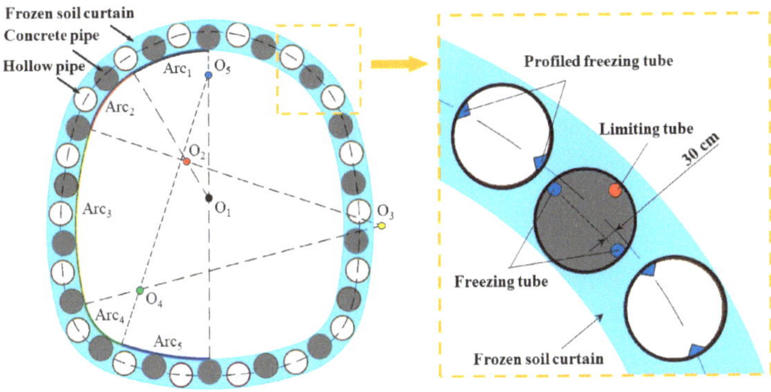

Figure 2. Layout of pipe roof and freezing tube in Gongbei Tunnel.

Based on the above assumptions, the model in Figure 2 is simplified, and freezing tubes as well as frozen soil curtains are selected as the main research objects [32]. Two types of freezing tubes, A and B, with radii R_0 are obtained, which are periodically arranged on the circumference line (R_2), and the spacing is set as the dislocation angle β. The inner and outer boundaries (R_1 and R_3) of the frozen soil curtain are circular (Figure 3):

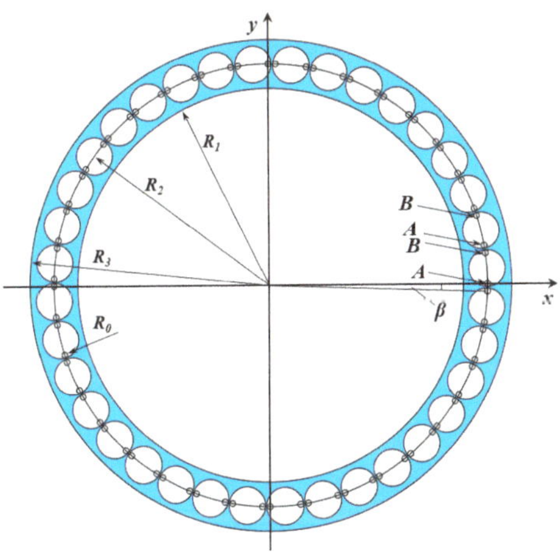

Figure 3. Simplified model of freezing temperature field of FSPR.

In Figure 3, the appropriate cartesian coordinate system is selected so that the center of a type A freezing tube is just on the positive semi-axis of the x-axis. Next, the corresponding mathematical expression of the two-dimensional steady-state temperature field is calculated using Equation group (1):

$$\begin{cases} \frac{\partial^2 T}{\partial r^2} + \frac{1}{r}\frac{\partial T}{\partial r} + \frac{\partial^2 T}{\partial \theta^2} = 0; \text{Two-dimensional steady-state heat conduction equation} \\ T(R_2 + R_0, k\frac{2\pi}{n}) = T_f; \text{Boundary condition for A-type freezing tubes} \\ T(R_2 + R_0, k\frac{2\pi}{n} - \beta) = T_f; \text{Boundary condition for B-type freezing tubes} \\ T(R_1, k\frac{2\pi}{n}) = T_0; \text{Inner boundary conditions for frozen soil curtain, } T_0 \neq 0 \\ T(R_3, k\frac{2\pi}{n}) = T_0; \text{Outer boundary conditions for frozen soil curtain, } T_0 \neq 0 \end{cases} \quad (1)$$

where:

r, θ – Polar diameter and polar angle
T_0, T_f – Frozen curtain boundary temperature and freezing tube wall temperature
n, R_0 – Number and radius of type A and B freezing tubes
k – Values from 0 to n − 1
R_1, R_3 – Inner and outer boundary radii of frozen curtain
R_2 – Freezing tube arrangement circle diameter
β – Dislocation angle between two adjacent freezing tubes

2.2. Conformal Mapping and Calculation Model Transformation

Considering that it is difficult to directly solve Equation group 1, conformal mapping should be considered during the conversion of circular boundary conditions of the model into the corresponding linear boundary conditions, thus, the logarithmic transformation function is introduced [33,34]:

$$\begin{cases} \zeta = i\ln\left(\frac{Z}{R_2}\right) \\ Z = re^{i\theta}; \text{Object plane} \\ \zeta = x + iy; \text{Image plane} \end{cases} \quad (2)$$

where by Z represents the object plane (i.e., the original plane in Figure 3), r and θ represent a point in the object plane, ζ represent the image plane, x and y represent a point in the image plane, we can obtain:

$$x + iy = -\theta + i\ln\left(\frac{r}{R_2}\right) \quad (3)$$

From Equations (2) and (3), we can convert the computational model in Figure 3 into the non-equidistant single-row tube with asymmetric development of the frozen curtain in the image plane, as shown in Figure 4:

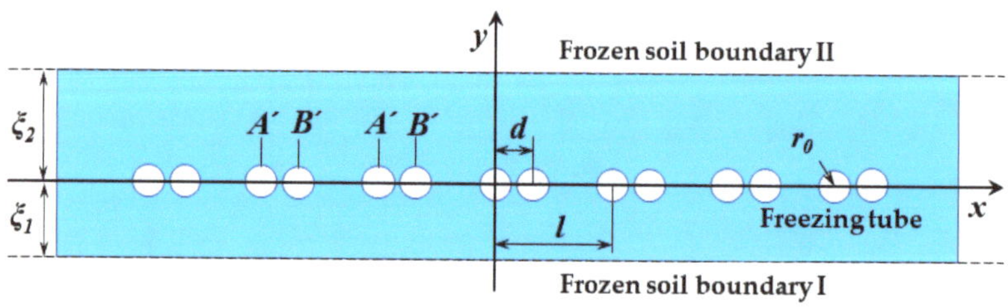

Figure 4. Freezing temperature field model in image plane of non-equidistant single-row tube with asymmetric development of frozen curtain.

This model can be calculated using Equation groups (4) and (5):

$$\begin{cases} x = -\theta, y = \ln \frac{r}{R_2} \\ \xi_1 = \ln \frac{R_2}{R_1}, \xi_2 = \ln \frac{R_3}{R_2} \\ l = \frac{2\pi}{n}, d = \beta, r_0 = \frac{R_0}{R_2} \end{cases} \tag{4}$$

$$\begin{cases} \frac{\partial^2 T}{\partial x^2} + \frac{\partial^2 T}{\partial y^2} = 0 \\ T(nl, r_0) = T_f; \text{Boundary condition for A}'\text{-type freezing tubes} \\ T(d + nl, r_0) = T_f; \text{Boundary condition for B}'\text{-type freezing tubes} \\ T(x, \xi_2) = T_0; \text{Frozen soil boundary II}, \ T_0 \neq 0 \\ T(x, -\xi_1) = T_0; \text{Frozen soil boundary I}, \ T_0 \neq 0 \end{cases} \tag{5}$$

where by, Equation group (4) denotes the conformal mapping function relationship between the models in Figures 3 and 4. Equation group (5) denotes the expressions of the model in Figure 4.

If the two types of freezing tubes A′ and B′ are separated, two linear single-row tubes' (equidistantly spaced) models of asymmetric frozen soil curtains can be obtained. Therefore, based on the separation variable solution method [35–37], the model in Figure 4 is regarded as a superposition of the two types of linear single-row tube equidistant arrangement models A′ and B′ (Figure 5):

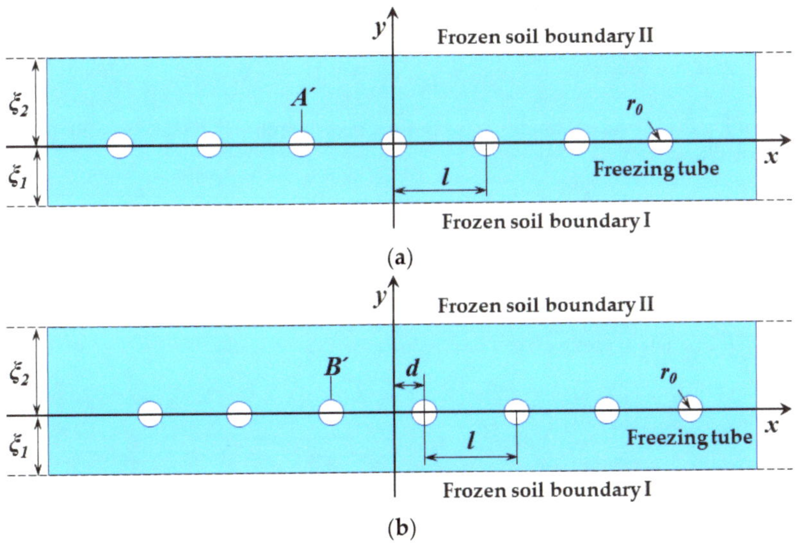

Figure 5. Freezing temperature field model in image plane of two kinds of linear single-row tubes: (**a**) A′-type linear single-row tube; (**b**) B′-type linear single-row tube.

Mathematical expressions of A′-type and B′-type linear single-row tube models are shown in Equation groups (6) and (7), respectively:

$$\begin{cases} \frac{\partial^2 T_1}{\partial x^2} + \frac{\partial^2 T_1}{\partial y^2} = 0 \\ T_1(nl, r_0) = T_f - a; \text{Boundary condition for A}'\text{-type freezing tubes} \\ T_1(d + nl, r_0) = b; \text{Boundary condition for B}'\text{-type freezing tubes} \\ T_1(nl, \xi_2) = T_0; \text{Frozen soil boundary II}, \ T_0 \neq 0 \\ T_1(nl, -\xi_1) = T_0; \text{Frozen soil boundary I}, \ T_0 \neq 0 \end{cases} \tag{6}$$

$$\begin{cases} \frac{\partial^2 T_2}{\partial x^2} + \frac{\partial^2 T_2}{\partial y^2} = 0 \\ T_2(nl, r_0) = a; \text{Boundary condition for A}'\text{-type freezing tubes} \\ T_2(d+nl, r_0) = T_f - b; \text{Boundary condition for B}'\text{-type freezing tubes} \\ T_2(nl, \xi_2) = T_0; \text{Frozen soil boundary II, } T_0 \neq 0 \\ T_2(nl, -\xi_1) = T_0; \text{Frozen soil boundary I, } T_0 \neq 0 \end{cases} \tag{7}$$

where the model temperature field T (Equation group (5)) is the superposition of temperature fields T_1 and T_2 of two different frozen soil boundary conditions, namely: $T = T_1 + T_2$. a and b are the coefficients to be solved.

2.3. Analytical Solution for Freezing Temperature Field Model in the Image Plane of Non-Equidistant Single-Row Tube with Asymmetric Development of Frozen Curtain

For the A′-type model in Figure 5a, according to Bakholdin's single-row tube freezing model theory and characteristics of asymmetric frozen soil curtains, the general form of its steady-state analytical solution can be obtained as:

$$T = \frac{T_f - T_0}{\ln\frac{2\pi r_0}{l} - \frac{\pi}{l}\cdot\frac{2\xi_1\xi_2}{\xi_1+\xi_2}}\left\{\frac{1}{2}\ln\left[2(\cosh\frac{2\pi y}{l} - \cos\frac{2\pi x}{l})\right] - \frac{\pi}{l}\cdot\frac{2\xi_1\xi_2}{\xi_1+\xi_2} + \frac{\pi}{l}\cdot\frac{\xi_1-\xi_2}{\xi_1+\xi_2}y\right\} + T_0 \tag{8}$$

Using Equation (8), the solution for Equation group (6) can be obtained as:

$$T_1 = \frac{T_f - a - T_0}{\ln\frac{2\pi r_0}{l} - \frac{\pi}{l}\cdot\frac{2\xi_1\xi_2}{\xi_1+\xi_2}}\left\{\frac{1}{2}\ln\left[2(\cosh\frac{2\pi y}{l} - \cos\frac{2\pi x}{l})\right] - \frac{\pi}{l}\cdot\frac{2\xi_1\xi_2}{\xi_1+\xi_2} + \frac{\pi}{l}\cdot\frac{\xi_1-\xi_2}{\xi_1+\xi_2}y\right\} + T_0 \tag{9}$$

Substituting the boundary condition for B′-type freezing tubes in Equation group (6) into Equation (9), we get:

$$b = \frac{T_f - a - T_0}{\ln\frac{2\pi r_0}{l} - \frac{\pi}{l}\cdot\frac{2\xi_1\xi_2}{\xi_1+\xi_2}}\left\{\frac{1}{2}\ln\left[2(\cosh\frac{2\pi r_0}{l} - \cos\frac{2\pi(d+nl)}{l})\right] - \frac{\pi}{l}\cdot\frac{2\xi_1\xi_2}{\xi_1+\xi_2} + \frac{\pi}{l}\cdot\frac{\xi_1-\xi_2}{\xi_1+\xi_2}r_0\right\} + T_0 \tag{10}$$

In the same way, the B′-type model in Figure 5b is equivalent to the A′-type model where each freezing tube is shifted to the right by a distance d. From Equation (8), the solution for Equation group (7) can be obtained as:

$$T_2 = \frac{T_f - b - T_0}{\ln\frac{2\pi r_0}{l} - \frac{\pi}{l}\cdot\frac{2\xi_1\xi_2}{\xi_1+\xi_2}}\left\{\frac{1}{2}\ln\left[2(\cosh\frac{2\pi y}{l} - \cos\frac{2\pi(x-d)}{l})\right] - \frac{\pi}{l}\cdot\frac{2\xi_1\xi_2}{\xi_1+\xi_2} + \frac{\pi}{l}\cdot\frac{\xi_1-\xi_2}{\xi_1+\xi_2}y\right\} + T_0 \tag{11}$$

Substituting the boundary condition for A′-type freezing tubes in Equation group (7) into Equation (11), we get:

$$a = \frac{T_f - b - T_0}{\ln\frac{2\pi r_0}{l} - \frac{\pi}{l}\cdot\frac{2\xi_1\xi_2}{\xi_1+\xi_2}}\left\{\frac{1}{2}\ln\left[2(\cosh\frac{2\pi r_0}{l} - \cos\frac{2\pi(nl-d)}{l})\right] - \frac{\pi}{l}\cdot\frac{2\xi_1\xi_2}{\xi_1+\xi_2} + \frac{\pi}{l}\cdot\frac{\xi_1-\xi_2}{\xi_1+\xi_2}r_0\right\} + T_0 \tag{12}$$

Since Equations (10) and (12) are relatively complex, they are simplified before the simultaneous solution:

$$\begin{cases} a = (T_f - b - T_0)\frac{\eta_\zeta}{\varphi_\zeta} + T_0 \\ b = (T_f - a - T_0)\frac{\eta_\zeta}{\varphi_\zeta} + T_0 \end{cases} \tag{13}$$

where:

$$\eta_\zeta = \frac{1}{2}\ln\left[2(\cosh\frac{2\pi r_0}{l} - \cos\frac{2\pi d}{l})\right] - \frac{\pi}{l}\cdot\frac{2\xi_1\xi_2}{\xi_1+\xi_2} + \frac{\pi}{l}\cdot\frac{\xi_1-\xi_2}{\xi_1+\xi_2}r_0$$

$$\varphi_\xi = \ln\frac{2\pi r_0}{l} - \frac{\pi}{l}\cdot\frac{2\xi_1\xi_2}{\xi_1+\xi_2}$$

Equation group (13) is solved to get:

$$a = b = \frac{T_f \frac{\eta_\zeta}{\varphi_\zeta} + T_0(1 - \frac{\eta_\zeta}{\varphi_\zeta})}{1 + \frac{\eta_\zeta}{\varphi_\zeta}} \tag{14}$$

Substitute Equation (14) into Equations (9) and (11) respectively, and according to $T = T_1 + T_2$, we get:

$$T = \frac{\gamma_\zeta}{\varphi_\zeta + \eta_\zeta} \cdot (T_f - 2T_0) + 2T_0 \tag{15}$$

where:

$$\gamma_\zeta = \frac{1}{2}\ln\left[2(\cosh\frac{2\pi y}{l} - \cos\frac{2\pi x}{l})\right] + \frac{1}{2}\ln\left[2(\cosh\frac{2\pi y}{l} - \cos\frac{2\pi(x-d)}{l})\right] - \frac{2\pi}{l}\cdot\frac{2\xi_1\xi_2}{\xi_1+\xi_2} + \frac{2\pi}{l}\cdot\frac{\xi_1-\xi_2}{\xi_1+\xi_2}y \tag{16}$$

Equation (16) is the analytical solution for the freezing temperature field model in the image plane of a non-equidistant single-row tube with asymmetric development of the frozen curtain shown in Figure 4.

2.4. Analytical Solution for Freezing Temperature Field Model in Object Plane of FSPR

Substituting Equation group (4) into Equation (16), we get:

$$T = \frac{\gamma_Z}{\varphi_Z + \eta_Z} \cdot (T_f - 2T_0) + 2T_0 \tag{17}$$

where:

$$\gamma_Z = \frac{1}{2}\ln\left[\left(\frac{r}{R_2}\right)^n + \left(\frac{R_2}{r}\right)^n - 2\cos n\theta\right] + \frac{1}{2}\ln\left[\left(\frac{r}{R_2}\right)^n + \left(\frac{R_2}{r}\right)^n - 2\cos n(\theta+\beta)\right] - \frac{2n\ln\frac{R_2}{R_1}\ln\frac{R_3}{R_2}}{\ln\frac{R_3}{R_1}} + \frac{n\ln\frac{R_2^2}{R_1R_3}}{\ln\frac{R_3}{R_1}}\cdot\ln\frac{r}{R_2}$$

$$\varphi_Z = \ln\frac{nR_0}{R_2} - \frac{n\ln\frac{R_2}{R_1}\ln\frac{R_3}{R_2}}{\ln\frac{R_3}{R_1}}$$

$$\eta_Z = \frac{1}{2}\ln\left(e^{\frac{nR_0}{R_2}} + e^{-\frac{nR_0}{R_2}} - 2\cos n\beta\right) - \frac{n\ln\frac{R_2}{R_1}\ln\frac{R_3}{R_2}}{\ln\frac{R_3}{R_1}} + \frac{n\ln\frac{R_2^2}{R_1R_3}}{2\ln\frac{R_3}{R_1}}\cdot\frac{R_0}{R_2}$$

Equation (17) is the analytical solution to steady-state temperature field of FSPR when considering different soil freezing points as shown in Figure 3.

3. Accuracy Verification of the Analytical Solution

3.1. Selection of Feature Parameters

For the calculation model of FSPR in Figure 3, since the freezing tubes are periodically arranged, take an A-type freezing tube and a B-type freezing tube adjacent to the x-axis as research objects. Taking the origin of coordinates as the center of the circle, select the fan-shaped area as shown in Figure 6 for calculation.

Select the x-axis direction (main section, $\theta = 0$), between two pipes (Section 1, $\theta = -\beta/2$), and the sector boundary (Section 2, $\theta = \pi/n - \beta/2$) as the three characteristic sections for temperature calculation, ξ_1 and ξ_2 are the inner and outer frozen curtain thicknesses, respectively.

Based on previous freezing engineering experience, six groups of freezing parameters are selected for calculation. The value range of freezing tube circle radius R_2 is 2.5~10.0 m; value range of jacking pipe diameter D is 0.8~2.0 m; value range of the frozen pipe radius R_0 is 0.06~0.16 m; value range of ξ_1/ξ_2 is 1~1.2; temperature of the freezing tube wall T_f is −30 °C; value range of the frozen soil boundary temperature is T_0 −1.5~0 °C; frozen tube parameters, such as n and β are shown in Table 1:

Table 1. Characteristic parameters for analytical solution.

Group	R_1/m	R_2/m	R_3/m	ξ_1/ξ_2	D/m	R_0/m	n	β/(°)	T_0/°C
1	6.0	7.0	8.0	1	1.46	0.06	20	4	0
2	6.0	7.0	8.0	1	1.02	0.06	25	4	0
3	8.0	9.0	10.0	1	1.62	0.06	36	2	−1.5
4	8.0	9.0	10.0	1	1.46	0.08	36	3	−1.5
5	7.9	9.0	10.0	1.1	1.59	0.08	36	2.6	−0.5
6	7.9	9.0	10.0	1.1	1.59	0.06	36	2.6	−0.5

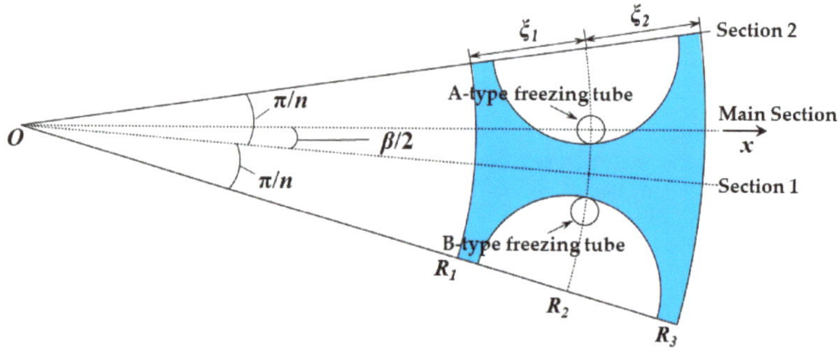

Figure 6. Computational periodic element model of the analytical solution.

3.2. Establishment of a Numerical Calculation Model

According to the model diagram in Figure 3 and characteristic parameters in Table 1, six two-dimensional steady-state temperature field numerical models are established using COMSOL Multiphysics [38]. The Heat Transfer in Porous Media module is used for calculation, and the correctness as well as the accuracy of the analytical solution are verified by comparing the results.

Taking the first group of parameters in Table 1 as an example, a sector of a period ($2\pi/n$ = 18°) is selected to build a model as shown in Figure 7a. The model is divided by triangular mesh elements. To ensure the calculation's accuracy, select the "Extra fine" option for the mesh element size, and increase the mesh density in the freezing tube area, as shown in Figure 7b.

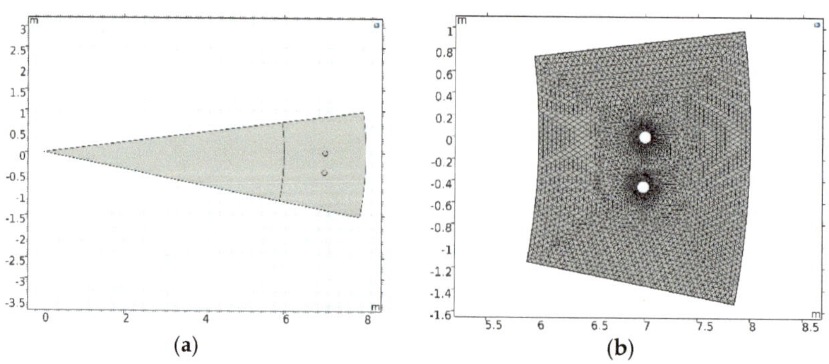

Figure 7. Numerical model of parameter group 1: (**a**) Model; (**b**) Computational domain grid.

3.3. Comparative Analysis of Calculation Results

Calculation results and the cloud map of the temperature field are compared as shown in Figures 8–13:

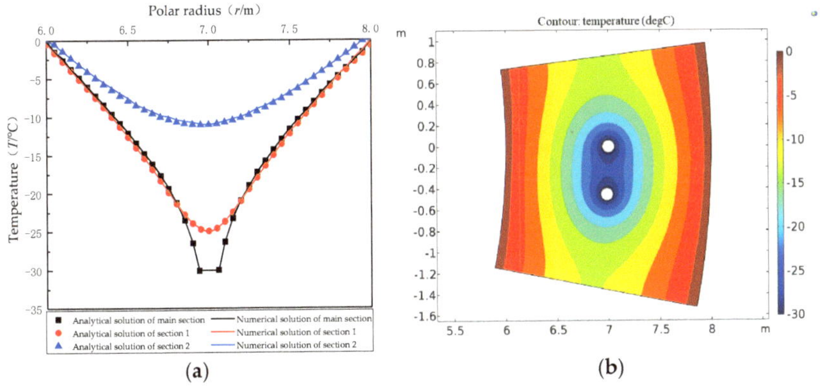

Figure 8. Comparison of the 1st group of calculation results: (**a**) Comparison curve between analytical and numerical solutions of section; (**b**) Steady-state temperature field.

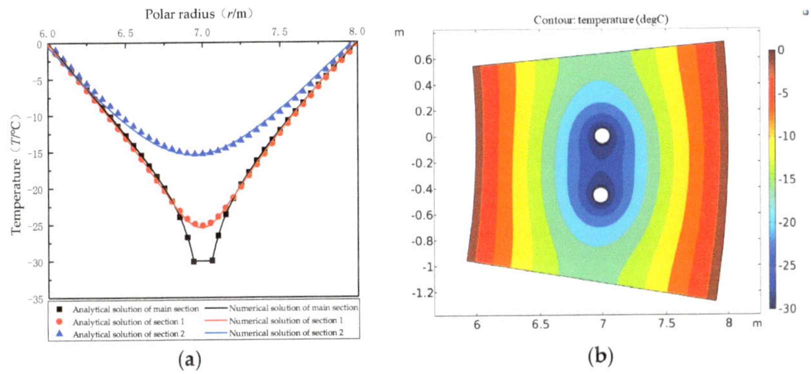

Figure 9. Comparison of the 2nd group of calculation results: (**a**) Comparison curve between analytical and numerical solutions of section; (**b**) Steady-state temperature field.

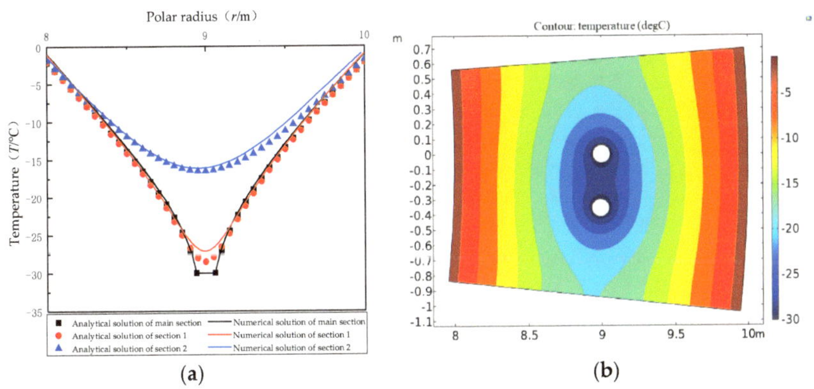

Figure 10. Comparison of the 3rd group of calculation results: (**a**) Comparison curve between analytical and numerical solutions of section; (**b**) Steady-state temperature field.

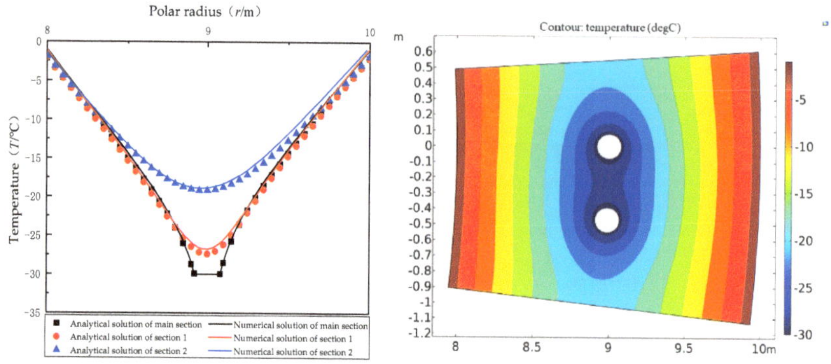

(a) (b)

Figure 11. Comparison of the 4th group of calculation results: (**a**) Comparison curve between analytical and numerical solutions of section; (**b**) Steady-state temperature field.

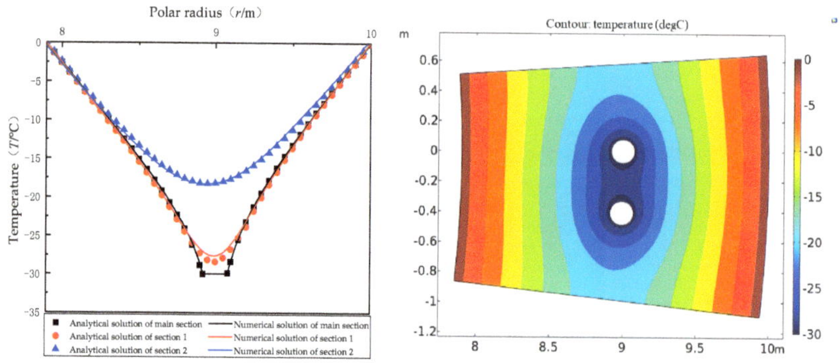

(a) (b)

Figure 12. Comparison of the 5th group of calculation results: (**a**) Comparison curve between analytical and numerical solutions of section; (**b**) Steady-state temperature field.

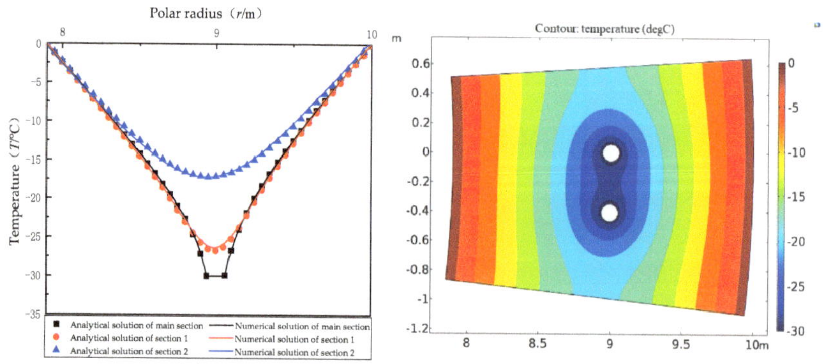

(a) (b)

Figure 13. Comparison of the 6th group of calculation results: (**a**) Comparison curve between analytical and numerical solutions of section; (**b**) Steady-state temperature field.

From Figures 8–13, the analytical solution and numerical solution curve for each model section under six groups of different parameters coincide, and maximum temperature differences at each point do not exceed 0.9 °C. The data indicate that the steady-state temperature field model of the complicated freezing tube layout in FSPR can be simplified, and conformal mapping and the separation-variable method can be employed to solve the problem. The derived analytical solution is highly accurate, and can accurately calculate the temperature value at any point in the model.

In the vicinity of the inner and outer boundaries of the model's frozen soil, temperatures of the three sections in each group are relatively close, corresponding to symmetrical and uniform temperature distributions on both sides of steady-state temperature field cloud map. In the area closer to the freezing tube, temperature differences among the three sections are greater, and the maximum temperature difference is located at the axial surface R_2 of the freezing tube. The temperature curve of Section 1 is located between Section 2 and the main section, and is closer to the latter.

In actual engineering, Section 1 is the position of the midline between the adjacent concrete pipe and hollow pipe, and is an important area for "freeze-sealing between pipes" in FSPR. The above figures show that the temperature range of this area within the size range of the pipe is −10 °C~−28 °C and the distribution is relatively uniform, implying that a reliable frozen soil curtain can be formed between the jacking pipes to ensure "freeze-sealing" effects and safety.

3.4. Discussion of Analytical Solution in FSPR

From the parameter selection of each group in Table 1, the smaller the dislocation angle β value and the larger the frozen tube radius R_0, the lower the temperature of the freezing tube area. In the FSPR method adopted in the Gongbei Tunnel, the approximate circle radius of the freezing tube ring is about 9 m, and the diameter of the jacking pipe is 1.62 m. The distance between adjacent jacking pipes is about 0.3 m, and the dislocation angle is about 2°. Taking the freezing parameters $R_1 = 7.9$ m, $R_2 = 9$ m, $R_3 = 10$ m, and $\xi_1/\xi_2 = 1.1$ at the end of the freezing stage, respectively, calculate the temperature difference of the three sections of the frozen soil curtain at the axial surface R_2 of the freezing tube, as shown in Table 2:

Table 2. Temperature differences of each section at the end of freezing.

	Main Section		Section 1		Section 2
Temperature/°C	−30		−28.51		−16.08
ΔT_1/°C		−1.49			——
ΔT_2/°C	——			−12.43	

At the end of freezing, the temperature difference between the main section and Section 1 is ΔT1 = −1.49 °C, while the temperature difference between sections 1 and 2 is ΔT2 = −12.43 °C. Section 2 is located at the axis of symmetry of the steel jacking pipe, which can provide effective water sealing capacity, therefore, only Section 1 should be considered. The temperature for Section 1 can drop to a lower temperature value at the end, which is very close to the temperature of the freezing tubes in the main section. Notably, frozen soils can be formed with sufficient strength and water sealing performance between the jacking pipes. Then, an effective frozen curtain is formed as a whole, which provides safe and reliable support for tunnel excavation.

Finally, the steady-state temperature field distribution cloud map of the overall model in FSPR is calculated and drawn in line with the analytical solution formula, (Figure 14). The analytical solution obtained in this article has sufficient accuracy to meet the calculation needs of different positions (pole radius r) in the FSPR model. The temperature field cloud map, drawn according to the calculation results, can intuitively reveal the overall temperature distribution and thickness of the frozen soil curtain, and realize visual processing. It is also proven that when the temperature field tends to steady-state in the latter stage of freezing, the shapes of the inner and outer boundaries of the frozen soil in this model can be considered to be a circular ring, and the wavy frozen soil boundary can be ignored.

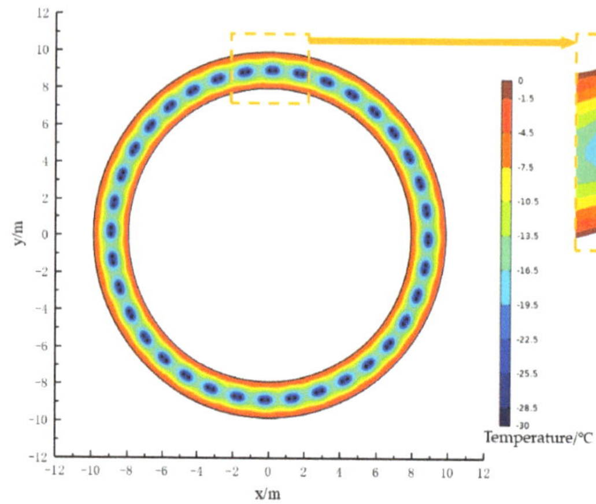

Figure 14. Two-dimensional steady-state temperature field based on analytical solution.

4. Conclusions

(1) During the freezing process of FSPR, formation of the frozen curtain is largely dependent on two types of freezing tubes to freeze the soil between the jacking pipes, and achieve the purpose of sealing water. Taking this as the main research object of the freezing steady-state temperature field, the model is assumed and simplified in combination with actual situation of the Gongbei tunnel project. Using the conformal mapping function and the separation-variable solution method, the analytical solution expression for the steady-state temperature field of FSPR under different soil freezing points is deduced, which is a quick calculation method that can be used by engineers and technicians during the designing stage and to evaluate the effect of on-site freezing construction.

(2) Different characteristic parameters and finite element software can be used to establish and solve the corresponding two-dimensional steady-state temperature field numerical calculation model. The correctness and accuracy of the analytical solution are verified by comparing the results. In this project, the calculation result is acceptable when the soil freezing point range is 0~−1.5 °C.

(3) Combined with the contour map of the steady-state temperature field, it is shown that when the number of frozen tubes is large, that is, the spacing between the freezing tubes is small, the shapes of the inner and outer boundaries of the frozen soil curtain can be approximately regarded as circular rings in the steady state.

(4) The temperature difference of the three sections is larger in the region closer to the freezing tube. Section 1 is the position of the midline between the adjacent concrete pipe and hollow pipe, and is an important area for "freeze-sealing between pipes" in FSPR. The calculated results show that the temperature range of this area within the size range of the pipe is −10 °C~−28 °C, implying that a reliable frozen soil curtain can be formed between the jacking pipes to ensure the effect of "freeze-sealing" and safety.

(5) How to adapt the analytical solution of temperature field to the operating state of various frozen tubes is a problem that requires further investigation. In addition, the actual tunnel section is closer to that of the ellipse, and considering these conditions, the analytical solution also deserves further exploration.

Author Contributions: Conceptualization, Y.D.; methodology, Y.D.; software, Y.D. and W.L.; validation, Y.D. and C.R.; writing—original draft preparation, Y.D.; writing—review and editing, Y.D. and X.H. All authors have read and agreed to the published version of the manuscript.

Funding: This research was funded by Scientific Research Startup Funding for Introducing Talents of Anhui University of Science and Technology, grant number 2022yjrc55; National Natural Science Foundation of China, grant number 51878005.

Data Availability Statement: The datasets generated and analyzed during the current study are available from the corresponding author upon reasonable request.

Acknowledgments: The authors would like to thank all the reviewers who participated in the review and MJEditor (www.mjeditor.com, accessed on 10 October 2022) for its linguistic assistance during the preparation of this manuscript.

Conflicts of Interest: The authors declare no conflict of interest.

References

1. Cheng, X.S. *Ground Freezing Method*, 1st ed.; China Communications Press: Beijing, China, 2013; pp. 3–4. (In Chinese)
2. Cheng, H. *Theory and Technology of Shaft Sinking by Freezing Method for Deep Alluvium*, 1st ed.; Science Press: Beijing, China, 2016; pp. 118–120. (In Chinese)
3. Wang, B.; Rong, C.X.; Cheng, H. Analytical Solution of Steady-state Temperature Field of Asymmetric Frozen Wall Induced by Directional Seepage. *Adv. Eng. Sci.* **2022**, *54*, 76–87. [CrossRef]
4. Taylor, R.N. *Geotechnical Centrifuge Technology*, 1st ed.; Blackie Academic & Professional: London, UK, 1994; pp. 273–280.
5. Cai, H.B.; Li, S.; Liang, Y.; Cheng, H. Model Test and Numerical Simulation of Frost Heave During Twin-tunnel Construction using Artificial Ground-freezing Technique. *Comput. Geotech.* **2019**, *115*, 103155. [CrossRef]
6. Huang, X.W.; Yao, Z.S.; Cai, H.B.; Li, X.; Chen, H. Performance evaluation of coaxial borehole heat exchangers considering ground non-uniformity based on analytical solutions. *Int. J. Therm. Sci.* **2021**, *170*, 107162. [CrossRef]
7. Li, K.Q.; Li, D.Q.; Liu, Y. Meso-scale investigations on the effective thermal conductivity of multi-phase materials using the finite element method. *Int. J. Heat Mass Transf.* **2020**, *151*, 119383. [CrossRef]
8. Hosseini, S.M.; Akhlaghi, M.; Shakeri, M. Transient Heat Conduction in Functionally Graded Thick Hollow Cylinders by Analytical Method. *Heat Mass Transf.* **2007**, *43*, 669–675. [CrossRef]
9. Jiang, B.S.; Shen, C.R.; Feng, Q. Analytical Formulation of Temperature Field of Single Freezing Pipe with Constant Outer Surface Temperature. *J. China Coal Soc.* **2010**, *35*, 923–927. [CrossRef]
10. Cai, H.B.; Xu, L.X.; Yang, Y.G.; Li, L. Analytical Solution and Numerical Simulation of the Liquid Nitrogen Freezing Temperature Field of a Single Pipe. *AIP Adv.* **2018**, *8*, 055119. [CrossRef]
11. Fang, T.; Hu, X.D. Generalized Analytical Solution to Steady-state Temperature Field of Single-row-piped Freezing. *J. China Coal Soc.* **2019**, *44*, 535–543. [CrossRef]
12. Sanger, F.J.; Sayles, F.H. Thermal and Rheological Computations for Artificially Frozen Ground Construction. *Eng. Geol.* **1979**, *13*, 311–337. [CrossRef]
13. Trupak, N.G. *Groud Freezing in Shaft Sinking;* Coal Technology Press: Moscow, Russia, 1954; pp. 20–65. (In Russian)
14. Bakholdin, B.V. *Selection of Optimized Mode of Ground Freezing for Construction Purpose;* State Construction Press: Moscow, Russia, 1963; pp. 21–27. (In Russian)
15. Hu, C.P.; Hu, X.D.; Zhu, H.H. Sensitivity Analysis of Control Parameters of Bakholdin Solution in Single-row-pipe Freezing. *J. China Coal Soc.* **2011**, *36*, 938–944. [CrossRef]
16. Yang, B.; Ding, W.Q.; Hu, X.D.; Liu, P. Sensitivity Analysis of Control Parameters for Average Temperature of Single-Row-Pipe Frozen Soil Wall. *Chin. J. Undergr. Space Eng.* **2012**, *8*, 1208–1214.
17. Hu, X.D.; Zhang, L.Y. Analytical Solution to Steady-State Temperature Field of Two Freezing Pipes with Diferent Temperatures. *J. Shanghai Jiaotong Univ. (Sci.)* **2013**, *18*, 706–711. [CrossRef]
18. Hu, X.D.; Han, Y.G.; Yu, X.F. Generalized Analytical Solution to Steady-state Temperature Field of Double-row-pipe Freezing. *J. Tongji Univ. (Nat. Sci.)* **2015**, *43*, 386–391. [CrossRef]
19. Hu, X.D.; Wang, Y. Analytical Solution of Three-row-piped Frozen Temperature Field by Means of Superposition of Potential Function. *Chin. J. Rock Mech. Eng.* **2012**, *31*, 1071–1080. [CrossRef]
20. Wang, B.; Rong, C.X.; Cheng, H. Elastic and Plastic Analysis of Heterogeneous Frozen Soil Wall of Triple-row Piped Freezing. *J. Yangtze River Sci. Res. Inst.* **2019**, *36*, 104–111. [CrossRef]
21. Hu, X.D.; Han, Y.G. General Analytical Solution to Steady-state Temperature Field of Single-circle-pipe Freezing. *J. Cent. South Univ. (Sci. Technol.)* **2015**, *46*, 2342–2349. [CrossRef]
22. Hu, X.D.; Fang, T.; Han, Y.G. Generalized Analytical Solution to Steady-state Temperature Field of Double-circle-piped Freezing. *J. China Coal Soc.* **2017**, *42*, 2287–2294. [CrossRef]
23. Hu, X.D.; Huang, F.; Bai, N. Models of Artificial Frozen Temperature Field Considering Soil Freezing Point. *J. China Univ. Min. Technol.* **2008**, *37*, 550–555.
24. Hong, Z.Q.; Shi, R.J.; Yue, F.T.; Han, L. Analytical Solution of Steady-state Temperature Field for Large-section Freezing with Rectangular Layout of Single-ring holes. *Chin. J. Geotech. Eng.* **2022**, *9*, 1–10. Available online: https://kns.cnki.net/kcms/detail/32.1124.TU.20220922.0943.006.html (accessed on 23 September 2022).

25. Qi, Y.; Zhang, J.; Yang, H.; Song, Y. Application of Artificial Ground Freezing Technology in Modern Urban Underground Engineering. *Adv. Mater. Sci. Eng.* **2020**, *2020*, 1619721. [CrossRef]
26. Liu, J.G.; Ma, B.S.; Cheng, Y. Design of the Gongbei Tunnel Using a Very Large Cross-section Pipe-roof and Soil Freezing Method. *Tunn. Undergr. Space Technol.* **2018**, *72*, 28–40. [CrossRef]
27. Kang, Y.S.; Liu, Q.S.; Cheng, Y.; Liu, X. Combined Freeze-sealing and New Tubular Roof Construction Methods for Seaside Urban Tunnel in Soft Ground. *Tunn. Undergr. Space Technol.* **2016**, *58*, 1–10. [CrossRef]
28. Zhang, D.M.; Pang, J.; Ren, H.; Han, L. Observed Deformation Behavior of Gongbei Tunnel of Hong Kong-Zhuhai-Macao Bridge During Construction. *Chin. J. Geotech. Eng.* **2020**, *42*, 1632–1641. [CrossRef]
29. Long, W.; Rong, C.X.; Duan, Y.; Guo, K. Numerical Calculation of Temperature Field of Freeze-sealing Pipe Roof Method in Gongbei Tunnel. *Coal Geol. Explor.* **2020**, *48*, 160–168. [CrossRef]
30. Hu, X.D.; Li, X.Y.; Wu, Y.H.; Han, L.; Zhang, C.B. Effect of Water-proofing in Gongbei Tunnel by Freeze-sealing Pipe Roof Method with Field Temperature Data. *Chin. J. Geotech. Eng.* **2019**, *41*, 2207–2214. [CrossRef]
31. Guo, K.; Duan, Y. Numerical Simulation of Temperature Field of Tube-Curtain Freezing Method for Shallow Buried Tunnel. *J. Anhui Univ. Sci. Technol. (Nat. Sci.)* **2019**, *39*, 63–68.
32. Hu, X.D.; Hong, Z.Q.; Fang, T. Analytical Solution to Steady-state Temperature Field with Typical Freezing Tube Layout Employed in Freeze-sealing Pipe Roof Method. *Tunn. Undergr. Space Technol.* **2018**, *79*, 336–345. [CrossRef]
33. Kong, X.Y. *Advanced Mechanics of Fluids in Porous Media*, 3rd ed.; USTC Press: Hefei, China, 2020; pp. 84–90. (In Chinese)
34. Li, H.; Xie, S.F. *Complex Variable Function & Integral Transform*, 5th ed.; Higher Education Press: Beijing, China, 2018; pp. 155–164. (In Chinese)
35. Wang, R.H.; Li, X.J. Superposition Calculation of Frozen Temperature Field and Its Computer Method. *J. Anhui Univ. Sci. Technol. (Nat. Sci.)* **2003**, *1*, 25–29. [CrossRef]
36. Hu, X.D.; Guo, W.; Zhang, L.Y. Mathematical Models of Steady-state Temperature Fields Produced by Multipiped Freezing. *J. Zhejiang Univ.-SCIENCE A (Appl. Phys. Eng.)* **2016**, *17*, 702–723. [CrossRef]
37. Chen, S.X.; Qin, T.H.; Zhou, Y. *Mathematical Physics Equation*, 1st ed.; Fudan University Press: Shanghai, China, 2003; pp. 61–63. (In Chinese)
38. COMSOL Multiphysics Reference Manual. 2018. Available online: https://doc.comsol.com/5.4/doc/com.comsol.help.comsol/COMSOL_ReferenceManual.pdf (accessed on 19 October 2022).

applied sciences

MDPI

Article

Research on Frost Heaving Distribution of Seepage Stratum in Tunnel Construction Using Horizontal Freezing Technique

Mengkai Li [1], Haibing Cai [1,*], Zheng Liu [1], Changqiang Pang [1] and Rongbao Hong [2]

1 School of Civil Engineering and Architecture, Anhui University of Science and Technology, Huainan 232001, China
2 School of Architectural Engineering, Nantong Vocational University, Nantong 226007, China
* Correspondence: haibingcai@163.com; Tel.: +86-13966450807

Abstract: During the horizontal freezing construction of a subway tunnel, the delay of the closure of the frozen wall occurs frequently due to the existence of groundwater seepage, which can be directly reflected by a freezing temperature field. Accordingly, the distribution of ground surface frost heaving displacement under seepage conditions will be different from that under hydrostatic conditions. In view of this, this paper uses COMSOL to realize the hydro–thermal coupling in frozen stratum under seepage conditions, then, the frost heaving distribution of seepage stratum in tunnel construction using horizontal freezing technique is researched considering the ice–water phase transition and orthotropic deformation characteristics of frozen–thawed soil by ABAQUS. The results show that the expansion speed of upstream frozen wall is obviously slower than that of the downstream frozen wall, and the freezing temperature field is symmetrical along the seepage direction. In addition, the ground frost heaving displacement field is asymmetrically distributed along the tunnel center line, which is manifested in that the vertical frost heaving displacement of the upstream stratum is less than that of the downstream stratum. The vertical frost heaving displacement of the ground surface decreases with the increase in tunnel buried depth, but the position of the maximum value remains unchanged as the tunnel buried depth increases. The numerical simulation method established in this paper can provide a theoretical basis and design reference for the construction of a subway tunnel in a water-rich stratum under different seepage using the artificial freezing technique.

Keywords: frost heaving; seepage stratum; horizontal freezing technique; COMSOL; ABAQUS

1. Introduction

With the acceleration of urbanization, there is more and more urgency to develop underground resources to expand the space for human activity [1,2]. Nevertheless, various complex geological conditions, such as water-rich sand, loose gravelly soil and silty clay with frozen–thawed sensitive characteristics, are encountered in underground engineering, and traditional geotechnical engineering methods face difficulties in solving such problems. Artificial ground freezing (AGF) is a method of artificial cooling that absorbs heat from the soil around the underground space to be excavated, gradually turning the natural soil into frozen soil and thus forming a dense frozen soil body. So, with its unique waterproof effect, adaptability to any complex geological conditions, flexible use and other characteristics, the geotechnical engineering community have paid more and more attention to AGF, which is now widely used in complex geotechnical engineering construction [3–5].

During the tunnel freezing period, ice intrusions such as ice crystals and interlayers formed in stratum frozen area will increase the volume of soil which is the frost heaving phenomenon of stratum. This phenomenon will be amplified when the stratum exits groundwater seepage, which brings difficulty to tunnel construction and damages the environment [6–8]. In view of this, domestic and foreign scholars have carried out much research by means of numerical simulation, model tests and field measurement.

Hou [9] represented the freezing characteristic function by combining the Young–Laplace law with the van Genuchten model. Then, a thermal-mechanical model of saturated media considering dynamic water–ice phase equilibrium was established and validated thorough model tests under various seepage velocities. Zhou [10] studied the development law of frozen wall in the process of artificial ground freezing under the construction of combined formation seepage by model test. Ji [11] studied the evolution process of frost heaving force under thermo–mechanical coupling through a series of freezing tests under different constraint stiffness and temperature gradient, and proposed a method to measure formation frost heaving force under constraint conditions. Tang et al. [12] proposed a prediction method with porosity as a variable using the finite difference method, and analyzed the physical changes of silty soil during freezing. Cai [13] established an optimized prediction method for ground surface frost heaving during the freezing construction of a subway tunnel using AGF based on stochastic medium theory. Lai et al. [14,15] proposed a hydro–thermal coupling model considering ice–water phase change based on heat transfer theory and seepage theory. Wang et al. [16] studied the evolution law of artificial frozen wall under seepage conditions in saturated gravel formations. Ahmed [17] studied the diachronic evolution law of temperature field using COMSOL finite element software considering different seepage velocities (0.5, 1.0, 1.5 m/d). Li et al. [18] proposed a moisture–heat coupling model considering ice–water phase transition to quantify the temperature distribution in soil during the freezing process. Huang et al. [19] adopted the same model to optimize the positions of freezing pipes around circular tunnel in seepage stratum with the velocity of 2.0 m/d. Yang et al. [20] further revealed the development law of the frozen wall upstream, and downstream of the tunnel under seepage conditions, using a large-scale model test.

The above researchers have made great contributions to the distribution laws of artificial freezing temperature and displacement fields; however, the research on the frost heaving laws of seepage strata during the freezing construction period is still insufficient. Therefore, in this paper, the finite element software COMSOL is selected to realize the hydro–thermal coupling in frozen stratum under seepage conditions. Then, the numerical simulation software ABAQUS is selected to realize the thermo-mechanical coupling, considering the ice–water phase transition and orthotropic deformation characteristics of frozen–thawed soil. The HyperMesh tool is used to divide the grid of models in COMSOL and ABAQUS, which ensures the consistency of the grid, and the calculation results of the former are directly input into the latter program. Finally, the formation and distribution laws of the displacement field and displacement field of the stratum under the groundwater seepage conditions are researched by taking tunnel horizontal freezing engineering as the research object.

2. Hydro–Thermal Coupling Mathematical Model

2.1. Basic Parameters

The volume content of each component in the soil is shown in Figure 1.

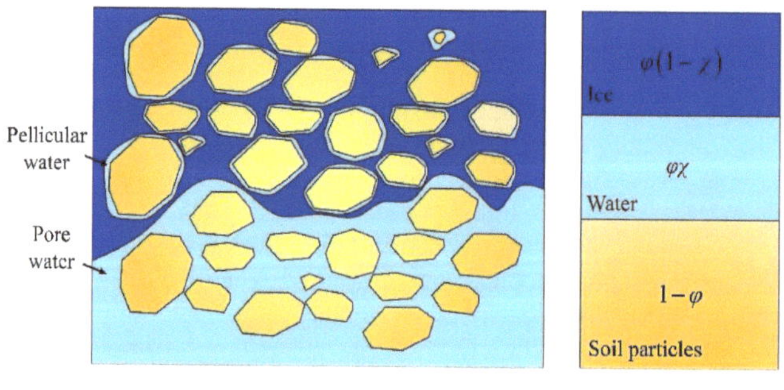

Figure 1. The volume content of each component in soil.

Where φ is the soil skeleton gap, χ is the volume content of water in the gap.

The unfrozen water content in frozen soil is always in dynamic equilibrium with the temperature of frozen soil [21,22], and content can be determined by Equation (1). The volume content of pore water can be expressed as

$$w_u = w_0 \left(\frac{T_f}{T} \right)^b \quad (1)$$

$$\chi = \begin{cases} 1, & T > T_f \\ \left(\frac{T_f}{T} \right)^b, & T \leq T_f \end{cases} \quad (2)$$

where w_0 denotes the initial water content of unfrozen soil; w_u denotes the free water content of frozen soil; T is soil temperature, °C; T_f is the initial freezing temperature of soil mass, °C; and b is the constant determined by soil properties.

2.2. Coupling of Temperature Field to Seepage Field

The temperature field during soil freezing is a transient heat transfer problem.it is assumed that ice, soil particles and water exist simultaneously in the freezing soil during the freezing process, and have different heat transfer characteristics.

According to the heat transfer and seepage principles [23], the control equations for the temperature and seepage fields considering the phase change are

$$\theta_i C_i \rho_i \frac{\partial T}{\partial t} + \varphi L \rho_l \frac{\partial \chi}{\partial T} \frac{\partial T}{\partial t} + \theta_l C_l \rho_l \frac{\partial T}{\partial t} + \rho_l C_l (\boldsymbol{u} \cdot \nabla T) + \theta_s C_s \rho_s \frac{\partial T}{\partial t} = \nabla[(\theta_i k_i + \theta_l k_l + \theta_s k_s) \nabla T] + Q \quad (3)$$

$$\rho_l [\alpha_l \varphi + \alpha_s (1 - \varphi)] \frac{\partial p}{\partial t} + \nabla \cdot (\rho_l \boldsymbol{u}) = \rho_l q \quad (4)$$

$$\boldsymbol{u} = -\frac{K\prime}{\rho_l g} \nabla p \quad (5)$$

$$K\prime = K_f + \left(K_0 - K_f \right) \chi \quad (6)$$

where t is time; C_i, C_l and C_s denote the specific heat capacity of water, ice and soil particles; K_i, K_l and K_s are their thermal conductivity; ρ_i, ρ_l and ρ_s are their density; L denotes ice–water phase change latent heat; $\boldsymbol{u}$ denotes the relative velocity of water; Q denotes the total amount of equivalent heat source; α_s is the basic equivalent compression ratio; φ is porosity; q denotes the volume of flow; p is the osmotic pressure; K' is hydraulic conductivity; and K_f and K_0 are the hydraulic conductivity of completely frozen area and unfrozen area respectively.

The change of absolute porosity during freezing can be expressed by the change of volume content of unfrozen water, therefore, Equation (4) can be expressed as

$$\rho_l [\alpha_l \theta_l + \alpha_s (1 - \theta_l)] \frac{\partial p}{\partial t} + \nabla \cdot (\rho_l \boldsymbol{u}) = \rho_l q \quad (7)$$

The temperature of the outer wall of the freezing tube is the freezing temperature, where the temperature boundary at the inlet of the freezing tube is the inlet temperature, and the convection condition at the outlet of the freezing tube is set, as shown

$$\boldsymbol{n}_d \cdot (-k \nabla T)|_{x=a_2} = 0 \quad (8)$$

where a_2 denotes the horizontal coordinate of the export of the freezing pipe, $\boldsymbol{n}_d$ denotes the axial direction vector at the export.

Setting the other boundaries to be adiabatic and not considering the heat transfer effect between the boundaries, the boundary conditions are expressed as

$$-\boldsymbol{n} \cdot (-k\nabla T)|_{y=b_1,b_2} = 0 \tag{9}$$

where b_1 and b_2 are the ordinates of the boundary, and $\boldsymbol{n}$ $\boldsymbol{n}$ denotes the direction vector of the normal direction within the boundary.

The temperature boundary considering other boundary heat transfer effects can be expressed as

$$T = T_0 \tag{10}$$

where T_0 is the temperature over time.

The initial conditions are

$$p|_{t=0} = p_0 \tag{11}$$

where p_0 denotes the soil pressure distribution in the initial state.

Therefore, the mathematical model of hydro–thermal coupling during freezing process can be composed of Equation (3), Equation (7) and the above boundary conditions.

3. Thermo–Mechanical Coupling Mathematical Model

3.1. Freezing Temperature Field

In the process of artificial ground freezing, the change of the temperature of the soil will cause the change of water migration and stress state, and the stress field will cause the redistribution of the temperature field. For this reason, the coupling of temperature field and stress field should be considered in frost heaving deformation. The controlling equation for the temperature field can be expressed as [24,25]

$$C^* \frac{\partial T}{\partial t} = \frac{\partial}{\partial x}\left(k^* \frac{\partial T}{\partial x}\right) + \frac{\partial}{\partial y}\left(k^* \frac{\partial T}{\partial y}\right) \tag{12}$$

where C^* is the equivalent volume specific heat; k^* is the equivalent thermal conductivity; and T is the soil temperature; and t is the time. C^* and k^* are given by

$$C^* = \begin{cases} C_f & T < T_d \\ \frac{C_f + C_u}{2} + \frac{L}{T_r - T_d} & T_d \le T \le T_r \\ C_u & T > T_r \end{cases} \tag{13}$$

$$k^* = \begin{cases} k_f & T < T_d \\ k_f + \frac{k_u - k_f}{T_r - T_d}(T - T_d) & T_d \le T \le T_r \\ k_u & T > T_r \end{cases} \tag{14}$$

where C_f and C_u denote the volumetric heat capacity of frozen soil and unfrozen soil (i.e., $C_f = \rho_f\, c_f$, $C_u = \rho_u\, c_u$), ρ_f and ρ_u denote the density of frozen soil and unfrozen soil, and c_f and c_u denote the mass specific heat of frozen soil and unfrozen soil. k_f and k_u denote the thermal conductivity of frozen soil and unfrozen soil, respectively. T_d denotes the freezing temperature of soil; T_r denotes the thawing temperature of soil.

The initial condition of temperature field is

$$T|_{t=0} = T_0 \tag{15}$$

where T_0 denotes the initial temperature of soil.

When the freezing pipe is transformed into one point, and the boundary condition of frozen pipe is expressed as

$$T|_{(x_p, y_p)} = T_c(t) \tag{16}$$

where (x_p, y_p) represents the central coordinate of the freezing pipe and T_c represents the brine temperature.

At the infinite distance of the frozen zone, the boundary condition is

$$T|_{(x=\infty,\ y=\infty)} = T_0 \tag{17}$$

3.2. Coupling of Stress Field to Temperature Field

For the plane strain problem, the equilibrium equation of the stress field can be expressed as

$$\begin{cases} \frac{\partial \sigma_x}{\partial x} + \frac{\partial \tau_{xy}}{\partial y} = 0 \\ \frac{\partial \tau_{xy}}{\partial x} + \frac{\partial \sigma_x}{\partial y} = -\gamma \end{cases} \tag{18}$$

where γ is the gravity of soil.

The geometric equation can be expressed as

$$\begin{cases} \varepsilon_x = \frac{\partial u}{\partial x} \\ \varepsilon_x = \frac{\partial u}{\partial x} \\ \varepsilon_{xy} = \frac{\partial u}{\partial y} + \frac{\partial v}{\partial x} \end{cases} \tag{19}$$

Due to the frozen soil being defined as a linear elastomer, the mechanical properties of frozen soil change with temperature, considering its orthotropic characteristics. So, the thermo–mechanical coupling mathematical model can be expressed as

$$\begin{cases} \varepsilon_x = \frac{1-\mu(T)^2}{E(T)}\left[\sigma_x - \frac{\mu(T)}{1-\mu(T)}\sigma_y\right] + \sigma_x^T \\ \varepsilon_y = \frac{1-\mu(T)^2}{E(T)}\left[\sigma_y - \frac{\mu(T)}{1-\mu(T)}\sigma_x\right] + \sigma_y^T \\ \varepsilon_{xy} = \frac{2[1-\mu(T)]}{E(T)}\tau_{xy} + \varepsilon_{xy}^T \end{cases} \tag{20}$$

where ε_x^T, ε_y^T and ε_{xy}^T are the instantaneous strain component caused by temperature.

To further effectively discuss the orthotropic deformation characteristics of frozen soil, generally, the frost heaving rate $\varepsilon_{f/r}^T$ is used to measure the strength of frozen soil and the parameter is defined as the instantaneous volumetric strain related to temperature. As shown in Figure 2, the local coordinate system is established in the direction of temperature gradient in three directions (1, 2 and 3).

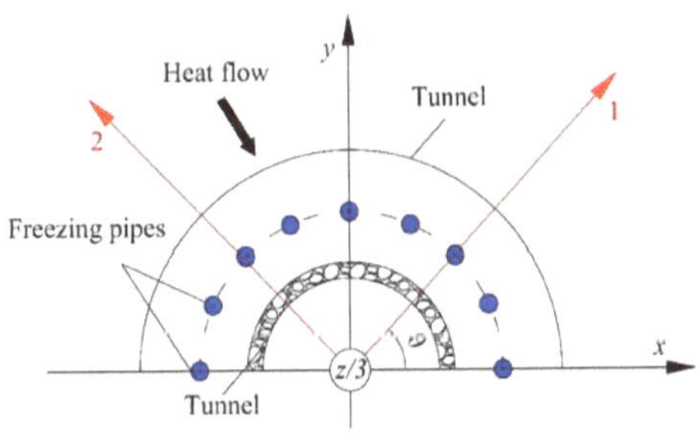

Figure 2. Local coordinate system.

According to the linear elastic theory of frozen soil, it can be known that the instantaneous volumetric strain ε_f^T and elastic strain ε^θ constitute the total strain ε of frozen soil when the soil begins to freeze and the instantaneous thermal strain component generated by temperature in the local coordinate system is expressed as

$$\begin{Bmatrix} \varepsilon_{11}^T & \varepsilon_{12}^T & \varepsilon_{13}^T \\ \varepsilon_{21}^T & \varepsilon_{22}^T & \varepsilon_{23}^T \\ \varepsilon_{31}^T & \varepsilon_{32}^T & \varepsilon_{33}^T \end{Bmatrix} = \begin{bmatrix} \eta & 0 & 0 \\ 0 & (1-\eta)/2 & 0 \\ 0 & 0 & (1-\eta)/2 \end{bmatrix} \varepsilon_f^T \tag{21}$$

where η is an orthotropic parameter. When $\eta = 1$, the frozen soil exhibits unidirectional deformation characteristics (i.e., in the direction of the temperature gradient).; when $\eta = 1/3$, the frozen soil exhibits isotropic deformation characterized. Therefore, whether unidirectional deformation or isotropic deformation, frozen soil has characteristics of orthotropic deformation.

Employing Hooke's law, the bulk strain component of frozen soil can be expressed as

$$\begin{cases} \varepsilon_{11} = \frac{1}{E}[\sigma_{11} - \mu(\sigma_{22} + \sigma_{33})] + \eta\varepsilon_f^T \\ \varepsilon_{22} = \frac{1}{E}[\sigma_{22} - \mu(\sigma_{11} + \sigma_{33})] + \frac{1}{2}(1-\eta)\varepsilon_f^T \\ \varepsilon_{33} = \frac{1}{E}[\sigma_{33} - \mu(\sigma_{11} + \sigma_{22})] + \frac{1}{2}(1-\eta)\varepsilon_f^T \\ \varepsilon_{12} = \frac{2(1+\mu)}{E}\tau_{12} \\ \varepsilon_{23} = \frac{2(1+\mu)}{E}\tau_{23} \\ \varepsilon_{13} = \frac{2(1+\mu)}{E}\tau_{13} \end{cases} \tag{22}$$

where E is the elastic modulus of frozen soil and μ is Poisson's ratio of frozen soil.

For the plane strain problem, $\varepsilon_{33} = \varepsilon_{13} = \varepsilon_{23} = 0$,

$$\sigma_{33} = \mu(\sigma_{11} + \sigma_{22}) - \frac{E}{2}(1-\eta)\varepsilon_f^T \tag{23}$$

Combined with Equations (22) and (23), bulk strain component can be expressed as

$$\begin{cases} \varepsilon_{11} = \frac{1-\mu^2}{E}\left[\sigma_{11} - \frac{\mu}{1-\mu}\sigma_{22}\right] + \left[\eta + \frac{\mu}{2}(1-\eta)\right]\varepsilon_f^T \\ \varepsilon_{22} = \frac{1-\mu^2}{E}\left[\sigma_{22} - \frac{\mu}{1-\mu}\sigma_{11}\right] + \frac{1}{2}(1-\eta)(1+\eta)\varepsilon_f^T \\ \varepsilon_{12} = \frac{2(1+\mu)}{E}\tau_{12} \end{cases} \tag{24}$$

From the above equation, the transiently thermal strain component is expressed as

$$\begin{cases} \varepsilon_{11}^T = \left[\eta + \frac{\mu}{2}(1-\eta)\right]\varepsilon_f^T \\ \varepsilon_{22}^T = \frac{1}{2}(1-\eta)(1+\eta)\varepsilon_f^T \\ \varepsilon_{12}^T = 0 \end{cases} \tag{25}$$

As shown in Figure 2, where θ represents the included angle between the coordinate axes (1 and x), and the transient thermal strain component is

$$\begin{Bmatrix} \varepsilon_x^T \\ \varepsilon_y^T \\ \varepsilon_{xy}^T \end{Bmatrix} = \begin{bmatrix} \cos^2\theta & \sin^2\theta & -2\sin\theta\cos\theta \\ \sin^2\theta & \cos^2\theta & 2\sin\theta\cos\theta \\ \sin\theta\cos\theta & -\sin\theta\cos\theta & \cos^2\theta - \sin^2\theta \end{bmatrix} \begin{Bmatrix} \varepsilon_{11}^T \\ \varepsilon_{22}^T \\ \varepsilon_{12}^T \end{Bmatrix} \tag{26}$$

Substituting Equation (25) into (26), we can get

$$\begin{cases} \varepsilon_x^T = \left\{\cos^2\theta\left[\eta + \frac{\mu}{2}(1-\eta)\right] + \frac{\sin^2\theta}{2}(1-\mu)(1+\mu)\right\}\varepsilon_f^T \\ \varepsilon_y^T = \left\{\sin^2\theta\left[\eta + \frac{\mu}{2}(1-\eta)\right] + \frac{\cos^2\theta}{2}(1-\mu)(1+\mu)\right\}\varepsilon_f^T \\ \varepsilon_{xy}^T = \frac{\sin\theta\cos\theta}{2}(3\eta - 1)\varepsilon_f^T \end{cases} \tag{27}$$

From this, the calculation formula of instantaneous temperature strain component considering the orthotropic deformation characteristics of freeze-thaw soil is obtained.

4. Finite Element Calculation Model

4.1. Calculation Model

The tunnel constructed by horizontal freezing method is selected as the calculation model. The cross-section of the tunnel is circular, the buried depth is 15.0 m, the initial temperature of the soil is set as 20 °C, the upper displacement boundary of the model is a free boundary, the horizontal displacement is limited on the left and right sides, and the bottom is a fixed boundary.

To import the temperature field here into ABAQUS as a predefined field, first, the temperature field in COMSOL is derived in the form of numerical nodes, then a predefined field of temperature is established in ABAQUS, and finally the exported node data are imported into the initial predefined field in the form of a mapping field to obtain the cloud diagram of temperature field. The specific numerical simulation analysis process is shown in the Figure 3 below.

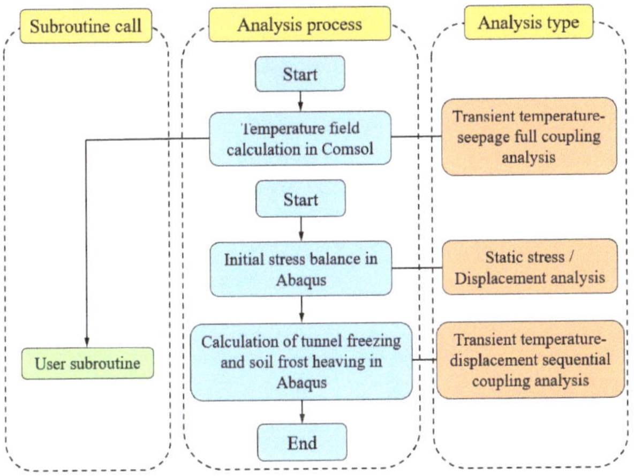

Figure 3. Flow chart of numerical simulation.

As shown in Figures 4 and 5, the HyperMesh tool is used to divide the grid of models in COMSOL and ABAQUS which ensures the consistency of the grid, and the calculation results of the former are directly input into the latter program.

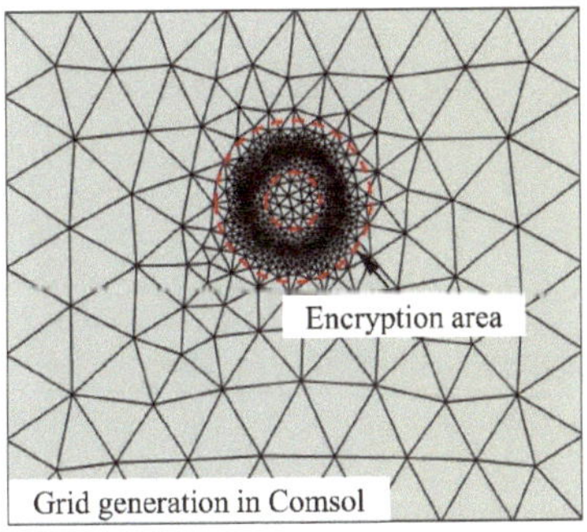

Figure 4. Grid generation in COMSOL.

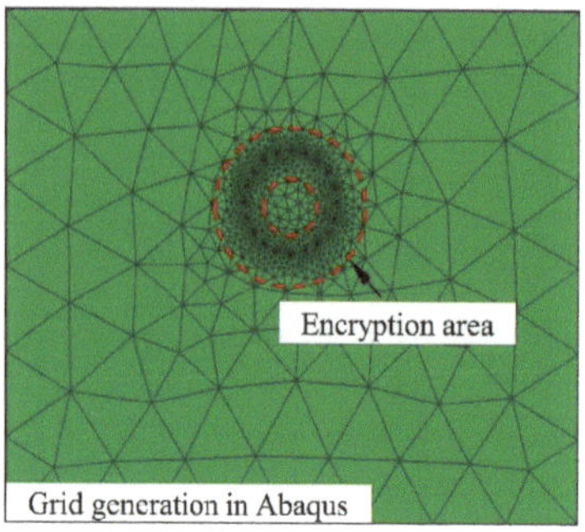

Figure 5. Grid generation in ABAQUS.

In order to combine the two finite element software more effectively, the temperature field obtained by hydro–thermal coupling in COMSOL is imported into ABAQUS as a predefined field (i.e., the temperature field results are transformed into nodal form). The cloud diagram of temperature field calculated by hydro–thermal coupling in COMSOL is shown in Figure 6, and that calculated by thermal–mechanical coupling in ABAQUS is shown in Figure 7. It can be seen from the above figure that the distribution of the temperature clouds calculated by both software are the same, and the freezing fronts show a non-uniform distribution along the flow direction due to the presence of the seepage field. The accuracy of the previously proposed thermal–hydraulic–mechanical coupled numerical simulation method is verified, thus realizing a more accurate and comprehensive study of the freezing and swelling law of seepage strata during the freezing construction period.

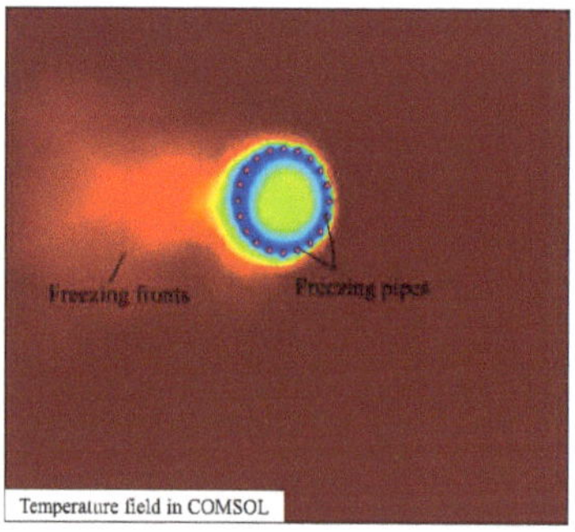

Figure 6. Temperature field in COMSOL.

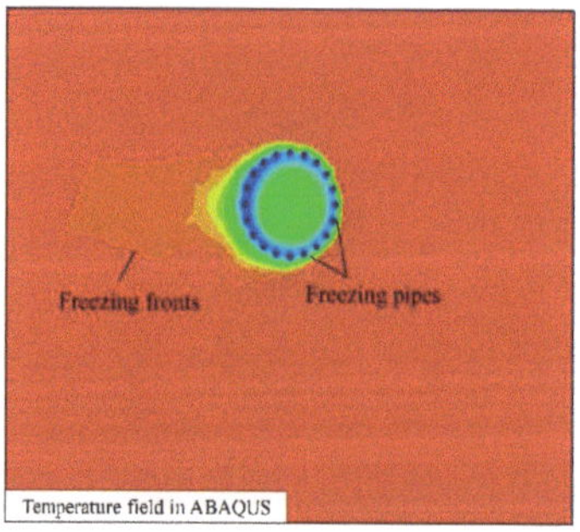

Figure 7. Temperature field in ABAQUS.

According to the orthogonal anisotropy of frozen soil, the freezing front will be formed only in the direction perpendicular to the temperature gradient during freezing process. The freezing front moves with the decrease in temperature. Thus, in this section, finite element software COMSOL and ABAQUS are selected to realize the hydro–thermal–mechanical coupling in frozen stratum under different seepage conditions. In addition, the ABAQUS frost heaving subroutine interface is introduced in the process of numerical analysis [26]. The process of calculating frost heaving deformation by subroutine interface is shown in Figure 8.

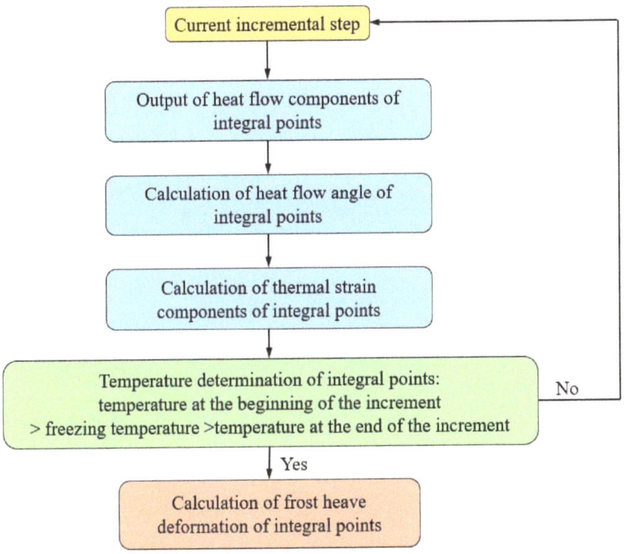

Figure 8. Flow diagram of subroutine programming.

As shown in Table 1, in order to analyze the influence of different seepage velocities and tunnel depths on surface deformation, five groundwater seepage velocities (0 m/d, 0.5 m/d, 1.0 m/d, 1.5 m/d and 2.0 m/d) and three tunnel depths (11 m, 13 m and 15 m) are selected.

Table 1. Model design.

Buried Depth	Seepage Velocities				
	0 m/d	0.5 m/d	1.0 m/d	1.5 m/d	2.0 m/d
11 m	Model 11-1	11-2	11-3	11-4	11-5
13 m	Model 13-1	13-2	13-3	13-4	13-5
15 m	Model 15-1	15-2	15-3	15-4	15-5

4.2. Model Parameters

In all calculation models, the radius of the horizontal freezing pipe arrangement circle (R_d) is 4.0 m, a total of 22 freezing pipes with a wall temperature of −25 °C are arranged, and the effective design freezing thickness is 2.0 m. On the upper surface of the model, the convective heat transfer value between the surface and the air is set as 175 kcal/(m^2·d·°C).

The thermal physical parameters and mechanical parameters of soil are shown in Tables 2 and 3. Considering the orthogonal anisotropy of soil, the deformation characteristic coefficient η is set as 0.9, and the frost heaving ratio is set as 1% [27].

Table 2. Physical parameters of soil.

Materials	Density/(kg·m^{-3})	Thermal Conductivity /(kcal·m^{-1}·h^{-1}·°C^{-1})	Latent Heat/(kcal·kg^{-1})	Specific Heat Capacity /(kcal·kg^{-1}·°C^{-1})	Permeability Coefficient /(m·d^{-1})
Soil	2110	1.55	-	0.301	-
Frozen soil	-	1.792	12.736	-	5.3×10^{-4}
Unfrozen soil	-	1.252	12.736	-	4.6×10^{-2}
Water	1000	0.56	-	0.495	-
Ice	918	2.24	-	0.998	-

In this numerical simulation, $T_{trans} = -1.0$ °C, $dT = 0.5$ °C.

Table 3. Mechanical parameters of soil.

Names	Elastic Modulus (MPa)	Poisson's Ratio	Cohesion (kPa)	Friction Angle (°)
Frozen soil	44.27 + 7.73 | T |	0.25	-	-
Unfrozen soil	24	0.35	15	20

T is soil temperature.

4.3. Initial Boundary Conditions

4.3.1. Boundary Conditions and Initial Conditions of Seepage Field

(1) Dirichlet boundary condition

The inlet and outlet are set as constant pressure boundaries, and the corresponding value is taken according to the initial pressure distribution.

$$p = p_0 \tag{28}$$

(2) Neumann boundary condition

$$-n\left(\vec{u}\,p\right)\Big|_{y=b_1,b_2} = 0 \tag{29}$$

where b_1 and b_2 are the ordinates of the boundary, and n represents the internal normal direction.

(3) Initial condition

It is assumed that the seepage velocity in initial state is only from right to left perpendicular to the inlet boundary. To form a stable seepage velocity in initial state, initial seepage velocity control method is as follows

$$p(x,y,t) = p_0(x,y,0) \tag{30}$$

4.3.2. Boundary Conditions and Initial Conditions of Temperature Field

(1) Dirichlet boundary condition

$$T = T_{a_1} \tag{31}$$

where a_1 is the abscissa of the inlet boundary and the upper and lower boundaries are set as adiabatic boundaries, which can be expressed as

$$n_d(-k\nabla T)|_{y=b_1,b_2} = 0 \tag{32}$$

where b_1 and b_2 are the ordinates of boundary, and n is the inner normal direction vector of the boundary.

(2) Neumann boundary condition

$$n_d(-k\nabla T)|_{x=a_2} = 0 \tag{33}$$

where a_2 represents the abscissa of outflow boundary condition.

(3) Initial condition

$$T(x,y,t) = T_0(x,y,0) \tag{34}$$

where T_0 is the initial temperature of soil layer.

5. Discussion

5.1. Effect of Groundwater Seepage on Freezing Temperature Field

The temperature range of cloud diagram is set as −26~−1 °C (i.e., the freezing temperature of soil is −1 °C) to characterize the thickness of frozen wall in numerical simulation. Then, it can be found that the development of frozen front is uneven due to the influence of seepage. Therefore, a unified measure is established in numerical simulation by defining the thickness of frozen wall at different locations AB, BC, DE and EF which correspond to L_{AB}, L_{BC}, L_{DE} and L_{EF}, respectively, when the construction requirements are met. As shown in Figure 9, L_{AB} is the thickness from the center of freezing pipe to the upstream outer front, L_{BC} is the thickness from the center of the freezing pipe to the upstream inner front, L_{DE} is the thickness from the center of the freezing pipe to the downstream inner front and L_{EF} is the thickness from the center of the freezing pipe to the upstream outer front.

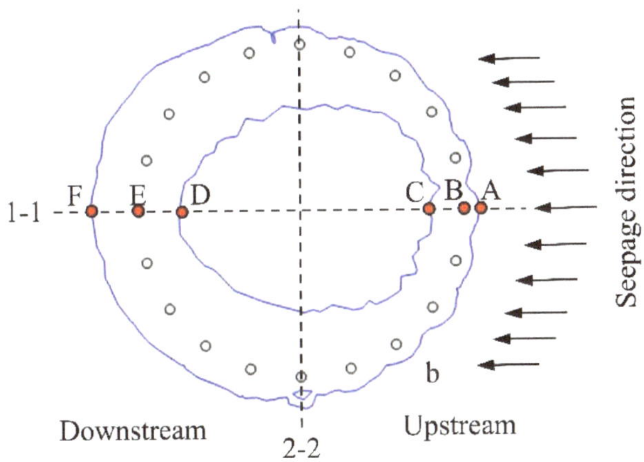

Figure 9. Custom frozen wall thickness.

The frozen wall formed at three groundwater seepage velocities for the given boundary conditions are shown in Figure 10. The closure time of frozen wall at 0 m/d is only 9 d, while it is 24 d at 1.0 m/d and 61 d at 2.0 m/d. The faster the seepage is, the more cooling capacity generated by the upstream freezing pipes is brought to the downstream, and the later the upstream frozen wall closes, that is, the longer the closure time of frozen wall is.

Similarly, the time required to reach design frozen wall thickness for construction at different seepage velocities is also different, as shown in Figure 10. The time required to reach design frozen wall thickness at 0 m/d, 1.0 m/d and 2.0 m/d is 41d, 83d and 121d, respectively. Therefore, it will take more time to complete the freezing engineering in the seepage stratum with a large flow velocity. The temperature field at different seepage velocities is shown in Figures 11 and 12.

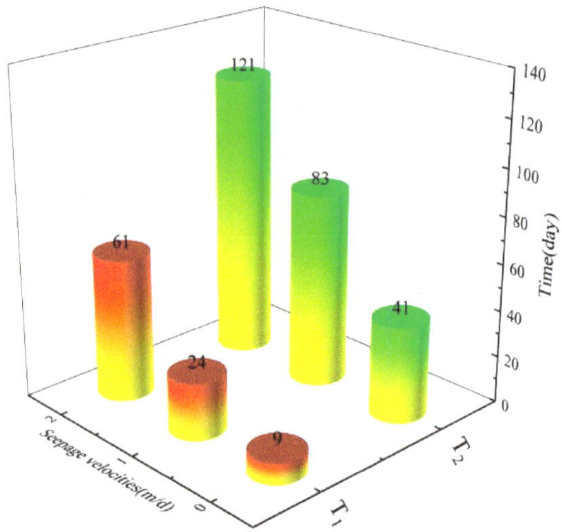

Figure 10. Freezing time of frozen wall under different states: T_1 is the time for initial closure of frozen wall, and T_2 is the time for complete closure of frozen wall.

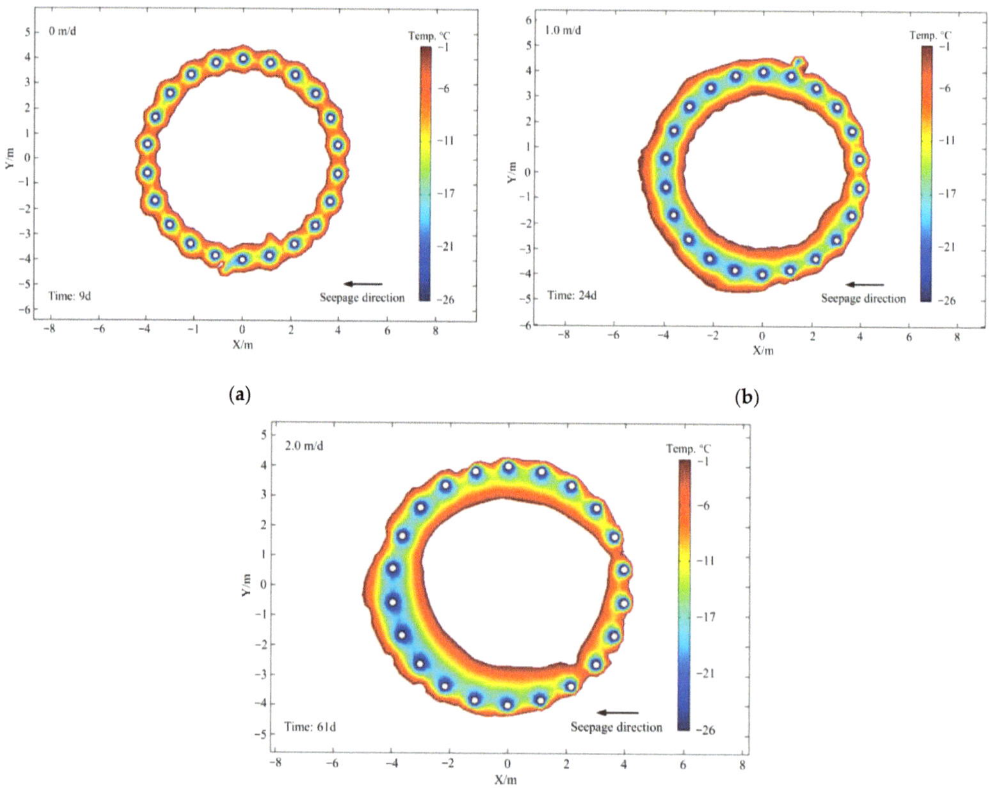

(a) (b) (c)

Figure 11. Initial closure of frozen wall: (**a**) 0 m/d; (**b**) 1.0 m/d; and (**c**) 2.0 m/d.

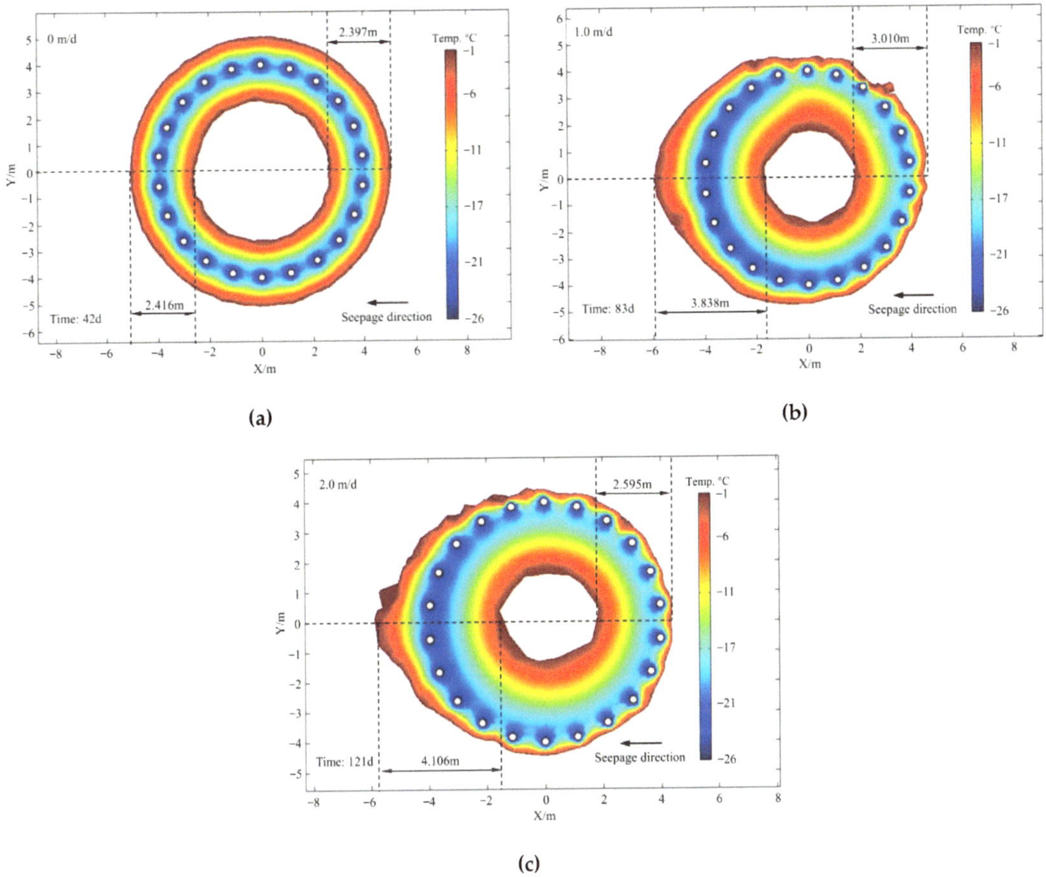

Figure 12. Complete closure of frozen wall: (**a**) 0 m/d; (**b**) 1.0 m/d; and (**c**) 2.0 m/d.

Table 4 shows that when the minimum thickness of the frozen wall is satisfied (i.e., the construction requirements are met), the thickness of the upstream frozen wall L_{AB} decreases correspondingly with the increase in seepage velocity, and the thickness of the downstream frozen wall L_{DE} and L_{EF} increases correspondingly. However, the thickness variation of the upstream frozen wall L_{BC} does not vary linearly. This is because with the greater seepage velocity, the heat at the inner front of the upstream frozen wall is carried away, resulting in a change in the thickness of the frozen wall L_{BC}. Therefore, in the actual project, when the seepage velocity is large enough, sufficient attention should be paid to the upstream freezing wall and corresponding measures should be taken.

Table 4. Relevant parameters of frozen wall at different flow velocity.

Seepage Velocity (m/d)	T_1 (d)	T_2 (d)	L_{AB} (m)	L_{BC} (m)	L_{DE} (m)	L_{EF} (m)
0	9	41	1.139	1.258	1.281	1.135
1.0	24	83	0.752	2.258	2.112	1.726
2.0	61	121	0.247	2.348	2.347	1.759

As shown in Figure 13, the development trend of frozen wall thickness at each position is similar at the same seepage velocity, the thickness L_{AB} is much smaller than L_{BC}, L_{DE}

and L_{EF} when groundwater seepage occurs, and this phenomenon change linearly with the seepage. The thickness L_{BC}, L_{DE} and L_{EF} reached 1.600 m, 1.754 m and 1.364 m, respectively, when freezing for 83 d at 2.0 m/d of seepage velocity, while L_{AB}, only 0.410 m, is less than one-third of the other places' thickness.

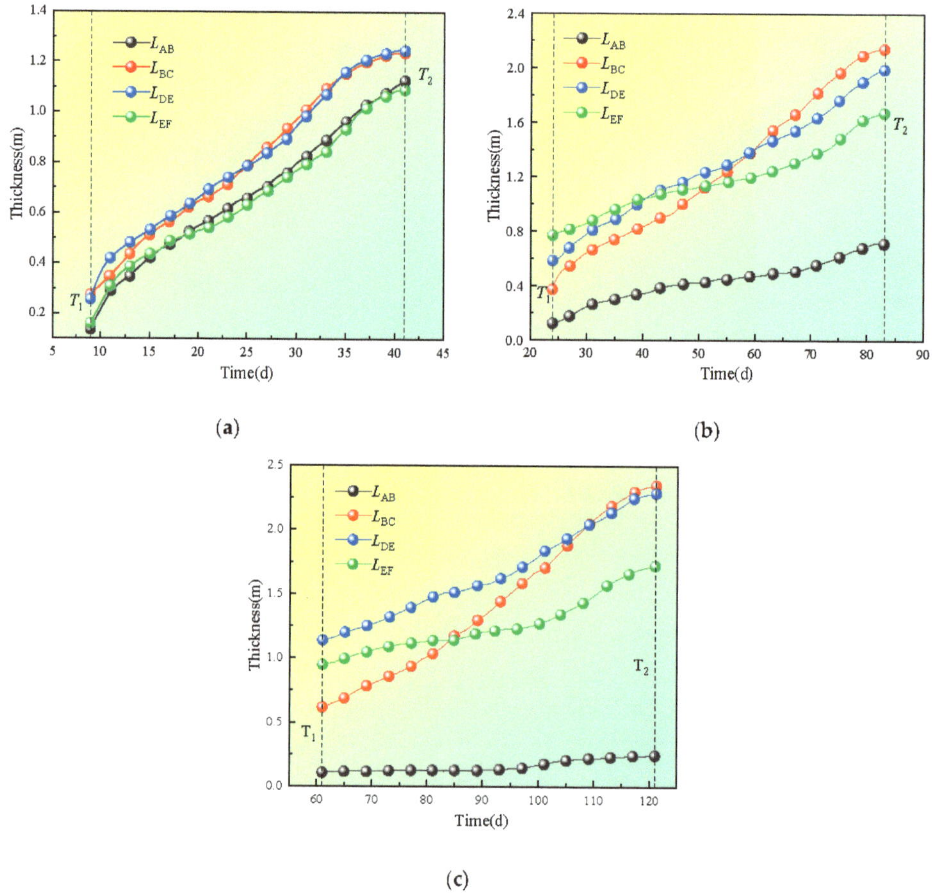

Figure 13. Variation of frozen wall thickness with time at different positions: (**a**) 0 m/d; (**b**) 1.0 m/d; (**c**) 2.0 m/d.

There are also great differences in the development speed of frozen wall thickness between the inner front and outer front. The development rates of frozen wall thickness L_{AB} and L_{EF} in the outer ring gradually decrease with freezing time. the thickness of frozen wall in upstream finally tends to be flat, and the phenomenon is related to the seepage velocity. When the thickness of frozen wall reached the construction standard, the development speed of the frozen wall thickness L_{AB} is 33.3 mm/d at 1.0 m/d of seepage, while the development rate of L_{BC} is only 0.5 mm/d. The frozen wall thickness L_{BC} and L_{DE} on the inner front increase linearly with freezing time, which is due to the closed area formed inside the freezing circle after the frozen wall intersects, which is basically no longer affected by the underground water.

5.2. Effect of Groundwater Seepage on Frost Heaving Displacement of Ground

In this section, taking the tunnel buried depth of 15 m as an example, five groundwater seepage velocities (0 m/d, 0.5 m/d, 1.0 m/d, 1.5 m/d and 2.0 m/d) are considered to explore the distribution law of stratum frost heaving displacement under the condition of groundwater seepage.

The development law of vertical displacement of the strata under different seepage velocities is shown in Figure 14. The vertical freezing expansion of the stratum is distributed asymmetrically along the center of the tunnel, which is due to the transfer of cold from the upstream freezing pipe to the downstream stratum under the action of seepage flow, resulting in the uneven thickness of the upstream and downstream freezing walls and the offset of the peak position of the stratum in the upper part of the stratum tunnel. When the effective frozen wall thickness is reached, the maximum vertical displacement of the upper stratum of the tunnel under different seepage velocity (0 m/d, 0.5 m/d, 1.0 m/d, 1.5 m/d and 2.0 m/d) is 43.1 mm, 45.2 mm, 48.2 mm, 51.0 mm and 53.5 mm.

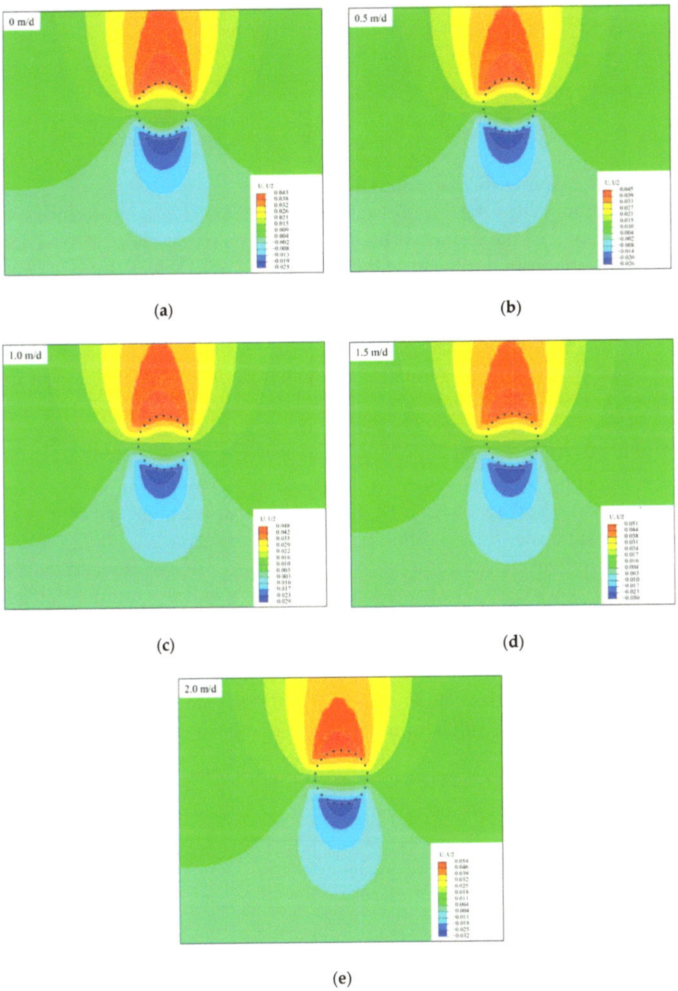

Figure 14. Distribution of stratum vertical displacement field of stratum under different seepage velocities: (**a**) 0 m/d; (**b**) 0.5 m/d; (**c**) 1.0 m/d (**d**) 1.5 m/d; and (**e**) 2.0 m/d.

The surface frost heaving displacement curve is shown in Figure 15, which is "normal distribution"; furthermore, the peak of the single peak locates directly above the center of the tunnel when the seepage velocity is 0 m/d, and the position of the peak will "shift" to the direction of water flow with the increase in seepage velocity.

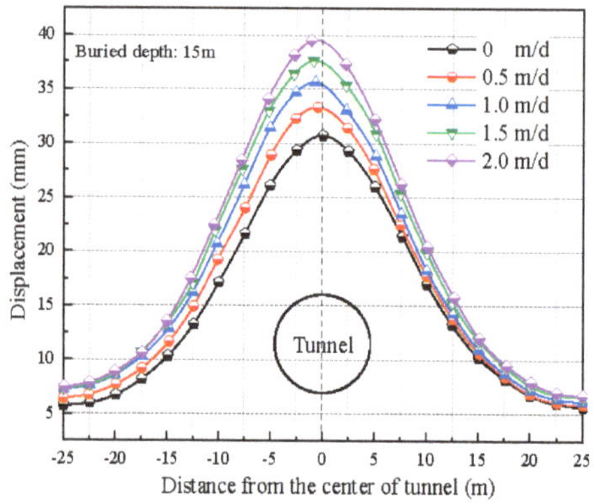

Figure 15. Vertical displacement distribution law of surface under different seepage velocities.

The maximum frost heaving displacement is located above the center of the tunnel under no seepage condition. When the flow velocity reaches 0.5 m/d, it is located 161 mm downstream from the center of the tunnel. When the flow velocity reaches 1.0 m/d, it is located 374 mm downstream from the center of the tunnel. When the flow velocity reaches 1.5 m/d, it is located 556 mm downstream from the center of the tunnel, and when the flow velocity reaches 2.0 m/d, it is located 968 mm downstream from the center of the tunnel. These data show that the existence of seepage leads to the thickness of frozen wall downstream being greater than that upstream, which leads to the deviation of the maximum frost heaving positive on the ground.

5.3. *Effect of Tunnel Buried Depth on Frost Heaving Displacement of Ground*

In this section, selecting the flow velocity of 1.0 m/d as the invariant, three buried depth of tunnel (11 m, 13 m, 15 m) are considered, to explore the distribution law of stratum frost heaving displacement under the condition of groundwater seepage.

The vertical displacement distribution law of seepage stratum and the vertical displacement curve of frost heaving on the surface after the frozen wall reaches the effective thickness are shown in Figures 16 and 17. On the one hand, it is not difficult to find that, under the same seepage velocity, the maximum frost heaving of stratum is inversely proportional to the buried depth of tunnel. On the other hand, when the buried depth is 11 m, the maximum frost heaving of surface is located at 368 mm from downstream to the center of tunnel; when the buried depth is 13 m, it is located at 384 mm from downstream to the center of tunnel, and the maximum is 38.1 mm; when the buried depth is 15 m, the vertical displacement of surface frost heaving reaches the maximum at 372 mm from downstream to the center of the tunnel, and the maximum is 35.2 mm.

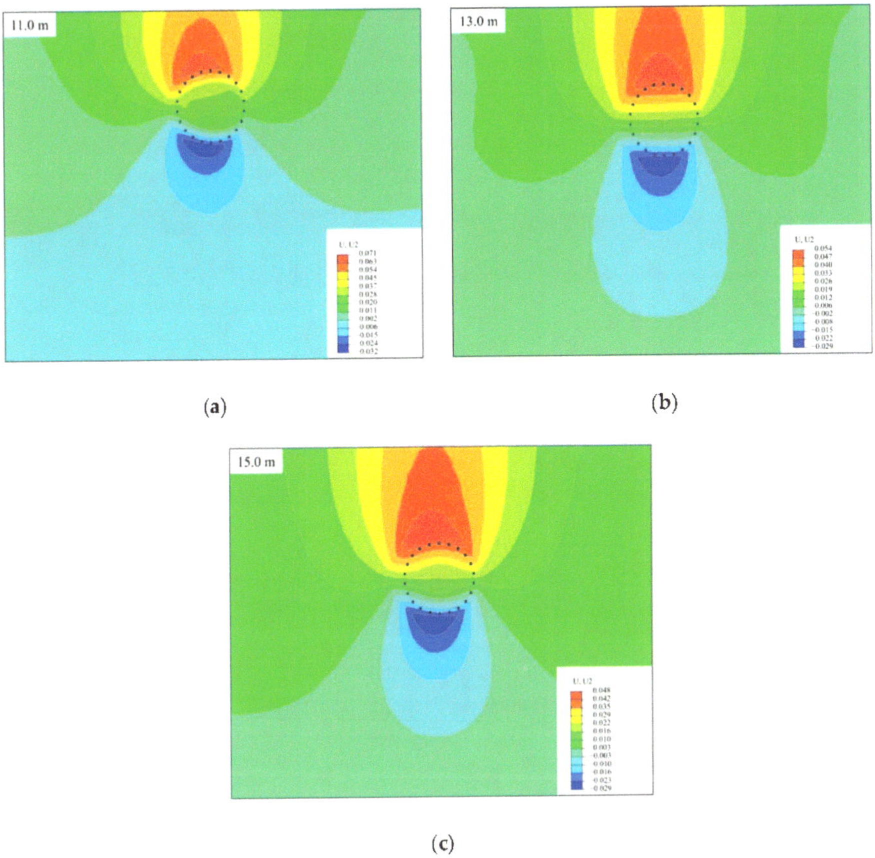

Figure 16. Distribution of stratum vertical displacement under different buried depth: (**a**) 11 m; (**b**) 13 m; and (**c**) 15 m.

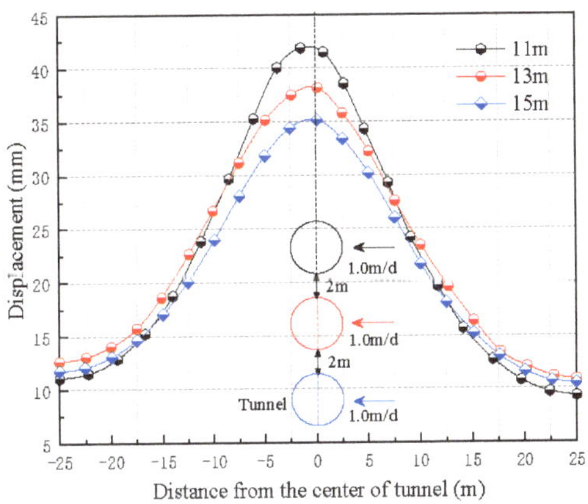

Figure 17. Vertical displacement distribution law of surface under different buried depths.

At the same time, the effect of tunnel buried depth on the vertical deformation of ground frost heaving under the action of seepage has the characteristics of regional differences. The frost heaving displacement of the surface is greatly affected by the buried depth of the tunnel within 9.2 m upstream and 12.1 m downstream of the tunnel center, and it decreases with the increase in buried depth.

6. Conclusions

In this paper, the development and distribution law of the freezing temperature field and frost heave displacement field of seepage stratum during the freezing construction period are studied by the numerical simulation method with a horizontal freezing engineering of subway tunnel as the research background. The main research contents and conclusions are as follows:

(1) A thermo–hydro–mechanical coupling numerical model is established using the finite element method. Specifically, the numerical simulation analysis software COMSOL is selected to realize the hydro–thermal coupling in frozen stratum under seepage condition. The numerical simulation analysis software ABAQUS is selected to realize the thermo–mechanical coupling, considering the ice–water phase transition and orthotropic deformation characteristics of frozen–thawed soil. The HyperMesh tool is used to divide the grid of models in COMSOL and ABAQUS to ensure the consistency of the grid, and the calculation results of the former are directly input into the latter program.

(2) Under the action of seepage, the frost heaving displacement of stratum is positively correlated with seepage velocity when the frozen wall reaches the effective thickness. It is worth point out that the maximum vertical displacement of frost heaving is 45.2 mm at the seepage velocity of 0.5 m/d, and the maximum vertical displacement of frost heaving is 53.5 mm at 2.0 m/d; furthermore, the frost heaving displacement of surface also increases from 30.8 mm to 39.6 mm.

Author Contributions: Conceptualization, M.L.; Methodology, R.H.; Software, C.P.; Formal analysis, Z.L.; Data curation, M.L.; Funding acquisition, H.C. All authors have read and agreed to the published version of the manuscript.

Funding: This research was funded by the National Natural Science Foundation of China, grant number 51778004; Research Activities Fund Project for Reserve Candidate of Academic and Technical Leaders of Anhui Province, China, grant number 2018H170; and Academic Funding for Top-notch Talents in University Disciplines (Majors) of Anhui Province, China, grant number gxbjZD10; and Research on Graduate Science Project in Anhui Province, China, grant number YJS20210383; and Graduate Innovation Fund Project of Anhui University of Science and Technology, grant number 2020CX2024.

Data Availability Statement: The datasets generated and analyzed during the current study are available from the corresponding author upon reasonable request.

Acknowledgments: The authors would like to thank Fangxing Yao, Zhe Yang, Heyin Wang for the numerical simulation data measurements.

Conflicts of Interest: The authors declare no conflict of interest.

References

1. Zhan, Y.X.; Lu, Z.; Yao, H. Numerical Analysis of Thermo-Hydro-Mechanical Coupling of Diversion Tunnels in a Seasonally Frozen Region. *J. Cold Reg. Eng.* **2020**, *34*, 04020018. [CrossRef]
2. Cui, J.Q.; Broere, W.; Lin, D. Underground space utilisation for urban renewal. *Tunn. Undergr. Space Technol.* **2021**, *108*, 103726. [CrossRef]
3. Alzoubi, M.; Xu, M.H.; Hassani, F.; Poncet, S.; Sasmito, A. Artificial ground freezing: A review of thermal and hydraulic aspects. *Tunn. Undergr. Space Technol. Inc. Trenchless Technol. Res.* **2020**, *104*, 103534. [CrossRef]
4. Pang, C.Q.; Cai, H.B.; Hong, R.B.; Li, M.K.; Yang, Z. Evolution Law of Three-Dimensional Non-Uniform Temperature Field of Tunnel Construction Using Local Horizontal Freezing Technique. *Appl. Sci.* **2022**, *12*, 8093. [CrossRef]
5. Alireza, A.; Hirokazu, A. Artificial ground freezing application in shield tunneling. *Jpn. Geotech. Soc. Spec. Publ.* **2015**, *3*, 71–75.
6. Lu, X.L.; Chen, X. Risk prevention and control of artificial ground freezing (AGF). *Chin. J. Geotech. Eng.* **2021**, *43*, 2308–2314.
7. Zhang, M.Y.; Zhang, X.Y.; Li, S.Y.; Lu, J.G.; Pei, W.S. Effect of Temperature Gradients on the Frost heaving of a Saturated Silty Clay with a Water Supply. *J. Cold Reg. Eng.* **2017**, *31*, 04017011. [CrossRef]

8. Zhang, M.Y.; Zhang, X.Y.; Lu, J.G.; Peng, W.S.; Wang, C. Analysis of volumetric unfrozen water contents in freezing soils. *Exp. Heat Transf.* **2019**, *32*, 426–438. [CrossRef]
9. Hou, S.S.; Yang, Y.G.; Cai, C.Z.; Chen, Y.; Li, F.L.; Lei, D.W. Modeling heat and mass transfer during artificial ground freezing considering the influence of water seepage. *Int. J. Heat Mass Transf.* **2022**, *2022 Pt 1*, 194. [CrossRef]
10. Zhou, J.; Guo, Z.Q.; Wang, C.H.; Li, Z.Y.; Zhou, H.D.; Pei, W.S. Analysis of Freeze–Thaw Response and Pore Characteristics of Artificially Frozen Soft Soil under Combined Formation Seepage. *Appl. Sci.* **2022**, *12*, 10687. [CrossRef]
11. Ji, Y.J.; Zhou, G.Q.; Hall, M.R. Frost heaving and frost heaving-induced pressure under various restraints and thermal gradients during the coupled thermal-hydro processes in freezing soil. *Bull. Eng. Geol. Environ.* **2019**, *78*, 3671–3683. [CrossRef]
12. Tang, Y.Q.; Xiao, S.Q.; Zhou, J. Deformation Prediction and Deformation Characteristics of Multilayers of Mucky Clay under Artificial Freezing Condition. *KSCE J. Civ. Eng.* **2019**, *23*, 1064–1076.
13. Cai, H.B.; Liu, Z.; Li, S. Improved analytical prediction of ground frost heaving during tunnel construction using artificial ground freezing technique. *Tunn. Undergr. Space Technol.* **2019**, *92*, 103050.
14. Lai, Y.M.; Wu, Z.; Zhu, Y.; Zhu, L. Nonlinear Analyses for the Couple Problem of Temperature, Seepage and Stress Field in Cold Region Tunnels. *Cold Reg. Sci. Technol.* **1999**, *29*, 89–96. [CrossRef]
15. Lai, Y.M.; Pei, W.S.; Zhang, M.Y. Study on theory model of hydro-thermal-mechanical interaction process in saturated freezing silty soil. *Int. J. Heat Mass Transf.* **2014**, *78*, 805–819.
16. Wang, T.L.; Song, H.F.; Shu, Y.; Zhang, F.; He, Y.M. Temperature field of artificially frozen gravel formation and optimization of freezing pipe layout parameters under seepage flow. *Geofluids* **2022**, *2022*, 1597645.
17. Marwan, A.; Zhou, M.M.; Abdelrehim, M.Z.; Guenther, M. Optimization of artificial ground freezing in tunneling in the presence of seepage flow. *Comput. Geotech.* **2016**, *75*, 112–125.
18. Li, Z.M.; Chen, J.; Sugimoto, M.; Ge, H.Y. Numerical simulation model of artificial ground freezing for tunneling under seepage flow conditions. *Tunn. Undergr. Space Technol.* **2019**, *92*, 103035. [CrossRef]
19. Huang, S.B.; Guo, Y.L.; Liu, Y.Z.; Ke, L.H.; Liu, G.F.; Chen, C. Study on the influence of water flow on temperature around freeze pipes and its distribution optimization during artificial ground freezing. *Appl. Therm. Eng.* **2018**, *135*, 435–445. [CrossRef]
20. Yang, X.; Ji, Z.Q.; Zhang, P.; Qi, J.L. Model test and numerical simulation on the development of artificially freezing wall in sandy layers considering water seepage. *Transp. Geotech.* **2019**, *21*, 100293. [CrossRef]
21. Zhang, M.L.; Guo, Z.Y.; Han, X.B.; Wang, B.; Wei, H.T.; Gao, Q. Analysis of coupled water and heat transfer in frozen soil based on mathematical module of COMSOL multiphysics. *Sci. Technol. Eng.* **2018**, *18*, 7–12.
22. Cheng, H.; Lin, J.; Wang, B.; Rong, C.X. Mathematical model and test verification of seepage freezing in saturated sand layer. *Sci. Technol. Eng.* **2018**, *18*, 38–44.
23. Wang, B.; Rong, C.X.; Cheng, H.; Cai, H.B. Experimental investigation on heat transfer law of multiple freezing pipes in permeable stratum with high seepage velocity. *Int. J. Heat Mass Transf.* **2022**, *182*, 121868. [CrossRef]
24. Cai, H.B.; Yao, F.X.; Hong, R.B.; Lin, J.; Zeng, K. Multi-loop pipe freezing optimization of deep shaft considering seepage effect. *Arab. J. Geosci.* **2022**, *15*, 153. [CrossRef]
25. Zheng, H.; Kanie, S.J.; Niu, F.J.; Akagawa, S.; Li, A.Y. Application of practical one-dimensional frost heaving estimation method in two-dimensional situation. *Soils Found.* **2016**, *56*, 904–914. [CrossRef]
26. Cai, H.B.; Hong, R.B.; Xu, L.X.; Wang, C.B.; Rong, C.X. Frost heaving and thawing settlement of the ground after using a freeze-sealing pipe-roof method in the construction of the Gongbei Tunnel. *Tunn. Undergr. Space Technol.* **2022**, *125*, 104503. [CrossRef]
27. Hong, R.B.; Cai, H.B.; Li, M.K. Factor's Influence Analysis of Frost heaving during the Twin-Tunnel Construction Using Artificial Horizontal Ground Freezing Method. *Adv. Civ. Eng.* **2020**, *2020*, 6650182.

applied sciences

MDPI

Article

Analysis of Elastoplastic Mechanical Properties of Non-Uniform Frozen Wall Considering Frost Heave

Bin Wang [1,2,3,*], Shenwei Liang [1], Yi Cao [1], Chuanxin Rong [1,3] and Shengmin Yu [1]

1 School of Civil Engineering and Architecture, Anhui University of Science and Technology, Huainan 232001, China
2 Postdoctoral Research Station, China Coal Mine Construction Group Co., Ltd., Hefei 230091, China
3 Engineering Research Center of Underground Mine Construction, Ministry of Education, Anhui University of Science and Technology, Huainan 232001, China
* Correspondence: wangbin@aust.edu.cn; Tel.: +86-182-9760-9818

Abstract: The aim of this study was to analyze the force and deformation law of an artificial frozen wall. Thus, the frost heave coefficient was introduced to describe the frost heave characteristics, and the frozen wall was regarded as a heterogeneous material whose material properties changed in a parabolic pattern with the radius. The elastoplastic stress and displacement formulas of a non-uniform frozen wall considering frost heave characteristics were derived according to different strength criteria. Consequently, the derived formulas were used to calculate and analyze the mechanical characteristics of the artificial frozen wall. The results showed that the radial stress of the frozen wall changed linearly, whereas the circumferential stress change followed a parabolic pattern after considering the non-uniform characteristics. Moreover, the displacement of the outer edge of the frozen wall was always greater than that of the inner edge, and the displacement of the inner edge changed with the increasing temperature, significantly greater than that of the outer edge. When the frozen wall was in the elastic state, its displacement caused by frost heave was constant. When the frozen wall entered the elastic–plastic state, the displacement of its inner edge caused by frost heave increased with the increase in the radius of the plastic zone, whereas the displacement of the outer edge caused by frost heave decreased with the increase in the radius of the plastic zone.

Keywords: artificial ground freezing method; nonuniform frozen wall; frost heave; elastoplastic analysis; analytical solution

1. Introduction

The AGF (artificial ground freezing method) has been widely used in underground engineering construction in water-rich strata because the resultant frozen wall has a good isolation effect on the groundwater [1–4], the principle of AGF is shown in Figure 1. In the construction of shaft freezing method, the depth of the frozen wall depends on the buried depth of mineral resources. For example, the freezing depth of Polby potash salt mine in the UK reached 930 m, and the freezing depth of the auxiliary shaft of Hutaoyu coal mine in China's Gansu Province reached 950 m [5]. The frozen soil undergoes frost heaving during the formation freezing, which is caused by the migration of unfrozen water and the expansion of frozen soil. The deformation of stratum is the main cause of the damage of an underground building structure [6–8], and the frost heaving is a major cause of the fracture of freezing pipes and the deformation of surrounding strata of underground buildings [9,10].

Currently, research on frost heave of stratum has been mainly focused on two aspects: the frost heave of frozen surrounding rock of tunnels in cold areas and the frost heave caused by the freezing of artificial formations. In the theoretical calculation of the frost heave force of tunnels in cold regions, Zhang et al. [11–13] used the unified strength theory as the basis for determining tunnels, considering the comprehensive influence of the intermediate principal stress and transverse isotropic frost heave on the strength of the surrounding rock. They established the unified plastic analytical solution of the stress and displacement of tunnels in cold areas. Xia et al. [14,15] found an analytical solution of the frost heave force of tunnels in cold areas. In this solution, the anisotropic frost heave coefficient k describes the transverse isotropic frost heave of the surrounding rock. The rationality of the solution was verified by comparing its results with field data. Feng [16] established a model for determining the elasticity and plasticity of the surrounding rock of tunnels in cold zones. Liu [17] divided the frozen surrounding rock into partitions according to support strength and support time, established a calculation model that considers the influence of frost heave, and examined the stress and deformation of tunnel lining and surrounding rock in cold areas through engineering cases. Zhang [18] established a three-dimensional temperature field model based on the test results of the temperature field of a tunnel in a cold region, obtaining the variation rule of the freezing depth of surrounding rock using Stephen's formula. In addition, the author established a new frost heave model based on the freezing depth of surrounding rock, combined with the frost heave model of a freeze–thaw zone and the frost heave model of regolith. Jiang et al. [19] established an elastic–plastic damage mechanics calculation model, considering the characteristics of the uneven frost heave of surrounding rock, verifying the theoretical solutions and analyzing the parameters.

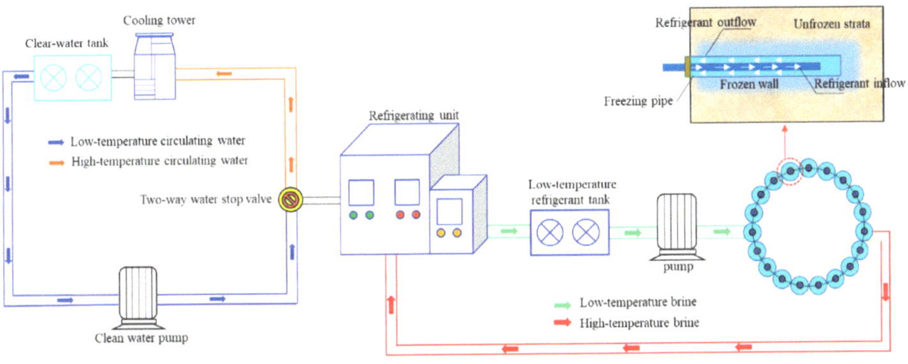

Figure 1. Schematic diagram of AGF.

Unlike the frozen surrounding rock of a tunnel in a cold zone, the artificial frozen wall is generated by freezing pipes. The heat conduction distance of freezing pipes and the direction of the temperature gradient influence the temperature within the frozen wall, which changes with the distance from the freezing pipes. Consequently, the internal strength of the frozen wall is non-uniform. The difference between the frozen surrounding rock of a tunnel and an artificial frozen wall is shown in Figure 2. If a frozen wall is treated as a homogeneous material to determine its mechanical properties, errors would be encountered in the calculation. Hu [20,21] derived a stress model for the frozen wall of single-row and double-row pipes using the M–C yield criterion. Wang et al. proposed a stress calculation formula for the heterogeneous frozen wall of three-row pipes [22] and derived a model for the directional-seepage-induced stress of the heterogeneous frozen wall by considering the influence of groundwater on the temperature distribution [23]. As a frozen wall does not exist in isolation in the soil, the unfrozen soil interacts with the frozen wall. Thus, soil excavation within the frozen wall influences the mechanical state of the wall. Yang et al. deduced the elastic–plastic stress solution of a frozen wall by considering the interaction of the wall with surrounding rock [24]. Wang developed a stress formula for the heterogeneous frozen wall, and the formula considers the unloading action and interaction with the unfrozen body [25].

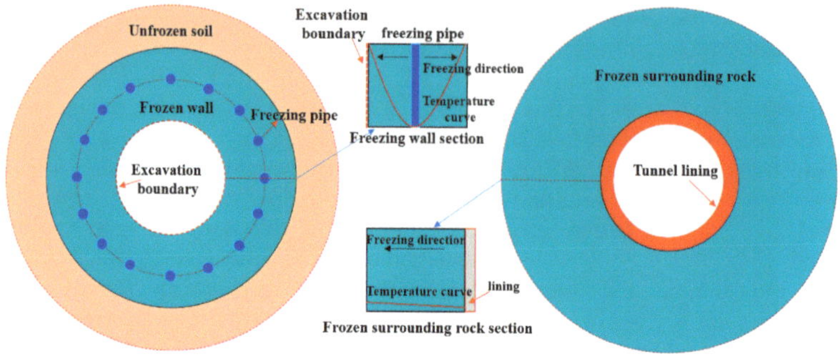

Figure 2. The difference between the frozen surrounding rock of a tunnel and an artificial frozen wall.

Currently, research on the theoretical calculation of the frost heave force of frozen surrounding rock of a tunnel in a cold area is relatively mature. Similarly, research on the mechanical mechanism of a heterogeneous frozen wall has advanced to appreciable levels. However, no relevant results have been obtained on the elastoplastic mechanical properties of a non-uniform frozen wall considering frost heave. The artificial frozen wall is different from frozen surrounding rock. The surrounding rock of a tunnel in a cold area is affected by unidirectional temperature gradient, which means it can be considered as a homogeneous material. However, the artificial frozen wall is formed using a freezing pipe; the temperature near the freezing pipe is low, whereas the temperature near the freezing front is high, resulting in the variation of the material properties of the frozen wall with the distance from the freezing pipe. In this case, the frost heave calculation theory of homogeneous frozen surrounding rock in a cold-region tunnel would not yield accurate calculation results of the stress and displacement of an artificial frozen wall. Therefore, the aim of this study is to introduce a frost heaving coefficient to describe the frost heaving characteristics of a frozen wall. The frozen wall is regarded as a functionally graded material whose material properties change parabolically with radius. The calculation formulas of the elastoplastic stress and displacement of a non-uniform frozen wall are derived using different strength criteria and considering frost heaving characteristics. This study is expected to provide a more accurate theoretical method for analyzing the mechanical properties of an artificial frozen wall.

2. Basic Assumptions and Frozen Soil Parameters

2.1. Basic Assumptions

1. Unfrozen soil is a homogeneous and an isotropic elastic medium [20–25]. The frozen wall is a heterogeneous material whose material properties vary with the wall radius [20–23];
2. The frost heave coefficient remains unchanged during freezing, and the frost heave rates of tangential and radial lines are unequal and remain unchanged during the freezing [14,15];
3. The Poisson's ratio μ and internal friction angle φ remain unchanged before and after freezing [20–23];
4. After entering the plastic state, the wall volume becomes incompressible [20–25];
5. The mechanical characteristics of the frozen wall are simplified as an axisymmetric plane strain problem [20–25].

2.2. Nonuniform Frost Heave Characteristics of Frozen Wall

An obvious temperature gradient exists between a freezing pipe and the surrounding soil during an artificial freezing process. The unfrozen water migrates to the freezing front along the temperature gradient, resulting in uneven frost heaving. The frost heave deformation along the direction of the temperature gradient $\alpha_{\parallel}$ is greater than that of the vertical temperature gradient $\alpha_{\perp}$.

The inhomogeneous frost heave coefficient k of soil is defined as

$$k = \alpha_{\parallel}/\alpha_{\perp}. \quad (1)$$

$k > 1$ represents the nonuniform frost heave characteristics of the frozen wall. The frost heave rate of the radial line of the frozen wall $\alpha_r = \alpha_{\parallel}$, and the circumferential frost heave rate $\alpha_\theta = \alpha_{\perp}$. As the axis direction of the frozen wall is perpendicular to the direction of the frozen soil, the linear frost heave rate along the axis is approximately equal to the circumferential linear frost heave rate. Thus, the volume frost heave rate of the frozen soil can be expressed as follows [11,14]:

$$\alpha_v = \alpha_r + \alpha_z + \alpha_\theta = \alpha_r + 2\alpha_\theta. \quad (2)$$

The relationship between α_v and k, α_r, and α_θ satisfies the following equations [11,14]:

$$\alpha_r = \frac{k\alpha_v}{k+2}, \quad (3)$$

$$\alpha_\theta = \frac{\alpha_v}{k+2}. \quad (4)$$

2.3. Equivalent Temperature Field of Frozen Wall

A linear relationship exists between the material properties of frozen soil and temperature [20], and the quadratic function curve reflects the temperature of the middle section of the adjacent freezing pipe within the range of freezing temperature commonly used in engineering practice [20]. Therefore, the elastic modulus and cohesion can be expressed in the following form:

$$E(r) = ar^2 + br + c \quad r \in [r_1, r_2], \quad (5)$$

$$c(r) = mr^2 + nr + l \quad r \in [r_1, r_2], \quad (6)$$

where a, b, c and m, l, n are the parameters related to the temperature distribution curve.

2.4. Strength Criteria for Frozen Soil

The strength criterion of frozen soil can be uniformly expressed as [24]

$$\sigma_\theta = M\sigma_r + Bc. \quad (7)$$

The values of M and B are presented in Table 1.

Table 1. Expressions of M and B.

Strength Criteria	M	B
Mohr Coulomb criteria	$\frac{1+\sin\varphi}{1-\sin\varphi}$	$-\frac{2\cos\varphi}{1-\sin\varphi}$
Druker–Prager criteria	$\frac{1+3\alpha}{1-3\alpha}$	$-\frac{2\kappa}{1-3\alpha}$
Tresca criteria	$\frac{1+2\sqrt{3}\alpha}{1-2\sqrt{3}\alpha}$	$-\frac{4\sqrt{3}\kappa}{3\left(1-2\sqrt{3}\alpha\right)}$
Twin shear unified failure criterion	$\frac{2(1+b)(1+\sin\varphi)-(1-\sin\varphi)b}{(1-\sin\varphi)(b+2)}$	$-\frac{2(1+b)}{(b+2)}\frac{2\cos\varphi}{1-\sin\varphi}$

where $\alpha = \frac{\sin\varphi}{\sqrt{3}\sqrt{3+\sin^2\varphi}}$ $\kappa = -\frac{\sqrt{3}\cos\varphi}{\sqrt{3+\sin^2\varphi}}$.

3. Analytical Solution of Stress and Displacement of Frozen Wall

The research flowchart of this study is depicted in Figure 3.

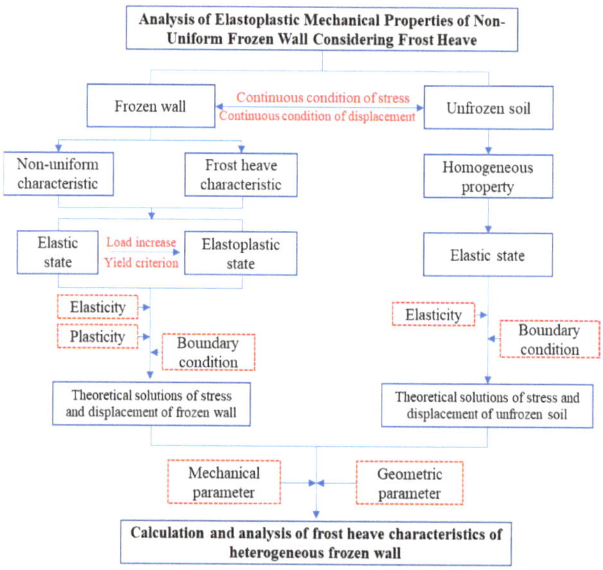

Figure 3. Research flowchart.

3.1. Mechanical Calculation Model of Frozen Wall

The mechanical model is depicted in Figure 4. In this model, P_0 is the formation pressure borne by the unfrozen body at infinity, P_1 is the external load exerted by the unfrozen body, and P_ρ is the interaction force at the junction of the plastic zone and the elastic zone. The inner and outer diameters of the frozen wall are denoted by R_1 and R_2, respectively, and the radius of any point on the section of the frozen wall is denoted by R. To facilitate the formula derivation, the relative radius $r = R/R_1$ is used to represent any point in the frozen wall section.

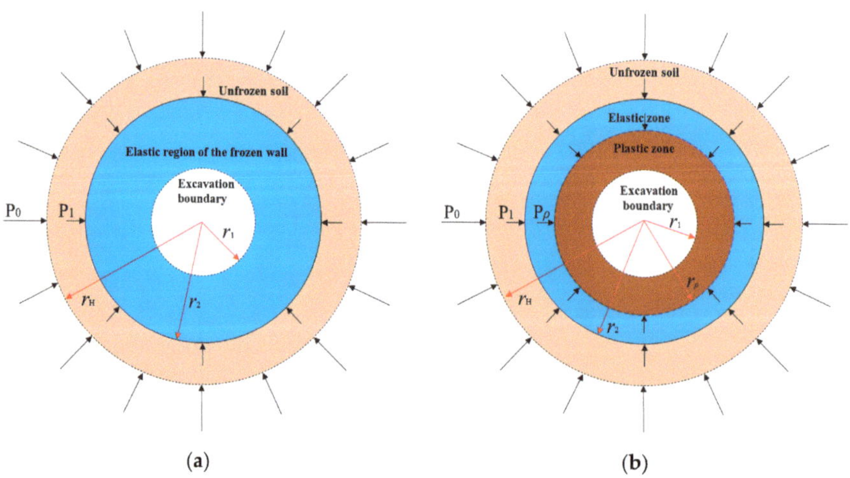

Figure 4. Mechanical calculation model of frozen wall. (**a**) Elastic model. (**b**) Elastoplastic model.

At the inner diameter: $R_1/R_1 = 1$
At the elastic–plastic boundary: $R_\rho/R_1 = r_\rho$
At the outer diameter: $R_2/R_1 = r_2$
At infinity: $R_\infty/R_1 = r_\infty$

When the external load is small, the frozen wall is in an elastic state. The mechanical model in this state is depicted in Figure 4a and comprises two parts: frozen wall $r \in [r_1, r_2]$ and unfrozen soil $r \in [r_2, r_\infty)$. When the external load is greater than its elastic ultimate bearing capacity, the frozen wall enters the elastoplastic state. The mechanical model in this state is depicted in Figure 4b. Here, the mechanical model mainly comprises three parts: plastic zone ($r \in [r_1, r_\rho]$), elastic zone ($r \in [r_\rho, r_2]$), and unfrozen soil ($r \in [r_2, r_\infty)$).

3.2. Calculation of Elastic Stress of Frozen Wall

When the frozen wall is in an elastic state, the calculation model shown in Figure 4a is used, and the boundary conditions of the model are as follows:

$$\begin{cases} \sigma_r^e = 0 \quad (r = r_1) \\ \sigma_r^e = P_1 \quad (r = r_2) \\ \sigma_r^s = P_0 \quad (r = r_\infty) \\ U_r^e = U_r^s \quad (r = r_2) \end{cases} \tag{8}$$

3.2.1. Unfrozen Area

The unfrozen soil is regarded as a uniform and elastic medium. Therefore, the formulas for calculating the elastic stress and displacement of the surrounding unfrozen soil are as follows [12,13]:

$$\sigma_r^s = \frac{r_2^2 r_\infty^2}{r_\infty^2 - r_2^2}\frac{P_1 - P_0}{r^2} + \frac{r_\infty^2 P_0 - r_2^2 P_1}{r_\infty^2 - r_2^2} = \left(1 - \frac{r_2^2}{r^2}\right)P_0 + \frac{r_2^2}{r^2}P_1 \tag{9}$$

$$\sigma_\theta^s = \frac{r_2^2 r_\infty^2}{r_\infty^2 - r_2^2}\frac{P_0 - P_1}{r^2} + \frac{r_\infty^2 P_0 - r_2^2 P_1}{r_\infty^2 - r_2^2} = \left(1 + \frac{r_2^2}{r^2}\right)P_0 - \frac{r_2^2}{r^2}P_1 \tag{10}$$

$$U^s = \frac{(1+\mu_3)(P_0 - P_1)r_2^2}{E_3 r} \tag{11}$$

where E_3 and μ_3 are the elastic modulus and Poisson's ratio of unfrozen soil, respectively, and the superscript s represents the unfrozen zone. The field measurement results show that the material properties of frozen soil and the surrounding unfrozen soil are the same; to simplify the calculation, let $E_3 = E(r_2)$ and $\mu_3 = \mu_2$.

3.2.2. Frozen Wall

When the frozen wall was in the elastic plane strain state, the following basic equations are satisfied:

Equation of equilibrium [24]:

$$\frac{d\sigma_r}{dr} + \frac{\sigma_r - \sigma_\theta}{r} = 0 \tag{12}$$

Geometric equation [24]:

$$\begin{cases} \varepsilon_r = du_r/dr \\ \varepsilon_\theta = u_r/r \end{cases} \tag{13}$$

Considering the condition of non-uniform frost heaving, the physical equation under plane strain in the elastic region is:

$$\begin{cases} \varepsilon_r^e = \frac{1-\mu_2{}^2}{E(r)}\sigma_r - \frac{\mu_2(1+\mu_2)}{E(r)}\sigma_\theta + (\alpha_r + \mu_2\alpha_\theta) \\ \varepsilon_\theta^e = \frac{1-\mu_2{}^2}{E(r)}\sigma_\theta - \frac{\mu_2(1+\mu_2)}{E(r)}\sigma_r + (\alpha_\theta + \mu_2\alpha_r) \end{cases} \tag{14}$$

Introducing the stress function φ, σ_r and σ_θ are represented by φ:

$$\begin{cases} \sigma_r = \varphi/r \\ \sigma_\theta = \dot{\varphi} \end{cases} \tag{15}$$

Combining (12)–(15) yields the following formula:

$$r^2\varphi'' + r\left(1 - \frac{E'(r)}{E(r)}r\right)\varphi' + \left(\frac{E'(r)}{E(r)}r - 1\right)\varphi = \frac{r(\alpha_r - \alpha_\theta)E(r)}{1+\mu_2} \tag{16}$$

As mentioned in Section 2.3, the elastic modulus changes with the relative radius r, and the expression of the elastic modulus $E(r)$ is substituted into Equation (16) to obtain

$$r^2\varphi'' - r\frac{ar^2 - c}{ar^2 + br + c}\varphi' + \frac{ar^2 - c}{ar^2 + br + c}\varphi = \frac{(k-1)\alpha_v(ar^2 + br + c)r}{(k+2)(1+\mu_2)} \tag{17}$$

Equation (17) is a second-order differential equation with variable coefficients, and the general solution is

$$\varphi = rC_0\left(a\ln r - b\frac{1}{r} - \frac{c}{2}\frac{1}{r^2}\right) + B_0 r + rA_0(ar^2 + 2br + 2c\ln r - c) \tag{18}$$

where $A_0 = \frac{\alpha_v(k-1)}{4(k+2)(\mu_2+1)}$.

By substituting the boundary condition in Equation (8) into Equation (18), the following is obtained:

$$C_0 = \frac{-P_1 - A_0(a(r_1^2 - r_2^2) + 2b(r_1 - r_2) + 2c\ln(r_1/r_2))}{a\ln(r_1/r_2) + b\left(\frac{1}{r_2} - \frac{1}{r_1}\right) + \frac{c}{2}\left(\frac{1}{r_2^2} - \frac{1}{r_1^2}\right)} \tag{19}$$

$$B_0 = -A_0(ar_1^2 + 2br_1 + 2c\ln r_1 - c) - C_0\left(a\ln r_1 - b\frac{1}{r_1} - \frac{c}{2}\frac{1}{r_1^2}\right) \tag{20}$$

$$\begin{cases} \sigma_r^e = \frac{A_0A_2(A_5 - A_3) + (P_1 + A_0A_1)(A_4 - A_6)}{A_2} \\ \sigma_\theta^e = \frac{A_0A_2(A_8 - A_3) + (P_1 + A_0A_1)(A_4 - A_7)}{A_2} \end{cases} \quad r \in [r_1, r_2] \tag{21}$$

where $A_1 = a(r_1^2 - r_2^2) + 2b(r_1 - r_2) + 2c\ln(r_1/r_2)$, $A_2 = a\ln(r_1/r_2) + b\left(\frac{1}{r_2} - \frac{1}{r_1}\right) + \frac{c}{2}\left(\frac{1}{r_2^2} - \frac{1}{r_1^2}\right)$, $A_3 = ar_1^2 + 2br_1 + 2c\ln r_1 - c$, $A_4 = a\ln r_1 - b\frac{1}{r_1} - \frac{c}{2}\frac{1}{r_1{}^2}$, $A_5 = ar^2 + 2br + 2c\ln r - c$, $A_6 = a\ln r - \frac{b}{r} - \frac{c}{2}\frac{1}{r^2}$, $A_7 = a\ln r + \frac{c}{2}\frac{1}{r^2} + a$, $A_8 = 3ar^2 + 4br + 2c\ln r + c$.

By substituting Equation (21) into physical Equation (14) and geometric Equation (13), the analytical solution of the elastic displacement of the heterogeneous frozen wall considering frost heave characteristics is obtained:

$$U^e = \frac{r\cdot(1-\mu_2^2)}{E(r)}\cdot\sigma_\theta^e - \frac{r\cdot\mu_2(1+\mu_2)}{E(r)}\cdot\sigma_r^e + \frac{(1+\mu_2 k)\alpha_v\cdot r}{k+2} \tag{22}$$

By substituting Equations (11) and (22) into the fourth expression in Equation (8) of interface displacement continuity, the expression of P_1 is obtained.

3.2.3. Determination of Elastoplastic State of Frozen Wall

The radial and circumferential stress of the frozen wall are obtained by substituting P_1 into Equation (21). The maximum elastic external load occurs at the inner diameter of the frozen wall. Let $r = r_1$ and substitute the stress into strength Equation (7) for judgment: if $\sigma^e_{\theta(r=r_1)} < M\sigma^e_{r(r=r_1)} + N$, the frozen wall is in the elastic state shown in Figure 4a; otherwise, it is in the elastic–plastic state shown in Figure 4b.

3.3. Mechanical Calculation Model of Frozen Wall

The mechanical model of the frozen wall in the elastic–plastic state is illustrated in Figure 4b. The model can be divided into the plastic zone ($r \in [r_1, r_\rho]$), elastic zone ($r \in [r_\rho, r_2]$), and unfrozen zone ($r \in [r_2, r_\infty)$), where $r = r_\rho$ is the interface between the plastic zone of the frozen wall and the elastic zone, and $r = r_2$ is the interface between the elastic zone of the frozen wall and the unfrozen soil. In this case, the stress displacement boundary conditions are as follows:

$$\begin{cases} \sigma^p_r = 0 & (r = r_1) \\ \sigma^p_r = P_\rho & (r = r_\rho) \\ \sigma^e_r = P_\rho & (r = r_\rho) \\ \sigma^e_r = P_1 & (r = r_2) \\ \sigma^s_r = P_0 & (r = r_\infty) \\ U^e_r = U^s_r & (r = r_2) \\ U^p_r = U^e_r & (r = r_\rho) \end{cases} . \tag{23}$$

3.3.1. Plastic Zone of the Frozen Wall

The stress balance equation in the plastic zone of the frozen wall is expressed as follows:

$$\frac{d\sigma^p_r}{dr} + \frac{\sigma^p_r - \sigma^p_\theta}{r} = 0. \tag{24}$$

By substituting Equation (7) into Equation (24), the following formula can be obtained:

$$\frac{d\sigma^p_r}{dr} + \frac{(1-M)\sigma^p_r}{r} - \frac{Bc(r)}{r} = 0. \tag{25}$$

By solving the first-order constant coefficient differential Equation (25), the general solution of radial stress in the plastic zone of the frozen wall can be expressed as follows:

$$\sigma^p_r = B_1 l r^2 + B_2 m r + B_3 n + C_1 \cdot r^{M-1}, \tag{26}$$

where $B_1 = \frac{B}{3-M}$, $B_2 = \frac{B}{2-M}$, and $B_3 = \frac{B}{1-M}$.

According to the stress boundary conditions expressed in Equation (23), Equation (26) can be re-expressed as follows:

$$C_1 = (-B_1 l r_1^2 - B_2 m r_1 - B_3 n) \cdot r_1^{1-M}. \tag{27}$$

Therefore, the radial stress in the plastic zone of the frozen wall is expressed as

$$\sigma^p_r = B_1 l r^2 + B_2 m r + B_3 n - (B_1 l r_1^2 + B_2 m r_1 + B_3 n) \cdot \left(\frac{r}{r_1}\right)^{M-1}. \tag{28}$$

According to the strength criterion of frozen soil expressed in Equation (7), the circumferential stress of the plastic zone can be obtained:

$$\sigma^p_\theta = M\sigma^p_r + Bc(r) \quad r \in [r_1, r_\rho]. \tag{29}$$

The volume of the frozen wall is incompressible when it is in the plastic state, and the average strain is 0 [20–25]:

$$\varepsilon_m = \frac{\varepsilon_r + \varepsilon_\theta + \varepsilon_z}{3} = 0. \tag{30}$$

In the axisymmetric plane strain problem, Equation (30) can be simplified as

$$\varepsilon_r + \varepsilon_\theta = 0. \tag{31}$$

By substituting the geometric Equation (13) into Equation (31), the following formula is obtained:

$$\frac{du_r}{dr} + \frac{u_r}{r} = 0. \tag{32}$$

By solving the first-order homogeneous differential Equation (32) with variable coefficients, the general displacement solution of the plastic zone is obtained:

$$u_r = \frac{C_2}{r}. \tag{33}$$

According to the stress boundary conditions expressed in Equation (23), the following expression is obtained:

$$C_2 = \frac{r_\rho^2(1-\mu_2^2)((M - \frac{\mu_2}{1-\mu_2})P_\rho + B \cdot c(r_\rho))}{ar_\rho^2 + br_\rho + c} + (\alpha_\theta + \mu_2\alpha_r)r_\rho^2. \tag{34}$$

By substituting Equation (34) into Equation (33), the analytical solution of the plastic location shift can be obtained:

$$U^\rho = \frac{r_\rho^2(1-\mu_2^2)((M - \frac{\mu_2}{1-\mu_2})P_\rho + B \cdot c(r_\rho))}{r(ar_\rho^2 + br_\rho + c)} + \frac{(\alpha_\theta + \mu_2\alpha_r)r_\rho^2}{r}. \tag{35}$$

3.3.2. Elastic Region of the Frozen Wall

By substituting $\sigma^e_{r(r=r_\rho)} = P_\rho$ and $\sigma^e_{r(r=r_2)} = P_1$ in Equation (23) into Equation (18) and then into Equation (15), the analytical solution of stress under the elastoplastic state is obtained:

$$\begin{cases} \sigma_r^e = \dfrac{A_0H_2(A_5 - H_3) + (P_\rho - P_1 - A_0H_1)(A_6 - H_4)}{H_2} \\ \sigma_\theta^e = \dfrac{A_0H_2(A_8 - H_3) + (P_\rho - P_1 - A_0H_1)(A_7 - H_4)}{H_2} \end{cases}, r \in [r_\rho, r_2], \tag{36}$$

where $H_1 = a\left(r_\rho^2 - r_2^2\right) + 2b(r_\rho - r_2) + 2c\ln\frac{r_\rho}{r_2}$; $H_2 = a\ln\frac{r_\rho}{r_2} + b\left(\frac{1}{r_2} - \frac{1}{r_\rho}\right) + \frac{c}{2}\left(\frac{1}{r_2^2} - \frac{1}{r_\rho^2}\right)$; $H_3 = ar_2^2 + 2br_2 + 2c\ln r_2 - c$; $H_4 = a\ln r_2 - b\frac{1}{r_2} - \frac{c}{2}\frac{1}{r_2{}^2}$; $A_5 = ar^2 + 2br + 2c\ln r - c$; $A_6 = a\ln r - \frac{b}{r} - \frac{c}{2}\frac{1}{r^2}$; $A_7 = a\ln r + \frac{c}{2}\frac{1}{r^2} + a$; and $A_8 = 3ar^2 + 4br + 2c\ln r + c$.

By substituting Equation (36) into geometric Equation (13) and physical Equation (14), the displacement solution of the elastic region is obtained:

$$U^e = \frac{(1-\mu_2)\cdot r}{E(r)}\left(\sigma_\theta^e - \frac{\mu_2}{1-\mu_2}\cdot\sigma_r^e\right) + \frac{(1+\mu_2 k)\alpha_v}{k+2}\cdot r. \tag{37}$$

According to the above formula, when $r_\rho = r_1$ and $r_\rho = r_2$, $P_{1-\max}$ is the elastic ultimate load and plastic ultimate load, respectively.

3.3.3. Unfrozen Zone

In the elastic–plastic state of the frozen wall, the expressions of stress and displacement of the unfrozen soil are as follows [12,13]:

$$\sigma_r^s = \left(1 - \frac{r_2^2}{r^2}\right) P_0 + \frac{r_2^2}{r^2} P_1, \tag{38}$$

$$\sigma_\theta^s = \left(1 + \frac{r_2^2}{r^2}\right) P_0 - \frac{r_2^2}{r^2} P_1, \tag{39}$$

$$U^s = \frac{(1 + \mu_3)(P_0 - P_1) r_2^2}{E_3 r}, \tag{40}$$

where E_3 and μ_3 are the elastic modulus and Poisson's ratio of unfrozen soil, respectively.

3.3.4. Radius of the Plastic Zone of the Frozen Wall

By combining Equations (23), (25), (28), (37) and (40), the following calculation formulas for P_1 and P_ρ are obtained:

$$P_\rho = B_1 l r_\rho^2 + B_2 m r_\rho + B_3 n - (B_1 l r_1^2 + B_2 m r_1 + B_3 n)\left(\frac{r_\rho}{r_1}\right)^{M-1}. \tag{41}$$

At the elastic–plastic junction, the stress in the elastic zone of the frozen wall meets the strength criterion of frozen soil:

$$\sigma^e_{r(r_\rho)} = M\sigma^e_{\theta(r_\rho)} + N_{(r_\rho)}. \tag{42}$$

By combining Equations (23), (36) and (42), the implicit equation of the radius of the plastic zone is obtained:

$$P_1 = \frac{P_0 H_2 (1 + \mu_3) r_2^2 + (\alpha_\theta + \mu_2 \alpha_r) H_2 E_3 r_2^2}{(1 + \mu_3)(1 - \mu_2)(2H_2 r_2^2 - a r_2^2 - b r_2 - c)} - \frac{(P_\rho - A_0 H_1 + 2A_0 H_2 r_2^2)(a r_2^2 + b r_2 + c) H_2 r_2^2}{H_2 (2H_2 r_2^2 - a r_2^2 - b r_2 - c) r_2^2}, \tag{43}$$

$$\frac{(1 - M) H_2 r_\rho^2 - a r_\rho^2 - b r_\rho - c}{H_2 r_\rho^2} P_\rho - \frac{(2A_0 H_2 r_\rho^2 - P_2 - A_0 H_1)(a r_\rho^2 + b r_\rho + c)}{H_2 r_\rho^2} = B(l r_\rho^2 + m r_\rho + n). \tag{44}$$

4. Mechanical Model of Frozen Wall

The inner and outer radii of the frozen wall of the new wind shaft considered in this study were 4 m and 10 m, respectively. The temperature at the freezing front was −3 °C. The thermal and physical parameters of the frozen soil are presented in Table 2. The coefficient of frost heave k was obtained by testing. Referring to the existing research results [11–15], k was set as 2.

Table 2. Thermophysical parameters of the frozen soil.

Elastic Modulus/MPa	Cohesive Force/MPa	Angle of Internal Friction	Poisson Ratio
−11.3T + 51.7	−0.26T + 1.17	10	0.35

The main external load of the frozen wall comes from formation pressure. Therefore, with the increase of the depth, the external load keeps increasing, and the stress state of the frozen wall changes accordingly. The stress distribution law of the frozen wall obtained by calculations at different depths is illustrated in Figure 5. According to the calculation results, the external load acting on the shaft wall increased continuously with an increase in depth; thus, the radial stress and circumferential stress on the section of the frozen wall also increased. As the non-uniform characteristics of frozen wall were considered in the derivation of the calculation formula, the calculated circumferential stress approximately mimicked a parabolic variation. The maximum circumferential stress appeared near $r = 1.5$.

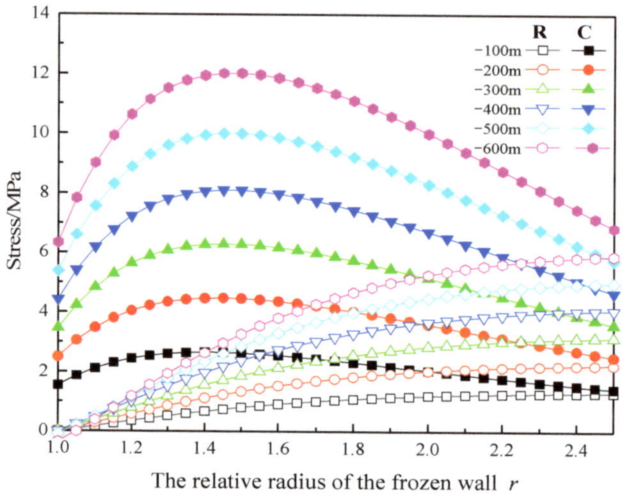

Figure 5. Stress distribution of the frozen wall at different depths. R represents the radial stress while C represents the circumferential stress.

In the process of the increase of the external load, the stress state of the frozen wall will go through three stages, namely, the elastic stage, the elastic–plastic stage, and the plastic stage. When the external load is small, the whole frozen wall is in the elastic state. When the external load increases to a certain critical value, the frozen wall enters the elastic–plastic state. The critical load from the elastic state into the plastic state is called the elastic limit load. When the frozen wall is in the elastic–plastic state, the frozen wall is divided into the plastic zone and the elastic zone from the inner edge to the outer edge. When the external load of the frozen wall increases further, the plastic zone expands gradually. When the whole frozen wall enters the plastic state, the corresponding external load is the plastic ultimate load. The stress state plays an important role in determining the stability of the frozen wall. The stress distribution of the frozen wall varies greatly under different stress states. Figure 6 illustrates the stress distribution of the frozen wall in the elastic limit state ($p = 5.90$ MPa), elastic–plastic state ($p = 11.14$ MPa), and plastic state ($p = 16.39$ MPa) according to the Mohr–Coulomb strength criterion. By comparison, the radial stress of the frozen wall increased with the increasing relative radius under different stress states, but the change law of circumferential stress was different. In the elastic limit state, elastoplastic state, and plastic limit state, the maximum circumferential stress of the frozen wall appeared at $r = 1.5$, the elastoplastic junction, and $r = 2.0$, respectively.

There are various yield criteria applicable to frozen soil. Currently, the commonly used ones are Mohr–Coulomb strength criteria, Druker–Prager strength criteria, Tresca strength criteria, and Twin shear unified failure criterion. Scholars usually choose one of these criteria as the basis for the calculation of mechanical properties of frozen wall [11–23]. In order to compare the difference of calculation results based on different criteria, the mechanical characteristics of frozen wall were calculated and analyzed with different yield criteria based on the same engineering condition in this study. Using different strength criteria, we calculated the elastic and plastic ultimate bearing capacity of the frozen wall at different average temperatures, as shown in Figures 7 and 8. The calculation results show that the bearing capacity of the frozen wall increased with a decrease in the average temperature. The calculation results of the frozen wall bearing capacity varied with criteria. The calculation results based on the Mohr–Coulomb criterion were close to those based on the Druker–Prager criterion, and the calculation results based on the twin shear strength criterion were greater than those based on other criteria. When the average temperature of the frozen wall was −12 °C, the elastic ultimate bearing capacity calculated according to the twin shear strength criterion, Tresca strength criterion, Mohr–Coulomb strength criterion, and Druker–Prager strength criterion was 5.35, 4.78, 4.06, and 4.03 MPa, respectively; the plastic bearing capacity was 15.79, 13.64, 11.11, and 11.03 MPa, respectively. When the average temperature of the frozen wall was −20 °C, the ultimate elastic bearing capacity calculated according to the twin shear strength criterion, Tresca strength criterion, Mohr–Coulomb strength criterion, and Druker–Prager strength criterion was 7.77, 6.94, 5.90, and 5.86 MPa, respectively; the plastic ultimate bearing capacity was 23.28, 20.14, 16.39, and 16.27 MPa, respectively.

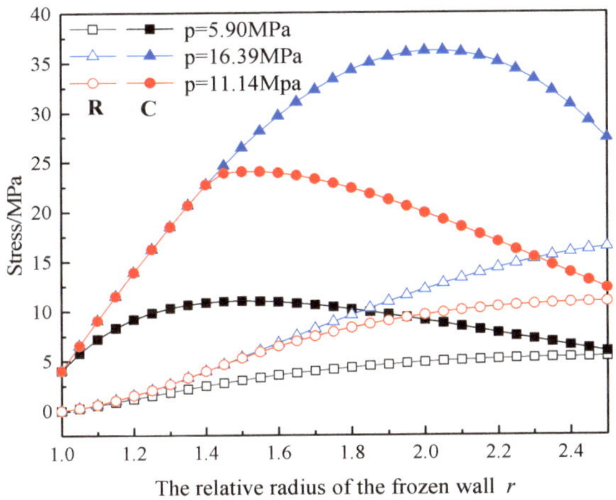

Figure 6. Stress distribution of the frozen wall under different stress states. R represents the radial stress while C represents the circumferential stress.

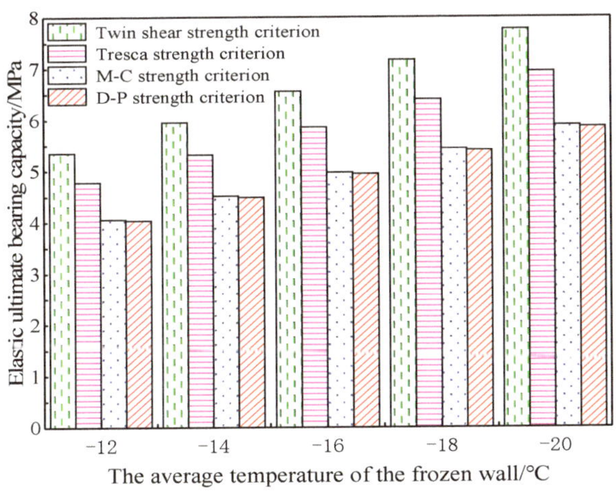

Figure 7. Calculation results of the elastic ultimate bearing capacity.

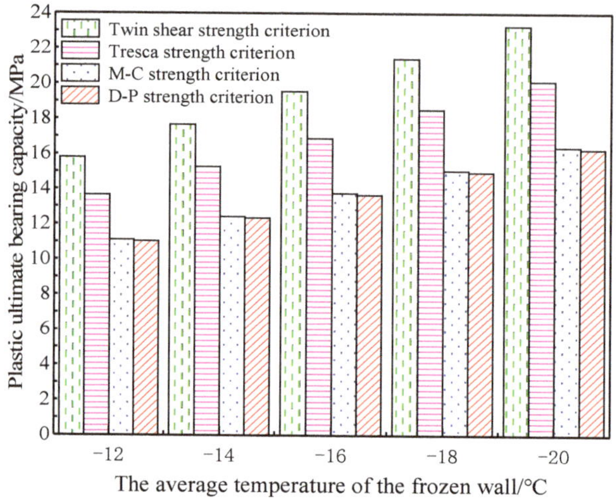

Figure 8. Calculation results of the plastic ultimate bearing capacity.

In the elastic limit and plastic limit states, the stress calculation results of the frozen wall based on different yield criteria are illustrated in Figures 9 and 10. A comparison shows that under the two stress states, the distribution law of the radial stress of the frozen wall was similar and increased with the increasing relative radius; the distribution law of the circumferential stress was significantly different. Although the circumferential stress of the frozen wall changed in an approximately parabolic shape under both conditions, the maximum value of the circumferential stress appeared at $r = 1.5$ in the elastic limit state and at $r = 2.0$ in the plastic limit state.

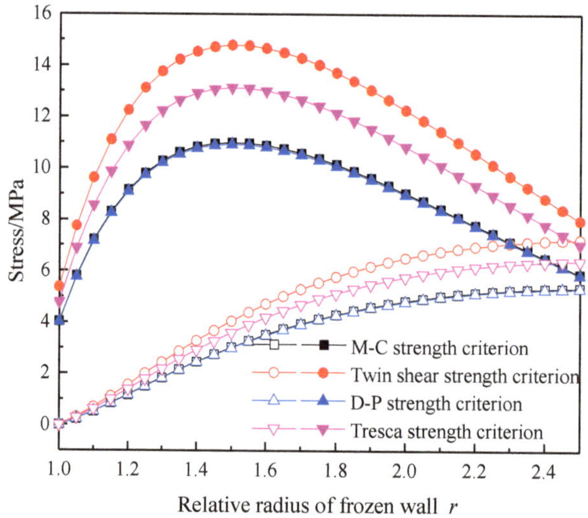

Figure 9. Stress distribution of the frozen wall in the elastic limit state.

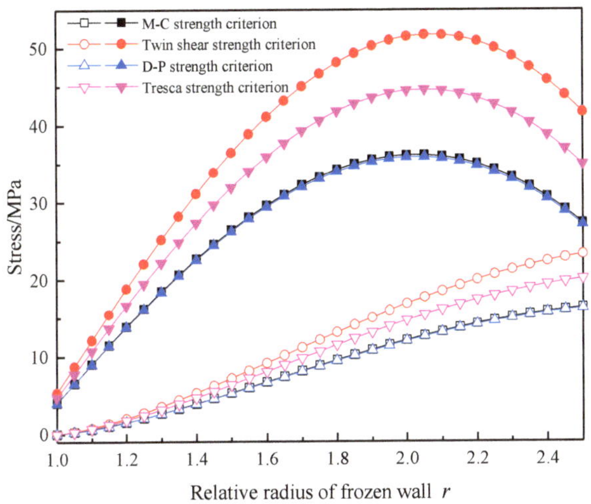

Figure 10. Stress distribution of the frozen wall in the plastic limit state.

The relationship between the bearing capacity of the frozen wall and the radius of the plastic zone was determined through calculations, and the results are illustrated in Figure 11. The results show that when the frozen wall entered the plastic state, its bearing capacity increased with the increase in the radius of the plastic zone, but the growth rate decreased gradually. Therefore, the larger the radius of the plastic zone was, the greater the damage risk of the frozen wall. The calculation results varied according to yield criteria, and the calculation results based on the Mohr–Coulomb and Druker–Prager criteria were mostly similar; the calculation results based on the twin shear strength criterion was the largest.

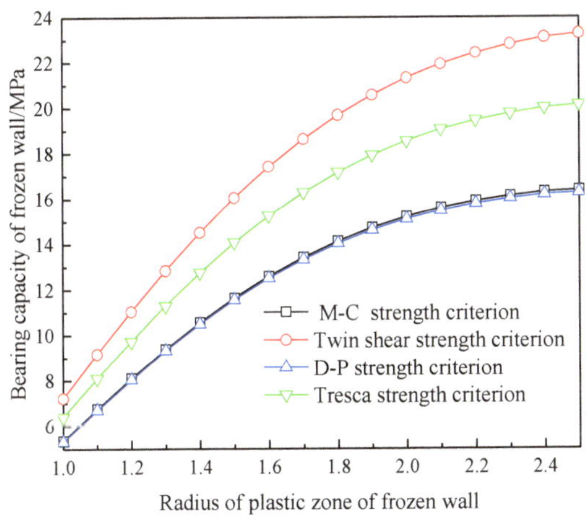

Figure 11. Relationship between plastic zone radius and bearing capacity of frozen wall.

The most common manifestation of frost heave in the freezing of a shaft is the displacement of the inner and outer edges of the frozen wall. The displacements of the inner and outer edges of the frozen wall in this study were calculated at different depths, as shown in Figure 12. As the depth increased, the external load of the frozen wall increased, and the displacement of the outer and inner edges of the frozen wall also increased. The lower the average temperature was, the greater the bearing capacity of the frozen wall; thus, the lower the average temperature was, the smaller the displacement of the frozen wall. The displacement of the outer edge of the frozen wall was always greater than that of the inner edge, which changed with temperature significantly more than that of the outer edge.

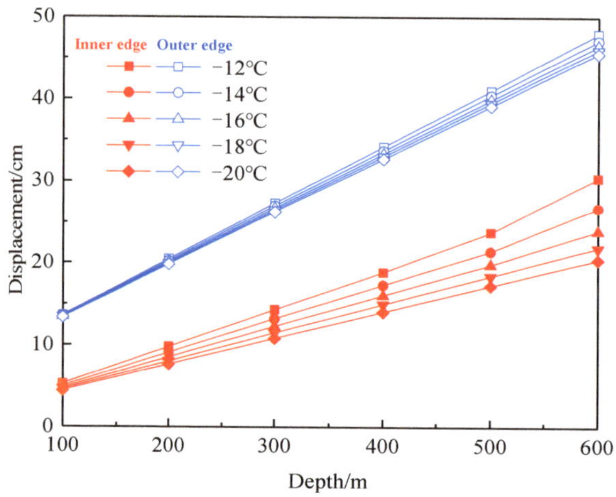

Figure 12. Variation of displacement of the inner edge and outer edge of the frozen wall.

When the average temperature was −20 °C, the relationship between the inner edge displacement and the radius of the plastic zone calculated based on different strength criteria was obtained, as illustrated in Figure 13. The results show that the displacement of increased rapidly with the increase in the radius of the plastic zone. When the relative radius *r* of the plastic zone reached 1.8, the calculated displacement results based on the twin shear strength criterion, Tresca strength criterion, Mohr–Coulomb strength criterion, and Druker–Prager strength criterion were 134.5, 118.2, 98.6, and 98.0 cm, respectively, which all exceed the allowable displacement range of the frozen wall. When a part of the frozen wall entered the plastic state, the displacement of the whole frozen wall increased considerably. Therefore, the bearing capacity of the frozen wall should be improved by increasing the thickness of the frozen wall or reducing the average temperature of the frozen wall, which would minimize the chances of the frozen wall entering the plastic state.

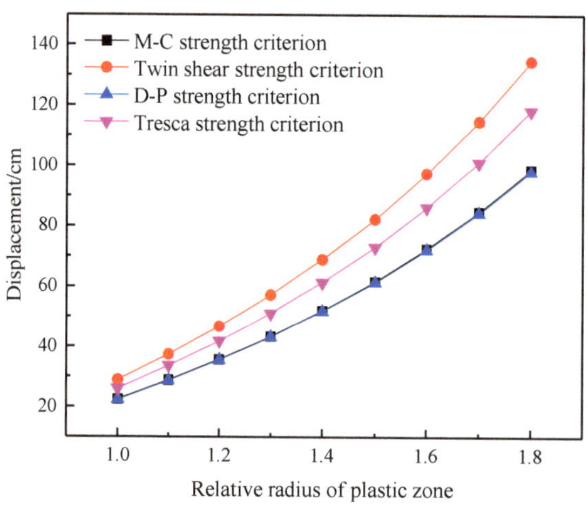

Figure 13. Variation of frozen wall displacement with plastic zone radius.

The results obtained using the heterogeneous frozen wall displacement formula derived in this paper, considering frost heave characteristics, were compared with the results of the frozen wall displacement formula without considering frost heave (converted from Ref. [20]). Figures 14 and 15 illustrate the comparison. The results show that the displacement mainly comprised two parts: the displacement generated by the frozen wall in the elastic or elastic–plastic state under the action of external load, and the displacement caused by soil frost heave. When the frost heave property was not considered, the calculated displacement of the frozen wall was smaller. Further analysis showed that the displacement of the inner and outer edges of the frozen wall due to frost heave remained unchanged when the buried depth was 100–300 m. In the depth range of 500–600 m, the frozen wall entered the elastoplastic state, and the displacement of the inner edge of the frozen wall caused by frost heave increased with the increase in the plastic zone. Meanwhile, the displacement of the outer edge caused by frost heave decreased with the increase in the plastic zone. A comparison of the displacements of the inner and outer edges showed that the total displacement of the outer edge of the frozen wall was greater than that of the inner edge at the same burial depth. Moreover, the displacement of frost heave on the outer edge was greater than that on the inner edge.

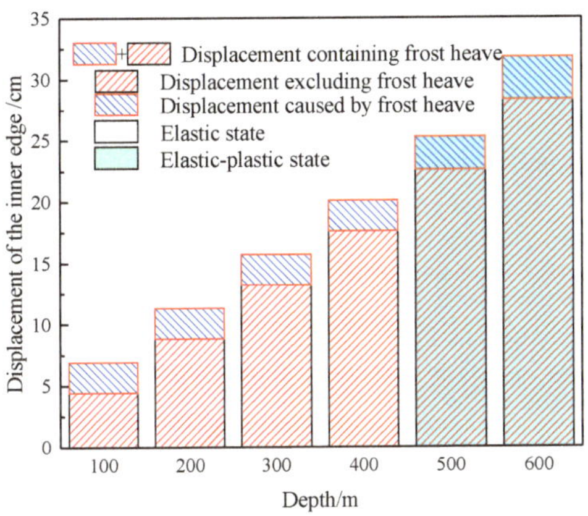

Figure 14. Comparison of the calculation results of the frozen wall inner edge displacement of the before and after frost heaving.

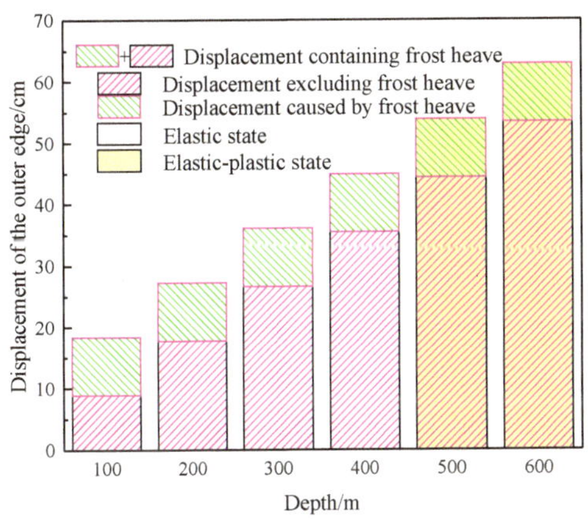

Figure 15. Comparison of the calculation results of the frozen wall outer edge displacement before and after considering frost heaving.

5. Conclusions

1. When the non-uniform characteristics were considered, the radial stress varied linearly with the relative radius r, and the circumferential stress of the frozen wall varied approximately in a parabolic shape. In this case, under the elastic limit state, the maximum value of the circumferential stress of the frozen wall appeared at $r = 1.5$. In the elastoplastic state, the maximum circumferential stress appeared at the elastoplastic junction. In the plastic limit state, the maximum circumferential stress appeared at $r = 2.0$;
2. The displacement of the outer and inner edges of the frozen wall increased with the increase in formation depth; the lower the average temperature of the frozen wall was, the smaller the displacement value. The displacement of the outer edge of the frozen wall was always greater than that of the inner edge, but the influence of temperature on the inner edge was greater than that of the outer edge;
3. When the frozen wall was in the elastic state, the displacement caused by frost heave was constant. However, when the frozen wall entered the elastic–plastic state, the displacement of the inner edge of the frozen wall caused by frost heave increased with the increase in the plastic zone, and the displacement of the outer edge of the frozen wall caused by frost heave decreased with the increase in the plastic zone;
4. When the frozen wall entered the plastic state, its bearing capacity increased with the increase in the radius of the plastic zone, but the growth rate decreased gradually. When a part of the frozen wall entered the plastic state, the displacement of the whole frozen wall increased considerably. Therefore, to improve the stability of the frozen wall, its bearing capacity should be enhanced by increasing its thickness or decreasing its average temperature, which would prevent the plastic state of the frozen wall.

Author Contributions: Conceptualization, B.W.; methodology, S.L.; software, Y.C.; validation, C.R., S.Y.; formal analysis, B.W.; investigation, B.W.; resources, B.W.; data curation, S.L.; writing—original draft preparation, B.W.; writing—review and editing, B.W.; visualization, S.L.; supervision, Y.C.; project administration, B.W.; funding acquisition, B.W. All authors have read and agreed to the published version of the manuscript.

Funding: This work was support by the Natural Science Foundation of Anhui Province (Grant No. 2108085QE251), China Postdoctoral Science Foundation (Grant No. 2021M703621), Anhui Postdoctoral Science Foundation (Grant No. 2022B635), Open Fund grants projects of Engineering Research Center of Underground Mine Construction of Ministry of Education (Grant No. JYBGCZX2022103), Key Project of Natural Science Research in Universities of Anhui Province (KJ2021A0425), Natural Science Foundation of Anhui University of Science and technology (xjzd2020-18), Talent Introduction Project of Anhui University of Science and Technology (13200403), Major Science and Technology Special Project of Anhui Province (202003c08020007), College student innovation and entrepreneurship program (S202210361048).

Institutional Review Board Statement: Not applicable.

Informed Consent Statement: Informed consent was obtained from all subjects involved in the study.

Data Availability Statement: The data used to support the findings of this study are available from the corresponding author upon request.

Conflicts of Interest: The authors declare that they have no known competing financial interests or personal relationships that could have appeared to influence the work reported in this paper.

References

1. Liu, Y.; Li, K.Q.; Li, D.Q.; Tang, X.S.; Gu, S.X. Coupled thermal–hydraulic modeling of artificial ground freezing with uncertainties in pipe inclination and thermal conductivity. *Acta Geotech.* **2022**, *17*, 257–274. [CrossRef]
2. Wang, B.; Rong, C.; Cheng, H.; Cai, H. Experimental investigation on heat transfer law of multiple freezing pipes in permeable stratum with high seepage velocity. *Int. J. Heat Mass Transf.* **2022**, *182*, 121868. [CrossRef]
3. Wang, B.; Rong, C.; Cheng, H. Analytical Solution of Steady-state Temperature Field of Asymmetric Frozen Wall Induced by Directional Seepage. *Adv. Eng. Sci.* **2022**, *54*, 76–87. (In Chinese)

4. Rong, C.; Wang, B.; Cheng, H.; Dong, Y.; Yang, F. Laboratory model test study on formation mechanisms of artificial frozen walls in permeable strata with high seepage velocity. *Chin. J. Rock Mech. Eng.* **2022**, *41*, 596–613. (In Chinese)
5. Yang, Z. Deformation Behaviour and Design Method of Icewall in Deep Soil Based on Co-bearing of Icewall and Stratum. Ph.D. Thesis, China University of Mining and Technology, Beijing, China, 2019. (In Chinese).
6. Ghiasi, V.; Omar, H.; Huat, B.B.K.; Muniandi, R.; Zainuddin, B. Risk management overview of tunnels using numerical modeling. *J. Eng. Des. Technol.* **2011**, *9*, 110–124. [CrossRef]
7. Ghiasi, V.; Koushki, M. Numerical and artificial neural network analyses of ground surface settlement of tunnel in saturated soil. *SN Appl. Sci.* **2020**, *2*, 939. [CrossRef]
8. Ghiasi, V.; Omar, H. Analysis of shotcrete lining of underground tunnels. *Pertanika J. Sci. Technol.* **2011**, *19*, 249–257.
9. Cao, G.; Cheng, H.; Rong, C. Theoretical solution to freezing pressure in Jurassic strata based on Kelvin damage model. *J. Min. Saf. Eng.* **2021**, *38*, 972–978. (In Chinese)
10. He, P.; Ma, W.; Mu, Y.; Dong, J.; Huang, Y. Elastic foundation beam model for frost heave damage of trapezoidal canal lining considering frost heave force and adfreeze force. *J. Cent. South Univ. Sci. Technol.* **2021**, *52*, 4148–4157. (In Chinese)
11. Zhang, C.; Gao, B.; Zhou, W.; Li, H. Plastic solutions of a cold region tunnel under freeze-thawcycles and non-uniform frost heave. *Chin. J. Theor. Appl. Mech.* **2022**, *54*, 252–262. (In Chinese)
12. Zhang, C.; Gao, B.; Li, T.; Shan, Y. An elastic-plastic solution for frost heaving force of cold region tunnels considering transversely isotropic frost heave and displacement release. *Rock Soil Mech.* **2021**, *42*, 2967–2976. (In Chinese)
13. Zhang, C.; Gao, B.; Shan, Y.; Li, Z. Unified plastic solution for stress and displacement of tunnels in cold regions considering transversely isotropic frost heave. *Chin. J. Geotech. Eng.* **2020**, *42*, 1825–1831. (In Chinese)
14. Xia, C.; Lv, Z.; Li, Q.; Huang, J.; Bai, X. Transversely isotropic frost heave of saturated rock under unidirectional freezing condition and induced frost heaving force in cold region tunnels. *Cold Reg. Sci. Technol.* **2018**, *152*, 48–58. [CrossRef]
15. Lv, Z.; Xia, C.; Wang, Y.; Luo, J. Analytical elasto-plastic solution of frost heaving force in cold region tunnels considering transversely isotropic frost heave of surrounding rock. *Cold Reg. Sci. Technol.* **2019**, *163*, 87–97. [CrossRef]
16. Feng, Q.; Jiang, B.; Zhang, Q.; Wang, L. Analytical elasto-plastic solution for stress and deformation of surrounding rock in cold region tunnels. *Cold Reg. Sci. Technol.* **2014**, *108*, 59–68. [CrossRef]
17. Liu, W.; Feng, Q.; Fu, S.; Wang, C. Elasto-plastic solution for cold-regional tunnels considering the compound effect of non-uniform frost heave, supporting strength and supporting time. *Tunn. Undergr. Space Technol.* **2018**, *82*, 293–302. [CrossRef]
18. Zhang, Y.; Xie, Y.; Li, Y.; Lai, J. A frost heave model based on space-time distribution of temperature field in cold region tunnels. *Rock Soil Mech.* **2018**, *39*, 1625–1632. (In Chinese)
19. Jiang, W.; Jiang, H.; Ma, Q.; Li, Y.; Li, Z. Unified elasto-plastic solution for cold regions tunnel considering damage and non-uniform frost heave. *J. South China Univ. Technol. Nat. Sci. Ed.* **2022**, *50*, 69–79+100. (In Chinese)
20. Hu, X.; Shu, C.; She, S. Elastic-plastic analytical solution for functionally graded material frozen soil wall with parabolic property under uniform load. *J. China Coal Soc.* **2012**, *37*, 379–384. (In Chinese)
21. Hu, X.; Shu, C. Stress field analysis of functionally graded material frozen soil wall in double-row-pipe shaft freezing. *Eng. Mech.* **2014**, *31*, 145–153. (In Chinese)
22. Wang, B.; Rong, C.; Cheng, H. Elastic and Plastic Analysis of Heterogeneous Frozen Soil Wall of Triple-row Piped Freezing. *J. Yangtze River Sci. Res. Inst.* **2019**, *36*, 104–111. (In Chinese)
23. Wang, B.; Rong, C.; Cheng, H.; Cai, H. Analysis of mechanical properties of heterogeneous frozen wall induced by directional seepage. *J. Glaciol. Geocryol.* **2022**, *44*, 1011–1020. (In Chinese)
24. Yang, W.; Yang, Z.; Bo, D. Elastic-plastic design theory of frozen soil wall based on interaction between frozen wall and surrounding rock. *Chin. J. Geotech.* **2013**, *35*, 175–180. (In Chinese)
25. Wang, B.; Rong, C.; Cheng, H. Stress analysis of heterogeneous frozen wall considering interaction with surrounding soil. *J. China Coal Soc.* **2017**, *42*, 354–361. (In Chinese)

applied sciences

MDPI

Article

Facies and Origin of Tufa Deposits from the Gostilje River Basin and the Sopotnica River Basin (SW Serbia)

Natalija Batoćanin [1,*], Wojciech Wróblewski [2], Ivana Carević [1], Uroš Durlević [1], Violeta Gajić [3] and Aleksandar Valjarević [1]

[1] Faculty of Geography, University of Belgrade, Studentski trg 3/3, 11000 Belgrade, Serbia
[2] Institute of Geological Sciences, Jagiellonian University, Gronostajowa 3a, 30-387 Krakow, Poland
[3] Faculty of Mining and Geology, University of Belgrade, Đušina 7, 11000 Belgrade, Serbia
* Correspondence: natalija.batocanin@gef.bg.ac.rs

Abstract: Tufa accumulations from the Gostilje River Basin and the Sopotnica River Basin in SW Serbia are represented by both active and fossil tufa precipitates. The aim of this study is to distinguish and describe different tufa facies and to determine the environmental conditions, based on stable isotope data. We also compare our analysis with other tufa deposits in Europe. Four facies are distinguished: moss tufa, algal tufa, stromatolitic laminated tufa, and phytoclastic tufa. The dominant constituent of all tufa samples is low Mg-calcite, whereas the presence of sylvite is noted in two samples from the Gostilje River Basin. The $\delta^{18}O$ values range from −9.07‰ to −10.79‰ (mean value: −9.81‰), while the $\delta^{13}C$ values range from −6.50‰ to −10.34‰ (mean values −9.01‰). The stable isotope values ($\delta^{13}C$ and $\delta^{18}O$) indicate that these tufa deposits were precipitated from cold, ambient water supported by CO_2 of an atmospheric origin. We emphasize that this is the first data about stable isotope analyses of tufa deposits from Serbia.

Keywords: tufa; facies; stable isotopes; Sopotnica River Basin; Gostilje River Basin; Serbia

1. Introduction

Tufa is a result of calcium carbonate precipitation under cool, ambient temperature in freshwater saturated in CaCO3. Therefore, tufa is found in fluvial, lacustrine, and palustrine environments [1]. Tufas are highly porous, poorly bedded [2], and composed of calcite precipitated from calcium bicarbonate water derived from the dissolution of carbonate rocks [3]. Tufa can be deposited both at springs and downstream at rapids or waterfalls, depending on the rate of CO_2 release. The first case occurs when dissolved CO_2 in groundwater, in equilibrium with soil CO_2 partial pressure, outgasses due to the lower partial pressure of CO_2 in the atmosphere. In the second case, there is a sharp increase in the outgassing rate due to the increased surface area between water and air caused by turbulence [4]. Tufa deposits usually contain abundant remains of micro- and macrophytes, invertebrates, and bacteria [5–8] and are characterized by a low or medium deposition rate (up to a few mm per year). Tufa deposition is related to specific, clearly defined geological, geomorphological, and environmental conditions, which, together, represent synonyms for "healthy" and preserved natural environments [9]. The distribution of tufa deposits depends on the availability of meteoric water and on local and climatic conditions. Since tufa often occurs at waterfalls, its distribution is also related to the microclimate in their immediate vicinity, which is characterized by cooler temperatures and higher humidity than in the surrounding area. Tufa deposits are a useful tool for deciphering long-term climatic changes during the Quaternary period, as they can be accurately dated using a variety of dating techniques [10]. The potential of fluvial tufa deposits as a valuable archive for paleoenvironmental and paleoclimate reconstructions has been investigated in several studies that focused on tufa oxygen and carbon isotopes and geochemical proxies [4,11–20].

The most used carbonate in paleoenvironmental reconstructions is calcite, as its precipitation is accompanied by the fractionation of stable carbon and oxygen isotopes [21]. A large number of tufa deposits is recorded on the territory of Serbia, mainly in eastern Serbia, but also in western Serbia, within the Dinaric Karst. They are located south of the Danube River and Sava River, usually at altitudes higher than 500 m.a.s.l. The bibliography of tufa accumulations in Serbia is insufficient. Most of the papers have been published in local languages and journals and are, therefore, not accessible to a wide geological community. In recent decades, more attention has been paid to individual smaller deposits in the context of environmental protection and geological heritage. Nevertheless, many authors have dealt with similar tufa deposits in the region [17,20,22–30]. The importance of tufa deposits is unambiguous. As it precipitates relatively fast, in specific conditions, it can serve as an applicable tool for the comparison of environmental conditions and regional correlations with similar deposits. Tufa deposits in Serbia were important in medieval time, as they were frequently used as a building material. However, nowadays, most of these deposits are under protection and exploitation is forbidden. In this work, tufa deposits from two sites were studied—tufa of the Sopotnica River Basin (further: SRB) and tufa of the Gostilje River Basin (further: GRB). Both sites include active and fossil tufa accumulations. The main goal of this study is to distinguish and describe different tufa facies and to determine environmental conditions based on stable isotope analysis. We also give a comparison with other tufa deposits in the region, as well as in Europe.

2. Study Sites

Two sites of active and fossil tufa accumulations were selected for this study. Tufa accumulations of the Sopotnica River appear in the village of the same name at the western slopes of the mountain Jadovnik, at about 15 km from the city of Prijepolje in southwestern Serbia. Tufa accumulations of the Gostiljska River are located approximately 25 km from the center of the Zlatibor Mountain. The geographical position of both sites is given in Figure 1.

SRB and GRB are located within the NW–SE trending Dinaridic ophiolite belt (DOB), which represents a remnant of an ocean-type basin. It comprises two essential components: Mélange, also known as the Dinaride olistostrome; and the ultramafic bodies, also known as the Dinaride ophiolites [31,32].

In Serbia, about 9% of the territory is made of carbonate rocks [33], which cover mostly the Western (Dinaric Karst) and Eastern (Carpatho-Balkanides) part of Serbia. Both sites of the SRB and GRB are part of the Dinaric Karst, which is the largest continuous karstland in Europe, covering approximately 60,000 km^2 [34]. The whole area is fed by karstic springs of Mesozoic limestones, which commonly exceed several hundred meters in thickness.

Tufa accumulations of the Sopotnica River appear in the village of the same name on the western slopes of the mountain Jadovnik, about 15 km from the city of Prijepolje in southwestern Serbia. The Sopotnica River is a short tributary of the Lim River, having a source in the Middle Triassic limestones in the village of the same name. At the contact of limestones and serpentinites, the underground river emerges and continues its flow as the Sopotnica River. The whole Jadovnik Mountain is intensively karstified and lacks of surface flows [9]. The Sopotnica River drains most of the mountain.

Tufa accumulations of the Gostiljska River are located approximately 25 km from the center of the Zlatibor Mountain, where twelve thermal springs occur in the Triassic limestones. Ten of them, including Gostilje, appear at the edge of the Zlatibor ultramafic massif. The Gostiljska River is a short tributary of the Katušnica River. The spring of the Gostiljska River is in the Triassic limestones, which cover an area of about 900 km^2 [31,35].

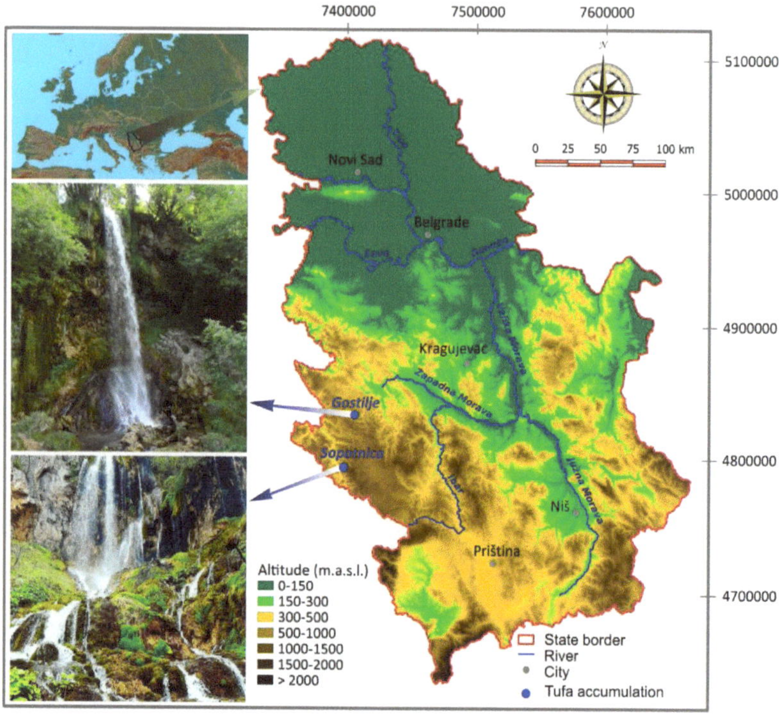

Figure 1. Geographical position of the studied sites.

3. Materials and Methods

Field work and sample selection was performed during the autumn months of 2020. A total of 59 tufa samples were collected at two locations—in the GRB and in the SRB. Tufa samples were collected from: active waterfalls, barrages, pools (recent tufa), inactive waterfalls, and fossil calcareous tufa. Facies are mainly distinguished based on field observations and confirmed afterward by detailed analyzes in thin sections. All samples were cleaned of weathered surfaces, before preparing thin sections, and then optically analyzed using a petrographic polarized microscope for transmitted light (Leica DMLSP), which was connected to a Leica DFC290 HD camera via the LAS V4.1 application. The mineralogical compositions of the twelve samples were determined by X-ray powder diffraction (XRD). Fifteen representative tufa samples were selected and examined using a JEOL JSM-6610LV scanning electron microscope connected to an X-Max energy dispersion spectrometer. Samples were covered with carbon using a BALTEC-SCD-005 Sputter coating device. Results were recorded under high vacuum conditions with an accelerating voltage of 20 kV and a beam current of 0.5–1.8 nA. All the above analyses were performed at the Department of Mineralogy, Crystallography, Petrology and Geochemistry, Faculty of Mining and Geology, University of Belgrade.

Stable isotope (oxygen and carbon) signatures in tufa have been used to clarify important climatic and environmental conditions in their formation [3,12,36–42]. In general, oxygen isotopes are indicators for paleotemperature, while carbon isotopes can point to a source of CO_2. A set of eight samples (four samples from SRB, as well as four samples from GRB) was selected for stable isotope analyzes. Stable oxygen and carbon isotope values ($\delta^{13}C$ and $\delta^{18}O$) of the carbonates were measured at the Jožef Stefan Institute in Ljubljana, Slovenia. The measurements were calibrated on the VPDB scale with the reference materials IAEA-CO8 and NBS-19. The measurements' precision was better than 0.1‰ for both the O and C values.

4. Results

4.1. Tufa Morphology and Petrology

Both localities are characterized by both active and fossil tufa precipitation, due to the changed direction of the river flow. Tufa is actively forming throughout the entire river flow at: waterfalls, cascades, and at the river bottom. Finally, deposition of the phytoclastic material led to the formation of dams and barrages, which locally surround pools. Dams and barrages are developed transverse to the watercourse. At those dams, barrages, and pools, there is also active tufa precipitation. The size of the barrages ranges from 10 cm to a few decimeters in height, and they are usually irregular or tongue shaped. They occur as single or as several piled-up barrages (Figure 2e).

Figure 2. (**a**) The tallest, multi-step waterfall at the River Sopotnica with active tufa precipitation; (**b**) Curtain-type waterfall at the River Sopotnica; (**c**) Fossil tufa, with visible traces of exploitation; (**d**) The main waterfall at Gostiljska River with active tufa precipitation; (**e**) Several tufa barrages surrounding swallow pools; (**f**) Fossil tufa outcrop; (**g**) Limestone fragments mixed with tufa debris; (**h**) Stromatolitic laminated fossil tufa; (**i**) Algae occupying the river bottom; (**j**) "Dry" waterfall; (**k**) Fossil tufa.

The pools vary in length and width from a few decimeters to 3 m, and water depth is around a few decimeters, up to one meter maximum. Tufa outcrops differ in size; sometimes they reach 12 m in height (Figure 2f). Outcrops are well preserved, and their lateral extension varies, but it usually is less than 100 m. Fossil tufa is white, very hard, compact (Figure 2k), and contains less organic matter. On the other side, recent tufa is brownish, more porous, and has many plants' remains. The thickness of tufa bodies is variable, sometimes reaching 20 m. Older, allochthonous tufa is sporadically present as debris. In some places, tufa debris is mixed with fragmented limestones (Figure 2g). On the riverbanks, fossil tufa sometimes displays well-developed lamination (Figure 2h).

Tufa accumulations of the Sopotnica River appear in the village of the same name at the western slopes of the mountain Jadovnik, at 1100 m.a.s.l. Although it flows shortly, at about 3.5 km, the Sopotnica River creates a series of waterfalls all the way down before its confluence with the Lim River, due to having constant stream throughout the year. The tallest waterfall is 25 m in height (Figure 2a), and, generally, it is a multi-step waterfall [43]. Locally, the waterfalls display features of both a segmented and curtain type (Figure 2b). The other waterfalls are on average 19–23 m in height. Some of the smaller waterfalls are active only during spring, while they are dry during other months (Figure 2j). The deposition of tufa starts beneath the limestone cliff Podstijenje at 1120 m.a.s.l. and continues down to 850 m.a.s.l., i.e., within 270 m in vertical directions and, for some, 500 m over mélange rocks. The deposition of tufa was disrupted at the middle of the river flow, due to exploitation (Figure 2c).

Tufa accumulations of the Gostiljska River are located approximately 25 km from the center of Zlatibor Mountain, at 867 m.a.s.l. Despite its short flow (about 2 km), the Gostiljska River creates tufa accumulations along the whole flow, as well as numerous waterfalls and small cascades. The biggest of them is 22 m high and it is created just before the end of the river flow, before its confluence with the Katušnica River (Figure 2d). Water drops abruptly from a 22 m high limestone cliff, forming further in the course of river, a several smaller tiered waterfalls and rapids yet to its confluence with the Katušnica River. Those waterfalls and rapids are always covered with hanging stems, mosses, and grass.

According to the results obtained by XRD investigations, the dominant constituent of all tufa samples is low-Mg calcite (LMC). The terrigenous component was too low to be identified with XRD, but SEM analysis confirmed the presence of sand- and silt-sized detrital quartz, as well as sylvite. The former occurs in isometric, oval-shaped grains, and its presence is noted in numerous samples and in all facies except the stromatolitic laminated tufa. The latter was found only in two samples of algal tufa from the Gostilje River basin. Even macroscopically tufa samples are very different, and four facies could be distinguished, ordered by their abundance: moss tufa, phytoclastic tufa, algal tufa, and stromatolitic laminated tufa. In general, micrite is the most abundant constituent in all facies, except for the stromatolitic laminated tufa, where sparite is dominant.

4.2. Tufa Facies

Four tufa facies are recognized herein. They range from moss, algal, stromatolitic laminated, and phytoclastic tufa facies.

Moss tufa is the most common facies at studied localities. Mosses are widespread, especially in the SRB, where they usually occur at the steep sides and waterfalls (Figure 2a,b). The moss filaments are generally randomly oriented (Figure 3b).

Moss tufa consists dominantly of micrite, but sparite and microsparite are also observed (Figure 4c). The size of the individual crystals varies, but, in general, they are less than 0.1 mm and, most commonly, isometric or elongated. Oval-shaped or irregular voids, which are usually empty, are the most abundant in moss tufa. Further magnification showed that moss tufa usually consists of rhombohedron calcite crystals (Figure 5c), as well as dagger-like, elongated crystals with sharp edges (Figure 5d).

Figure 3. Samples of the most representative tufa samples. (**a**) Algal tufa; (**b**) Moss tufa; (**c**) Phytoclastic tufa; (**d**) Stromatolitic laminated tufa.

Figure 4. Optical photomicrographs of thin sections. (**a**) Algal tufa. Note regular oval-shaped voids, surrounded by spherical calcite crystals (white arrows); (**b**) Algal tufa. White arrow points to void, filled by sparry calcite crystal; (**c**) Moss tufa built by micrite and sparite; (**d**) Phytoclastic tufa dominantly built by micrite; (**e**) Fan-shaped sparry calcite crystals in stromatolitic laminated tufa; (**f**) Micritic laminae (white arrow) alternates with sparitic laminae (pink arrow) in stromatolitic laminated tufa.

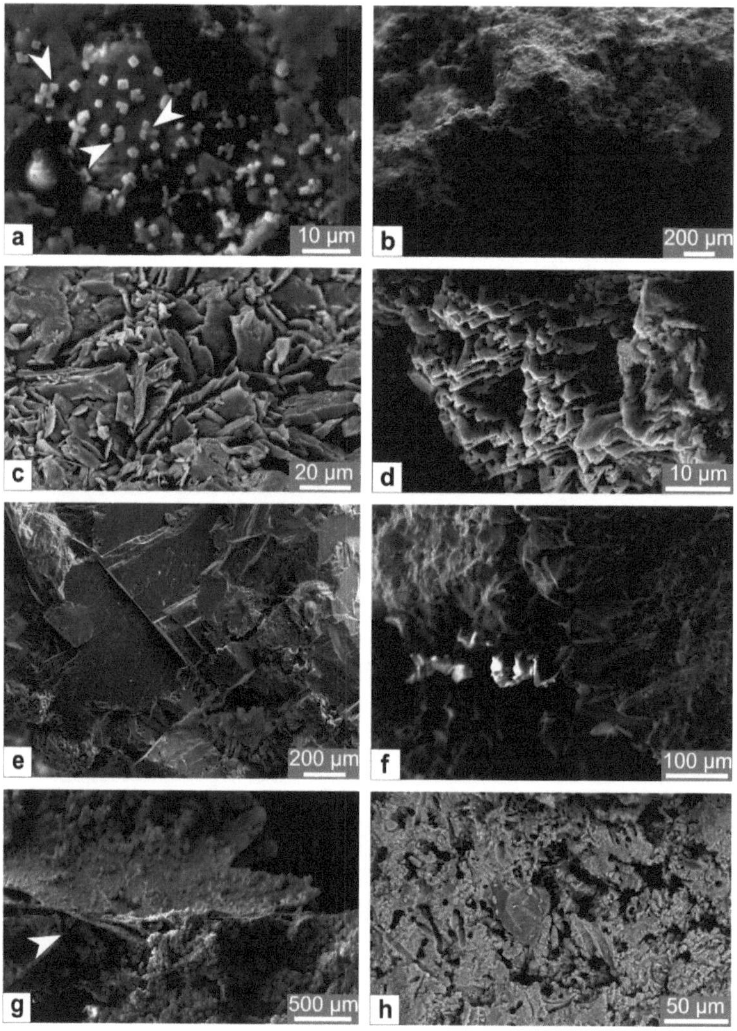

Figure 5. SEM images of tufa samples. (**a**) White arrows point to sylvite crystals; (**b**) Massive calcite aggregate in algal tufa; (**c**) Crystals of a primitive rhombohedron in moss tufa; (**d**) Dagger-like calcite crystals in moss tufa; (**e**) Massive calcite aggregate with well-developed cleavage; (**f**) Scalenohedron calcite crystals; (**g**) Phytoclastic tufa. Note the leaf imprint (white arrow); (**h**) Tubular cavities in calcite left by loss of bacterial filaments.

Algal tufa is usually found at the river bottom, which is intensively occupied by algae (Figure 2i). At some places, it occurs as a thin crust above the host rock. Algal tufa is brownish and, usually, very porous and friable (Figure 3a). It is dominantly built by clotted micrite, with scarce traces of sparite and microsparite, which locally surround micritic grains, forming a rim. Secondary sparry calcite crystals are usually found inside voids of different thicknesses (Figure 4b). Calcite grains are about 0.1 mm in size, and, in this facies, they are generally equal in size and shape. Regular oval-shaped voids are commonly present, surrounded by spherical calcite crystals (Figure 4a). It is observed that algal tufa usually displays massive or clotted textures and does not contain well-developed crystals (Figure 5b).

Stromatolitic laminated tufa is exclusively fossil tufa, usually found in a lower altitude. Stromatolitic laminated tufa is non-porous, very compact, and consists of about 1.5 mm thick light sparitic and dark micritic laminas (Figures 3d and 4f). Usually, light and dark laminas are of equal thickness, but some variations were noted. Respectively, sporadically light sparitic laminas are thicker than the micritic ones. Lamination is horizontal or wavy. In these facies, columnar and fan-shaped or needle-shaped sparite crystals are found in laminas (Figure 4e). Calcite crystals are perpendicular to lamination and occasionally reach 1 mm in length. SEM analyses showed that stromatolitic laminated tufa consists of coarse scalenohedron calcite crystals (Figure 5f) with well-developed cleavage (Figure 5e).

Phytoclastic tufa is formed among the whole river flow, especially at the barrages and small dams. It is more friable and more porous than other facies. These facies contain allochthonous plant fragments, such as leaves and twigs (Figures 3c and 5g). Phytoclastic tufa is almost completely built by clotted micrite, which surrounds numerous irregular and elongated fissures (Figure 4d). The size of the individual crystals is about 5μ. Rarely, secondary calcite crystals are found inside voids. Investigation in scanning electron microscope showed that there are also nanoscopic-sized holes, equal in shape and size, considering the loss of bacterial filaments (Figure 5h).

SEM analysis confirmed the presence of sylvite in two tufa samples from GRB. In one of those samples, sylvite occurs as dendritic aggregates while sylvite crystals appeared in a form of hexahedron in the other cubic (isometric) (Figure 5a). Sylvite crystals are of small dimensions that are usually about 2 μm, with a maximum of up to 3 μm.

4.3. Stable Isotope Composition

Both recent and fossil tufa deposits show similar and relative values of $\delta^{18}O$ and $\delta^{13}C$. The $\delta^{18}O$ values are between −9.07‰ and −10.79‰ (mean value: −9.81‰),whereas $\delta^{13}C$ values range between −6.50‰ and −10.34‰ (mean values −9.01‰). The highest values of $\delta^{18}O$ display fossil tufas from GRB, while the lowest values were recorded in fossil tufa from SRB. The highest values of $\delta^{13}C$ display fossil tufa from SRB, while the lowest values of $\delta^{13}C$ are represented by fossil tufa from GRB. There were no significant differences in the stable isotope composition between the different types and facies of analyzed tufa samples, except Sample Five, which has a significantly higher $\delta^{13}C$ value than the other tufa samples. The results of stable isotope analyses are given in Table 1.

Table 1. Stable isotope values in different tufa facies.

Sample	Site	Facies	$\delta^{18}O$ (‰ VPDB)	$\delta^{13}C$ (‰ VPDB)	Recent vs. Fossil Tufa
1	GRB	Algal tufa	−9.27 ± 0.05	−9.98 ± 0.04	Recent tufa
2	GRB	Phytoclastic tufa	−9.30 ± 0.08	−9.69 ± 0.02	Recent tufa
3	GRB	Stromatolitic laminated tufa	−9.07 ± 0.03	−8.84 ± 0.04	Fossil tufa
4	GRB	Moss tufa	−9.49 ± 0.02	−10.34 ± 0.02	Fossil tufa
5	SRB	Moss tufa	−9.48 ± 0.05	−6.50 ± 0.07	Fossil tufa
6	SRB	Algal tufa	−10.66 ± 0.09	−8.63 ± 0.03	Recent tufa
7	SRB	Stromatolitic laminated tufa	−10.79 ± 0.03	−8.53 ± 0.03	Fossil tufa
8	SRB	Phytoclastic tufa	−10.44 ± 0.01	−9.59 ± 0.02	Recent tufa

5. Discussion

The differences between tufa samples were obvious even macroscopically, especially between recent and fossil tufa samples, as well as evident different facies. Differences between tufa morphologies were influenced by "micro" variations in local conditions and, thus, by the diversity of the flora. In fact, the morphology of the tufa was controlled by a substrate, over which the spring water flows. The distribution of tufa facies points to a clear relationship between facies and local environmental conditions [44].

Moss tufa was, undoubtedly, the dominant facies at the waterfalls (recent tufa), but also in the fossil tufa samples. Algal tufa was mostly found in calmer conditions, such as river bottoms, i.e., recent tufas. Phytoclastic tufas are also recent but were mostly found among the whole river flow, especially at the barrages. Finally, the least present facies was stromatolitic tufa, and all of those tufa samples were considered fossil ones. In general, moss tufa was dominantly found in high energy conditions, whereas algal tufa and phytoclastic tufa were found in lower energy environments. The biggest difference between fossil and recent samples was that fossil tufa samples were harder and more compact than recent ones. This was expected, because the diagenesis of the tufa also included changes in the primary structural and textural characteristics of it, leading to more compact tufa by filling primary voids [7]. Additionally, the abundance of sparite/micrite, as well as porosity, varied in different facies, as could be seen in provided results.

The appearance of sylvite in two tufa samples was unusual, thus, it requires some explanation. As sylvite is the evaporite mineral that precipitates among the last ones out of solution, it is only found in a very dry saline areas. Nevertheless, supersaturation of certain ions results in salt crystallization, which is also assisted by higher porosity. It is well-known that sylvite is mostly used as a raw material for the production of fertilizers [45]. Thus, we assume that sylvite is most probably the result of anthropogenic contamination.

Origin of Tufa Deposits

Freshwater carbonates such as tufa can be useful tools for environmental conditions, assuming isotopic equilibrium during calcite precipitation. However, equilibrium between water and precipitated calcite is often disturbed due to different factors.

The changes in water temperature are the most responsible for the variations in $\delta^{18}O$. Increasing water temperature is related to decreasing $\delta^{18}O$ values [3]. Thus, we can assume that differences in $\delta^{18}O$ can be related to seasonal changes. Generally, oxygen isotopic composition suggests that tufa precipitated from cold, freshwaters enriched in calcium bicarbonate. Carbon isotope composition suggests that CO_2 is dominantly of an atmospheric origin, with various amounts of soil (organic) CO_2. Sample Five displays a very high carbon isotope value (−6.50‰). The carbon isotopes of carbonate rocks seem to be more complex than oxygen isotopes, which may have resulted from the many potential agents (e.g., abiotic and photosynthetic), degassing of CO_2, deposition of calcite, and CO_2 exchange with the atmosphere [46]. A higher $\delta^{13}C$ value in this sample can point to the input of heavier carbon isotope from dissolved limestones.

In general, no attention is paid to stable isotope analyses in tufas in Serbia. Nevertheless, tufa deposits have been studied worldwide as their isotopic composition can point to environmental conditions. A number of studies dealt with tufa deposits from an environmental point of view in the Mediterranean region [17,22,26,39,42,47–52], Central Europe [21,38,53], and Great Britain [54–56]. Stable isotope data of the studied sites were compared with some well-known tufa deposits in Europe at Figure 6.

It is evident that most of the tufa deposits have $\delta^{18}O$ values between −9.5‰ and −6‰. The highest $\delta^{18}O$ values are recorded in France [42], as well as in Great Britain [54]. It is interesting that stable oxygen values from SRB and GRB are generally lower than in mentioned studies. The most similar $\delta^{18}O$ signatures to SRB and GRB have tufa deposits from Turkey [50], Poland [38], and a part of tufa from Dinaric Karst [17,22], but they are, still, generally higher than tufa from SRB and GRB.

The situation with the $\delta^{13}C$ isotope data is even more complicated, i.e., more variations in different studies are noted. Most of the tufa deposits have $\delta^{13}C$ values between −10‰ and −6‰. Tufa samples from Serbia display more similarity with other deposits in $\delta^{13}C$ values than in $\delta^{18}O$ values. However, $\delta^{13}C$ values in this study are generally lower than in other sites in Europe. In many studies, $\delta^{13}C$ values in some samples are very high, usually near 0, or even positive [39,47,49,50]. Only data from France [42], and a part of data from Great Britain [54] have lower $\delta^{13}C$ (minimum −12.1‰) than the samples from SRB and GRB.

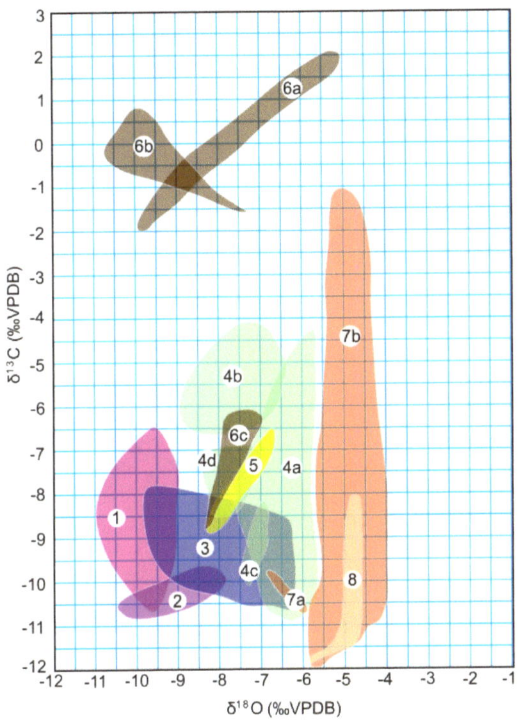

Figure 6. Comparison of stable isotope data in this study with other deposits in Europe. 1. Serbia, present study; 2. Poland [38]; 3. Dinaric Karst [17,22]; 4. Spain: a [47], b [39], c [48], d [51]; 5. Slovakia [53]; 6. Turkey: a [49], b [50], c [52]; 7. Britain: a [55], b [54]; 8. France [42].

Compared to stable isotope analyses of tufa in Dinaric Karst [17,20,22,26], tufa from SRB and GRB have similar $\delta^{13}C$ values and lower $\delta^{18}O$ values to other tufa accumulations in the region. The comparison between the stable isotope compositions from GRB and SRB to other tufa sites in Dinaric Karst is given in Figure 7. Mean $\delta^{13}C$ values for SRB and GRB are −9.01‰, whereas the mean value is −9.60‰ in the Zrmanja River Canyon, Krka River, and Krupa River [17,22,26]. In the Plitvice lakes, mean $\delta^{13}C$ values are around −8.1‰ [17]. On the other side, $\delta^{18}O$ values are a bit lower for SRB and GRB than in other tufa sites in the region. Mean $\delta^{18}O$ values for SRB and GRB is −9.81‰, whereas it is −7.02‰ in the Zrmanja River Canyon [22], −8.08‰ in the Krka River [26], and −7.8‰ in the Krupa River [17]. The most similar to this study are the $\delta^{18}O$ values for the Plitvice lakes—with a mean value of −9.7‰ [17].

Generally, $\delta^{13}C$ values in tufas from Serbia are lower, but still very similar with the compared regions in Europe. In contrast, the slight differences in the $\delta^{18}O$ values suggest that the analyzed tufas could have been formed in slightly different weather or climatic conditions. Most probably, lower values of $\delta^{18}O$ relative to other sites in Europe reflect differences in water temperature during tufa crystallization, which is commonly observed in tufas in this part of Europe [57]. Additionally, lower values of $\delta^{18}O$ can be caused by the altitude or continental effect on isotopic fractionation in the rainwater that supplies springs, within which tufas are formed [12,58,59]. SRB and GRB are both at higher elevations than the above-mentioned localities (SRB is at 1100 m.a.s.l. and GRB is at 867 m.a.s.l.), whereas the altitude of other sites is usually less than 500 m.a.s.l. According to the observations of [60], the differences in $\delta^{13}C$ and $\delta^{18}O$ values between analyzed tufa and other sites in the Dinaric region and Europe may result from seasonal differences in seasonal atmospheric precipitation rates and the degrees of CO_2 degassing along the course of the stream. Comparing only the samples from this study, values are, in general, very similar. Furthermore, the CO_2 source did not change, being mostly atmospheric, and the water temperature was similar all the time during tufa precipitation.

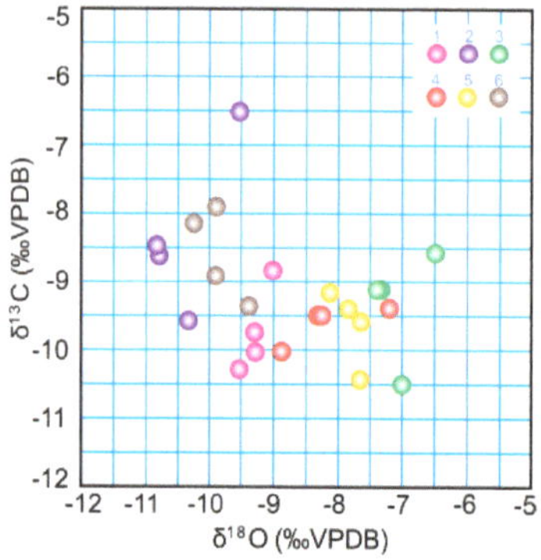

Figure 7. Comparison of stable isotope data in this study with other deposits in Dinaric Karst. 1—Gostilje River Basin; 2—Sopotnica River Basin; 3—Zrmanja River Basin; 4—Krka River Basin; 5—Krupa River Basin; 6—Plitvice Lakes.

6. Conclusions

The investigations of tufa accumulations at SRB and GRB led to the following conclusions:

(1) Four tufa facies are recognized, ordered by their abundance: moss tufa, phytoclastic tufa, algal tufa, and stromatolitic laminated tufa. Different facies display different environmental conditions.

(2) Stable isotope ($\delta^{13}C$ and $\delta^{18}O$) data from the tufa of SRB and GRB are similar in all tufa facies. Oxygen isotope composition suggests that the tufa precipitated from cold, freshwaters enriched in calcium bicarbonate. Carbon isotope composition suggests that CO_2 is dominantly of atmospheric origin, with various amount of soil (organic) CO_2.

(3) Differences in the $\delta^{13}C$ values between the tufa samples from Serbia and from the other sites in Europe are negligible. Lower $\delta^{18}O$ values in the tufa from SRB and GRB than in other sites in Europe indicate that these tufa deposits in Serbia precipitated in a slightly different condition. Lower $\delta^{13}C$ values in tufa from SRB and GRB than in other sites in Europe indicate that in tufa from Serbia, while the influence of isotopically heavier ^{13}C from dissolved limestones was less.

(4) Well-developed sylvite crystals in two samples from GRB suggest a strong anthropogenic influence.

Author Contributions: Conceptualization, N.B.; Methodology, N.B. and V.G.; Software, U.D. and A.V.; Investigation, V.G.; Data curation, I.C. and V.G.; Writing—review & editing, N.B. and I.C.; Visualization, N.B., W.W., I.C. and U.D.; Supervision, W.W. All authors have read and agreed to the published version of the manuscript.

Funding: The study was supported by the Ministry of Education, Science and Technological Development of the Republic of Serbia (Contract numbers 451-03-68/2022-14/200126 and 451-03-68/2022-14/200091).

Institutional Review Board Statement: Not applicable.

Informed Consent Statement: Not applicable.

Data Availability Statement: The data presented in this study are available on request from the corresponding author.

Conflicts of Interest: The authors declare no conflict of interest.

References

1. Roche, A.; Vennin, E.; Bundeleva, I.; Bouton, A.; Payandi-Rolland, D.; Amiotte-Suchet, P.; Gaucher, E.C.; Courvoisier, H.; Visscher, P.T. The Role of the Substrate on the Mineralization Potential of Microbial Mats in a Modern Freshwater River (Paris Basin, France). *Minerals* **2019**, *9*, 359. [CrossRef]
2. Gandin, A.; Capezzuoli, E. Travertine vs. calcareous tufa: Distinctive petrologic features and stable isotope signatures. *Ital. J. Quat. Sci.* **2008**, *21*, 125–136.
3. Andrews, J.E. Palaeoclimatic records from stable isotopes in riverine tufas: Synthesis and review. *Earth Sci. Rev.* **2006**, *75*, 85–104. [CrossRef]
4. Hamdan, M.A.; Brook, G.A. Timing and characteristics of Late Pleistocene and Holocene wetter periods in the Eastern Desert and Sinai of Egypt, based on ^{14}C dating and stable isotope analysis of spring tufa deposits. *Quat. Sci. Rev.* **2015**, *130*, 168–188. [CrossRef]
5. Brogi, A.; Capezzuoli, E.; Buracchi, E.; Branca, M. Tectonic control on travertine and calcareous tufa deposition in a low-temperature geothermal system (Sarteano, Central Italy). *J. Geol. Soc. Lond.* **2012**, *169*, 461–476. [CrossRef]
6. Capezzuoli, E.; Gandin, A.; Pedley, M. Decoding tufa and travertine (fresh water carbonates) in the sedimentary record: The state of the art. *Sedimentology* **2014**, *61*, 1–21. [CrossRef]
7. Ford, T.D.; Pedley, H.M. A review of tufa and travertine deposits of the world. *Earth Sci. Rev.* **1996**, *41*, 117–175. [CrossRef]
8. Liu, L. Factors Affecting Tufa Degradation in Jiuzhaigou National Nature Reserve, Sichuan, China. *Water* **2017**, *9*, 702. [CrossRef]
9. Đurović, P. *Bigar—Značajna prirodna vrednost krasa u Srbiji, Zaštita prirode [Tufa—significant natural value of Serbian karst. Nature conservation]*; Zavod za zaštitu prirode: Beograd, Srbije, 1998; pp. 163–170.
10. Sancho, C.; Arenas, C.; Vázquez-Urbez, M.; Pardo, G.; Lozano, V.M.; Peña-Monné, L.J.; Hellstrom, C.J.; Ortiz, E.J.; Osácar, C.M.; Auqué, L.; et al. Climatic implications of the Quaternary fluvial tufa record in the NE Iberian Peninsula over the last 500 ka. *Quat. Res.* **2015**, *84*, 398–414. [CrossRef]
11. Andrews, J.E.; Riding, R.; Dennis, F.P. The stable isotope record of environmental and climatic signals in modern terrestrial microbial carbonates from Europe. *Palaeogeogr. Palaeoclimatol. Palaeoecol.* **1997**, *129*, 171–189. [CrossRef]
12. Andrews, J.E.; Pedley, M.; Dennis, F.P. Palaeoenvironmental records in Holocene Spanish tufas: A stable isotope approach in search of reliable climatic archives. *Sedimentology* **2000**, *47*, 961–978. [CrossRef]
13. Brook, G.A.; Cherkinsky, A.; Railsback, L.B.; Marais, E.; Hipondoka, M.H.T. ^{14}C dating of organic residue and carbonate from stromatolites in Etosha Pan. Namibia: ^{14}C reservoir effect: Correction of published ages and evidence of >8-m-deep lake during the late Pleistocene. *Radiocarbon* **2013**, *55*, 1156–1163. [CrossRef]
14. Domínguez-Villar, D.; Vázquez-Navarro, J.A.; Cheng, H.; Edwards, R.L. Freshwater tufa record from Spain supports evidence for the past interglacial being wetter than the Holocene in the Mediterranean region. *Glob. Planet. Chang.* **2011**, *77*, 129–141. [CrossRef]
15. Dabkowski, J. The late-Holocene tufa decline in Europe: Myth or reality? *Quat. Sci. Rev.* **2020**, *230*, 106141. [CrossRef]
16. Hasan, O.; Miko, S.; Brunović, D.; Papatheodorou, G.; Christodolou, D.; Ilijanić, N.; Geraga, M. Geomorphology of Canyon Outlets in Zrmanja River Estuary and Its Effect on the Holocene Flooding of Semi-enclosed Basins (the Novigrad and Karin Seas, Eastern Adriatic). *Water* **2020**, *12*, 2807. [CrossRef]
17. Horvatinčić, N.; KrajcarBronić, I.; Obelić, B. Diferences in the ^{14}C age, $\delta^{13}C$ and $\delta^{18}O$ of Holocene tufa and speleothem in the Dinaric Karst. *Palaeogeogr. Palaeoclimatol. Palaeoecol.* **2003**, *193*, 139–157. [CrossRef]
18. Pentecost, A. *Travertine*; Springer: Berlin/Heidelberg, Germany, 2005; p. 445.
19. Sherriff, J.E.; Schreve, D.C.; Candy, I.; Palmer, A.P.; White, T.S. Environments of the climatic optimum of MIS 11 in Britain:Evidence from the tufa sequence at Hitchin, southeast England. *J. Quat. Sci.* **2021**, *36*, 508–525. [CrossRef]
20. Zavadlav, S.; Rožič, B.; Dolenec, M.; Lojen, S. Stable isotopic and elemental characteristics of recent tufa from a karstic Krka River (south-east Slovenia): Useful environmental proxies? *Sedimentology* **2016**, *64*, 808–831. [CrossRef]
21. Gruszczyński, T.; Małecki, J.J.; Romanova, A.; Ziułkiewicz, M. Reconstruction of Thermal Conditions in the Subboreal Inferred from Isotopic Studies of Groundwater and Calcareous Tufa from the Spring Mire Cupola in Wardzyń (Central Poland). *Water* **2019**, *11*, 1945. [CrossRef]
22. Barešić, J.; Faivre, S.; Sironić, A.; Borković, D.; Lovrenčić Mikelić, I.; Drysdale, R.N.; Krajcar Bronić, I. The Potential of Tufa as a Tool for Paleoenvironmental Research—A Study of Tufa from the Zrmanja River Canyon, Croatia. *Geosciences* **2021**, *11*, 376. [CrossRef]
23. Horvatinčić, N.; Ćalić, R.; Geyh, M.A. Interglacial Growth of Tufa in Croatia. *Quat. Res.* **2000**, *53*, 185–195. [CrossRef]
24. Lojen, S.; Dolenec, T.; Vokal, B.; Cukrov, N.; Mihelčić, G.; Papesch, W. C and O stable isotope variability in recent freshwater carbonates (River Krka, Croatia). *Sedimentology* **2004**, *51*, 361–375. [CrossRef]
25. Lojen, S.; Trkov, A.; Ščančar, J.; Vázquez-Navarro, J.A.; Cukrov, N. Continuous 60-year stable isotopic and earth-alkali element records in a modern laminated tufa (Jaruga, river Krka, Croatia): Implications for climate reconstruction. *Chem. Geol.* **2009**, *258*, 242–250. [CrossRef]
26. Rovan, L.; Zuliani, T.; Horvat, B.; Kanduč, T.; Vreča, P.; Jamil, Q.; Čermelj, B.; Bura-Nakić, E.; Cukrov, N.; Štrok, M.; et al. Uranium isotopes as a possible tracer of terrestrial authigenic carbonate. *Sci. Total Environ.* **2021**, *797*, 149103. [CrossRef] [PubMed]

27. Srdoč, D.; Horvatincic, N.; Obelic, B.; Krajcar, I.; Sliepcevic, A. Procesi taloženja kalcita u krškim vodama s posebnim osvrtom na Plitvicka jezera (Calcite deposition processes in karst waters with special emphasis on the Plitvice Lakes, Yugoslavia). *KRS Jugosl* **1985**, *11*, 1–104.
28. Horvatinčić, N.; Barešić, J.; Babinka, S.; Obelić, B.; Krajcar Bronić, I.; Vreča, P.; Suckow, A. Towards a deeper understanding how carbonate isotopes (^{14}C, ^{13}C, ^{18}O) reflect environmental changes: A study with recent ^{210}Pb dated sediments of the Plitvice lakes Croatia. *Radiocarbon* **2008**, *50*, 233–253. [CrossRef]
29. Srdoč, D.; Obelić, B.; Horvatinčić, N. Radiocarbon dating of calcareous tufa: How reliable data can we expect? *Radiocarbon* **1980** *22*, 858–862. [CrossRef]
30. Srdoč, D.; Krajcar Bronić, I. The application of stable and radioactive isotopes in karst water research. *NašKrš* **1986**, *12*, 37–47.
31. Karamata, S. The geodynamic framework of the Balkan Peninsula: Its origin due to the approach, collision and compression of Gondwanian and Eurasian units. In *Tectonic Development of the Eastern Mediterranean Region*; Robertson, A.H.F., Mountrakis, D., Eds. Geological Society: London, UK, 2006; Volume 260, pp. 155–178.
32. Robertson, A.; Karamata, S.; Šarić, K. Overview of ophiolites and related units in the Late Paleozoic—Early Cenozoic magmatic and tectonic development of Tethys in the northern part of the Balkan region. *Lithos* **2009**, *108*, 1–36. [CrossRef]
33. Gavrilović, D. *The Karst of Serbia*; Memories of the Serbian Geographical Society 13: Belgrade, Serbia, 1976; pp. 1–28.
34. Kranjc, A. Dinaric Karst. In *Encyclopedia of Caves and Karst Science*; Gunn, J., Ed.; Fitzroy Dearborn: New York, NY, USA; London UK, 2004; pp. 287–289.
35. Milivojević, M. Hydrogeothermal system of Zlatibor. In *Geologija Zlatibora (Geology of Zlatibor)*; Dimitrijević, M.D., Ed.; Geoinistitut Beograd, Srbija, 1996; Volume 18, pp. 121–130.
36. Oliveira, E.C.; Rossetti, D.F.; Utida, G. Paleoenvironmental Evolution of Continental Carbonates in West-Central Brazil. *An. Acad Bras. Ciênc.* **2017**, *89*. [CrossRef]
37. Andrews, J.E.; Brasier, A.T. Seasonal records of climatic change in annually laminated tufas: Short review and future prospects *J. Quat. Sci.* **2005**, *20*, 411–421. [CrossRef]
38. Pazdur, A.; Pazdur, M.F. Stable isotopes of Holocene calcareous tufa in southern Poland as paleoclimatic indicators. *Quat. Res* **1988**, *30*, 177–189. [CrossRef]
39. Soriano, M.C.O.; Abad, C.A.; Marcén, C.A.; Tirapu, G.P.; Bello, L.M. Stable-isotope changes in tufa stromatolites of the Quaternary Añamaza fluvial system (Iberian Ranges, Spain). *Geogaceta* **2017**, *61*, 167–170.
40. Pedley, H.M. Tufas and travertines of the Mediterranean region: A testing ground for freshwater carbonate concepts and developments. *Sedimentology* **2009**, *56*, 221–246. [CrossRef]
41. Martín-Algarra, A.; Martín-Martín, M.; Andreo, B.; Juliá, R.; González-Gómez, C. Sedimentary patterns in perched spring travertines near Granada (Spain) as indicators of the paleohydrological and paleoclimatological evolution of a karst massif *Sediment Geol.* **2003**, *161*, 217–228. [CrossRef]
42. Dabkowski, J.; Limondin-Lotouet, N.; Antoine, P.; Andrews, J.E.; Marca-Bell, A.; Robert, V. Climatic variations in MIS 11 recorded by stable isotopes and trace elements in a French tufa (La Celle, Seine Valley). *J. Quatern. Sci.* **2012**, *27*, 790–799. [CrossRef]
43. Beisel, R.H., Jr. *International Waterfall Classification System*; Outskirts Press: Parker, CO, USA, 2006; p. 294.
44. Gradzinski, M. Factors controlling growth of modern tufa: Results of a field experiment. In *Tufas and Speleothems*; Pedley, H.M., Rogerson, M., Eds.; Geological Society of London: London, UK, 2010; Volume 336, pp. 143–191. [CrossRef]
45. Castro-Suarez, J.R.; Colpas-Castillo, F.; Taron-Dunoyer, A. Chemical and Morphologic Characterization of Sylvite (KCl) Mineral from Different Deposits Used in the Production of Fertilizers. *Agronomy* **2023**, *13*, 52. [CrossRef]
46. Matsuoka, J.; Kano, A.; Oba, T.; Watanabe, T.; Sakai, S.; Seto, K. Seasonal variation of stable isotopic compositions recorded in a laminated tufa, SW Japan. *Earth Planet. Sci. Lett.* **2001**, *192*, 31–44. [CrossRef]
47. Ortiz, J.E.; Torres, T.; Delgado, A.; Reyes, E.; Diaz-Bautista, A. A review of the Tagus river tufa deposits (central Spain): Age and palaeoenvironmental record. *Quat. Sci. Rev.* **2009**, *28*, 947–963. [CrossRef]
48. Melon, P.; Alonso-Zarza, A.M. The Villaviciosa tufa: A scale model for an active cool water tufa system, Guadalajara (Spain). *Facies* **2018**, *64*, 5. [CrossRef]
49. Kosun, E. Facies characteristics and depositional environments of Quaternary tufa deposits, Antalya, SW Turkey. *Carbonates Evaporites* **2012**, *27*, 269–289. [CrossRef]
50. Toker, E. Quaternary fluvial tufas of Sarıkavak area, southwestern Turkey: Facies and depositional systems. *Quat. Int.* **2017**, *437*, 37–50. [CrossRef]
51. Arenas, C.; Osacar, C.; Sancho, C.; Vasquez-urbez, M.; Auque, L.; Pardo, G. Seasonal record from recent fluvial tufa deposits (Monasterio de Piedra, NE Spain): Sedimentological and stable isotope data. In *Tufas and Speleothems*; Pedley, H.M., Rogerson, M., Eds.; Geological Society of London: London, UK, 2010; Volume 336, pp. 119–142.
52. Ozkul, M.; Gokgoz, A.; Horvatincic, N. Depositional properties and geochemistry of Holocene perched springline tufa deposits and associated spring waters: A case study from the Denizli Province, Western Turkey. In *Tufas and Speleothems*; Pedley, H.M., Rogerson, M., Eds.; Geological Society of London: London, UK, 2010; Volume 336, pp. 245–262.
53. Dabkowski, J.; Frodlova, J.; Hajek, M.; Hajkova, P.; Petr, L.; Fiorillo, D.; Dudova, L.; Horsak, M. A complete Holocene climate and environment record for the Western Carpathians (Slovakia) derived from a tufa deposit. *Holocene* **2018**, *9*, 493–504. [CrossRef]
54. Andrews, J.E.; Riding, R.; Dennis, P.F. Stable isotopic compositions of recent freshwater cyanobacterial carbonates from the British Isles: Local and regional environmental controls. *Sedimentology* **1993**, *40*, 303–314. [CrossRef]

55. Andrews, J.E.; Pedley, H.M.; Dennis, P.F. Stable isotope record of palaeoclimatic change in a British Holocene tufa. *Holocene* **1994**, *4*, 349–355. [CrossRef]
56. Garnett, E.R.; Andrews, J.E.; Preece, R.C.; Dennis, P.F. Climatic change recorded by stable isotopes and trace elements in a British Holocene tufa. *J. Quat. Sci.* **2004**, *19*, 251–262. [CrossRef]
57. Bódai, B.; Czuppon, G.; Fórizs, I.; Kele, S. Seasonal study of calcite-water oxygen isotope fractionation at recent freshwater tufa sites in Hungary. *Geol. Carpathica* **2022**, *73*, 485–496. [CrossRef]
58. Kern, Z.; Hatvani, I.G.; Czuppon, G.; Fórizs, I.; Erdélyi, D.; Kanduč, T.; Palcsu, L.; Vreča, P. Isotopic 'altitude' effect and 'continental' effect in modern precipitation across the Adriatic–Pannonian region. *Water* **2020**, *12*, 1797. [CrossRef]
59. Gonfiantini, R.; Roche, M.; Olivry, J.; Fontes, J.; Zuppi, G. The altitude effect on the isotopic composition of tropical rains. *Chem. Geol.* **2001**, *181*, 147–167. [CrossRef]
60. Hori, M.; Kawai, T.; Matsuoka, J.; Kano, A. Intra-annual perturbations of stable isotopes in tufas: Effects of hydrological processes. *Geochim. Cosmochim. Acta* **2009**, *73*, 1684–1695. [CrossRef]

applied sciences

MDPI

Article

Mechanical Properties and Constitutive Relationship of Cretaceous Frozen Sandstone under Low Temperature

Siyuan Shu [1], Zhishu Yao [1,2,*], Yongjie Xu [1], Chen Wang [1] and Kun Hu [1]

1 Engineering Research Center of Underground Mine Engineering of Ministry of Education, Anhui University of Science and Technology, Huainan 232001, China

2 School of Civil Engineering and Architecture, Anhui University of Science and Technology, Huainan 232001, China

* Correspondence: zsyao@aust.edu.cn

Abstract: During the construction of coal mine shafts through Cretaceous water-rich stratum using the freezing method, the frozen shaft lining can break and lose stability. Hence, it is necessary to examine the mechanical properties and constitutive relationship of Cretaceous water-rich sandstone under the effect of surrounding rocks. To this end, in this work, the mechanical properties of red sandstone at different confining pressures and freezing temperatures were examined by using a ZTCR-2000 low-temperature triaxial testing system, wherein the 415–418 m deep red sandstone in the Lijiagou air-return shaft of Wenjiapo Mine was taken as the research object. The test results indicated that the stress–strain curves of rock under triaxial compression and uniaxial compression presented four stages: pore compaction, elastic compression, plastic yield, and post-peak deformation. The difference between the two cases was that the post-peak curve of the former was abrupt, while the latter exhibited a post-peak strain softening section. As the freezing temperature was constant, with the raise in the confining pressure, the elastic modulus and peak strength of the rock rose linearly, while the Poisson's ratio decreased quadratically. As the control confining pressure was constant, the elastic modulus and rock's peak strength increased with the decrease in the temperature, and under the condition of negative temperature, the two parameters were linearly correlated with the temperature, while the Poisson's ratio showed the opposite trend. The two-part Hooke's model and the statistical damage model based on Drucker–Prager (D-P) yield criterion were used to establish the stress–strain relationship models before and after the rock yield point, optimize the model parameters, and optimize the junction of the two models. The results revealed that the optimized model curve was in good agreement with the experimental curve, which suggests that the proposed model can accurately describe the stress–strain characteristics of rock under three-dimensional stress. This verified the feasibility and rationality of the proposed model for examining the constitutive relationship of rocks.

Keywords: Cretaceous red sandstone; triaxial compression; low-temperature freezing; mechanical properties; constitutive model

1. Introduction

Coal is still the largest and most reliable source of energy in the world. Since the majority of China's coal resources are located in the western region, the coal mining operations are being gradually shifted to this region [1–4]. During the construction of vertical shafts in the western region, the Cretaceous water-rich bedrock section contains a large number of sandstones with low strength and weak cementation, which often leads to the fracture of frozen pipe due to the instability or excessive deformation of surrounding rocks, causing serious engineering accidents [5,6]. Hence, it is of great significance to examine the mechanical properties and constitutive relationship of Cretaceous water-rich sandstone under the effect of surrounding rocks.

To study the mechanical properties of ultra-deep sandstone reservoir rocks, Li Qinghui et al. [7] used a variety of laboratory tests to examine the relationships between the compressive strength, internal friction angle, shear strength, and other mechanical parameters of sandstone and the confining pressure and temperature and compared them with those of shallow sandstone. Tan Wenhui et al. [8] studied the mechanical properties of granite under uniaxial compression by using computerized tomography (CT) scanning technology because of the significant influence of joints and fissures and deduced the constitutive relationship closer to the actual stress state of the sample. Meng Zhaoping et al. [9] studied the impact of lateral pressure on porosity, permeability, deformation, the mechanical properties, and failure mechanism and obtained the fitting equation for the relation between the physical and mechanical properties of sandstone and confining pressure. Although several studies have been conducted on the mechanical properties of rocks under different conditions [10–13], the mechanical properties of the Cretaceous water-rich sandstone in the western region of China have been rarely examined, and the stress state of sandstone under complex stress conditions needs to be further studied.

The stress state of rock under the crustal stress is very complex, and the rock mechanics basically involves studying the complete stress–strain relationship to characterize the deformation and failure of rocks, i.e., establishing the constitutive relationship of rock [14]. The early studies on the constitutive relationship of rocks were mainly based on the classical continuum mechanics theory. For example, Li Wenting et al. [15] characterized the post-peak elastic modulus obtained during the compressive failure of rock as a function of strain and considered the internal friction angle as the intermediate variable to establish the post-peak nonlinear constitutive relationship on the basis of the Mohr–Coulomb strength criterion. The obtained model accurately described the post-peak mechanical behavior of marble under different confining pressures. Zhang Chunhui et al. [16] introduced the post-peak strength decline index to describe the influence of confining pressure on the post-peak residual strength and secant modulus and established a post-peak strain softening mechanical model of rock by considering the influence of confining pressure. The rationality of the established model was verified by comparing the experimental and numerical simulation results. However, rock is a coherent aggregate of one or more minerals and has complex mechanical properties. On the basis of the continuum theory, the precision of the constitutive model is restricted to the actual working conditions [17].

Subsequently, it was found that the damage constitutive model established by introducing damage mechanics can more accurately describe the stress state of rocks. Cao Wengui et al. [18,19] derived a three-dimensional (3D) damage evolution equation and a damage-softening constitutive equation for rock considering that the micro-element strength of rock obeyed Weibull distribution and normal distribution. Huang Haifeng [20] assumed that the micro-element strength of rock obeyed the improved Harris probability density distribution, established a statistical damage softening model of rock considering the modification of damage variables, and compared the modeling results with the experimental data, proving the applicability and rationality of the proposed model. Considering that there is a compaction section before the peak of weakly cemented sandstone and a strain softening section after the peak, Zhang Weizhong et al. [21] established a parabolic curve linear elastic Duncan hyperbola plastic softening section residual ideal plasticity five-segment model for rock triaxial stress. Combined with the test curve, it was shown that the model could accurately reflect the deformation characteristics of sandstone under triaxial stress, but the multi-segment model had many nodes and complex forms, which is not suitable for engineering applications.

Although the mechanical properties and constitutive relationship of rocks have been extensively examined, there are few reports on the western Cretaceous water-rich sandstone under complex stress conditions. To this end, in this study, the freezing project of the Lijiagou air-return shaft in Wenjiapo Mine was taken as the research background, and a low-temperature triaxial test system was used to investigate the mechanical properties of red sandstone under different confining pressures and different freezing temperatures.

On this basis, to accurately describe the obvious crack compaction and strain softening sections of the Cretaceous red sandstone under the action of 3D stress, the stress–strain relationship was established on the basis of the two-part Hooke's model (TPHM) and the statistical damage model. The findings of this study provide useful insights on the mechanical properties of the Cretaceous sandstone and can serve as a theoretical reference for the constitutive modeling of western water-rich sandstone.

2. Test Plan

2.1. Sample Preparation

The rock samples in the test were taken from the Lower Cretaceous Luohe Formation of Lijiagou air-return shaft in Wenjiapo Mine, Shaanxi Province, and the burial depth of the rock samples was 415–418 m. A drilling sampler was used to sample the large rocks obtained from the site. After sampling, the rock samples were cut and polished. The diameter and height of the prepared samples were 50 and 100 mm, respectively. The verticality and flatness of the samples were checked on the basis of the relevant specifications of the International Society for Rock Mechanics [22].

The scanning electron microscopy (SEM) images of the rock sample are presented in Figure 1. It is clear from Figure 1a that there were numerous internal cracks in the rock. The rock could be easily broken, and the joints and fissures were relatively developed. When the rock sample was immersed in water, it disintegrated into granules, as depicted in Figure 1b. The particle size of rock was approximately 0.15–0.5 mm, and the particle size distribution was relatively uniform, but there were several pores and cracks in the rock, which corresponded to medium-grained rock characteristics. The rock primarily consisted of cement and mineral particles, wherein the mineral particles were wrapped by cement and connected with adjacent particles, as described in Figure 1c. The microscopic image of the outer cement is depicted in Figure 1d.

Figure 1. SEM images of rock: (**a**) 100 μm; (**b**) 200 μm; (**c**) 20 μm; (**d**) 2 μm.

Through X-ray diffraction (XRD) analysis, it was found that the rock contained minerals such as quartz, calcite, pyrite, K-feldspar, and clay minerals. Figure 2 shows the X-ray diffraction (XRD) results of the test soil sample, with the main crystal in the soil sample being SiO_2. According to the rock classification standard, the rock was red sandstone. The basic physical parameters of the sandstone are shown in Table 1.

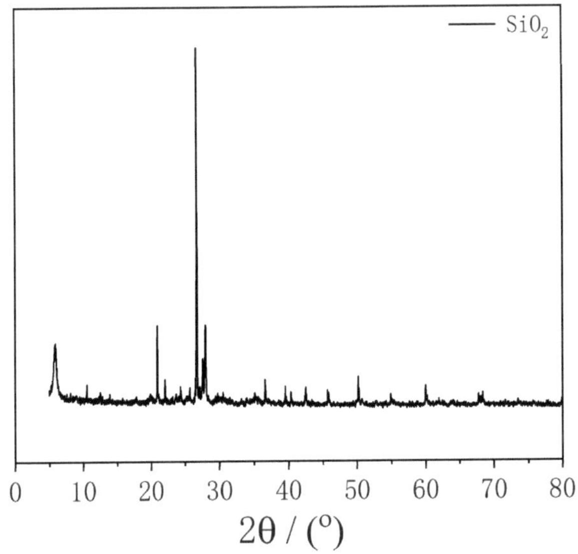

Figure 2. XRD of sandstone.

Table 1. Basic physical parameters of sandstone.

Natural Moisture Content/%	Saturated Water Content/%	Dry Density/g·cm^{-3}	Saturation Density/g·cm^{-3}	Porosity/%
5.18	8.69	2.12	2.23	23.61

2.2. Test Equipment

The experiment equipment mainly included sample preparation and characterization instruments, including a rock drilling sampler (ZS-100), rock cutting machine (DJ-1), double face grinder (SHM-200), electric blast drying oven (DHG9075A), vacuum saturator (NEL-VJH), and non-metallic ultrasonic testing analyzer (NM-4B).

A low-temperature rock triaxial testing system (ZTCR-2000) at the Anhui University of Architecture and Construction was used as the main testing equipment, which included a servo oil source, temperature control system, confining pressure system, axial pressure system, and computer. Presently, it is the most advanced rock testing system in China. The sample installation and working diagram of the rock triaxial compression testing system are shown in Figure 3.

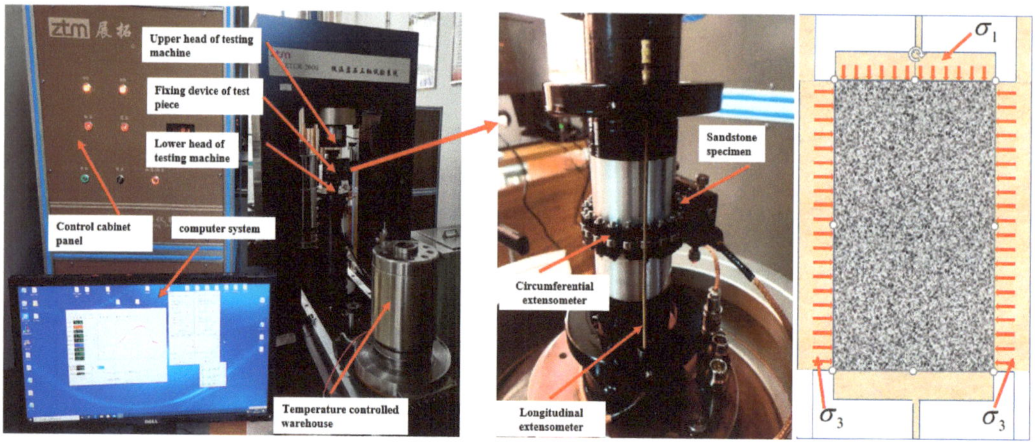

Figure 3. Stress diagram of the triaxial test system and test piece.

2.3. Test Design

According to the designed burial depth and freezing temperature of rock samples, the test temperatures were 25, −5, −10, and −15 °C, and four confining pressures were considered: 0 (no confining pressure), 2, 4, and 6 MPa. To mark the tested rock samples, the temperature and confining pressure were abbreviated as T and W, respectively. For example, W0T25 implies that the sample was tested at 0 MPa and 25 °C. The specific test scheme is shown in Table 2.

Table 2. Test scheme and number.

Experimental Temperature	Confining Pressure			
	0 MPa (No Confining Pressure)	**2 MPa**	**4 MPa**	**6 MPa**
25 °C	W0T25	W2T25	W4T25	W6T25
−5 °C	W0T-5	W2T-5	W4T-5	W6T-5
−10 °C	W0T-10	W2T-10	W4T-10	W6T-10
−15 °C	W0T-15	W2T-15	W4T-15	W6T-15

The selected rock samples were placed in a curing box for water retention, and then they were frozen at low temperature. The loading test was only able to be conducted when the temperature of the samples reached the design value and remained stable for at least 24 h. Before the loading test, the test box needed to be cooled to ensure that the ambient temperature of the samples met the test temperature requirements.

3. Mechanical Test Results of Cretaceous Red Sandstone

3.1. Uniaxial Compression Test of Cretaceous Red Sandstone

To examine the mechanical properties of Cretaceous red sandstone under complex crustal stress and different low temperature environments, firstly, uniaxial compression tests were conducted under different ambient temperatures [23,24]. The stress–strain curve of red sandstone under uniaxial compression is depicted in Figure 4. It can be seen that under different temperatures, the stress–strain curve of sandstone included four stages: pore compaction, elastic compression, plastic yield, and post-peak deformation, which were essentially consistent with the typical stress–strain curve of rocks. The difference was that at room temperature (25 °C), the sandstone was destroyed beyond the peak strength point, and the bearing capacity changed suddenly. Under the condition of negative temperature, the sandstone had a residual strength after reaching the peak strength through plastic deformation, which is called the "strain softening stage." Comparing the stress–strain curves of rock samples under room temperature and three negative temperatures, it was found that with the decrease in the temperature, the elastic modulus and peak strength of the rock increased, but the length of the crack compaction section decreased.

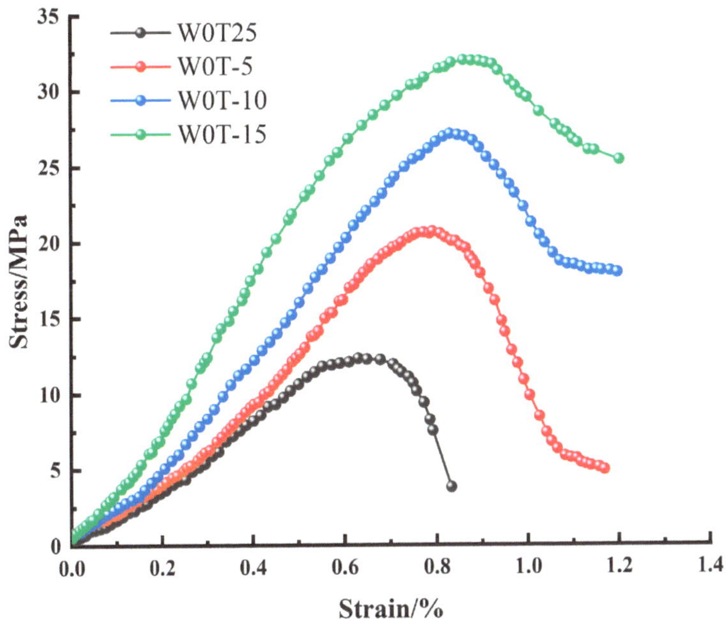

Figure 4. Stress–strain curve of red sandstone under uniaxial compression.

3.2. Triaxial Compression Test of Cretaceous Red Sandstone

The stress–strain curves of red sandstone under triaxial compression at different negative temperatures are provided in Figure 5, and the red sandstone sample under triaxial compression at room temperature was considered the control group. It was evident that under different temperature conditions, the stress–strain curve of red sandstone under triaxial compression was essentially consistent with that under uniaxial compression, conforming to the typical stress–strain curve of rock. Figure 4a demonstrates that at room temperature, with the increase in the confining pressure, the elastic modulus and peak strength were improved. The rock did not have residual strength after the peak strength point, but exhibited a sudden change in the strength, causing rock failure. Under the condition of negative temperature, the triaxial compression strength of sandstone increased with the raise in the confining pressure. Furthermore, with the raise in the confining pressure and the decrease in the temperature, the length of the compaction section decreased. This was because the increase in the confining pressure limited the compression effect of the pores. With the decrease in the temperature, the content of unfrozen water in the rock decreased, and the water in the pores condensed into ice, reducing the pores in the rock.

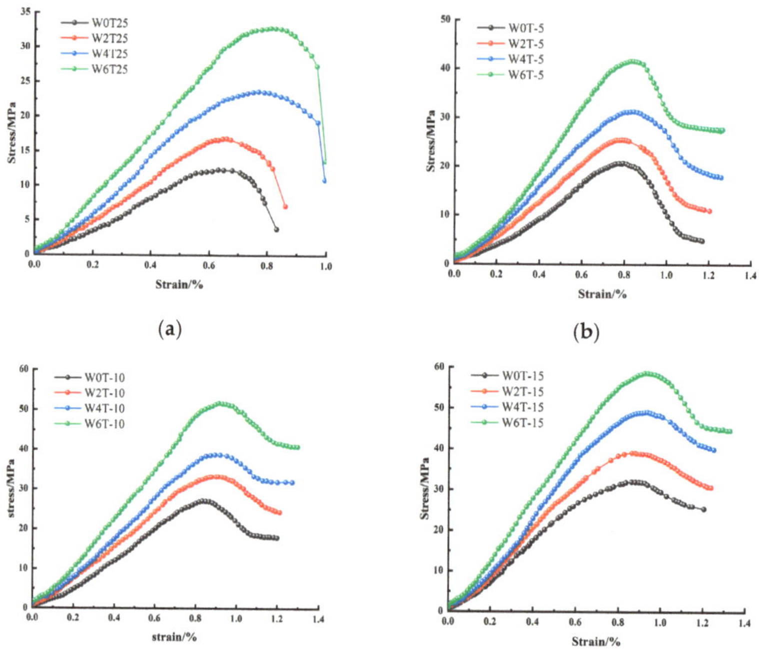

(a) (b) (c) (d)

Figure 5. Stress–strain curve of red sandstone under triaxial compression. (**a**) T = 25 °C; (**b**) T = −5 °C; (**c**) T = −10 °C; (**d**) T = −15 °C.

3.3. Analysis of Mechanical Properties of Rock

Generally, the ground stress of rock in the stratum is relatively complex, usually in the state of three-way loading, and obvious differences also exist due to the influence of temperature during freezing construction. On the basis of the Mohr–Coulomb strength criterion, the peak strength of frozen sandstone under different low temperatures and confining pressures can be obtained from triaxial tests using the values of σ_{max}, Poisson's ratio μ, and rock's elastic modulus E. The test results are shown in Table 3.

Table 3. Test results.

Specimen Number	σ_{max}/MPa	E/MPa	μ
W0T25	12.242	2407.36	0.249
W0T-5	20.689	2973.21	0.239
W0T-10	27.122	3728.39	0.230
W0T-15	31.812	4947.37	0.210
W2T25	16.827	3026.18	0.234
W2T-5	25.557	3802.13	0.226
W2T-10	33.277	4507.64	0.216
W2T-15	38.981	5807.88	0.197
W4T25	23.611	3713.07	0.222
W4T-5	31.276	4478.78	0.216
W4T-10	38.704	5292.02	0.204
W4T-15	49.109	6723.26	0.187
W6T25	32.784	4376.91	0.213
W6T-5	41.582	5439.73	0.207
W6T-10	51.706	6390.49	0.196
W6T-15	58.642	7507.74	0.180

3.3.1. Relationship between the Freezing Temperature/Confining Pressure and the Peak Intensity of Sandstone

The variation in the sandstone peak intensity as a function of the temperature and confining pressure is depicted in Figure 6. It can be seen in Figure 6a that the peak strength of sandstone was linearly proportional to the confining pressure. At room temperature, as the confining pressure of sandstone rose from 0 MPa to 2, 4, and 6 MPa, the corresponding peak strengths increased by 37.45%, 92.87%, and 167.80%, respectively. When the rock temperature was −15 °C, as the confining pressure increased from 0 MPa to 2, 4, and 6 MPa, the peak strengths of sandstone rose by 22.54%, 54.37%, and 84.34%, respectively. When the temperatures of sandstone were 25, −5, 10, and −15 °C, and the applied confining pressure increased from 0 to 6 MPa, the peak strengths increased by 167.80%, 96.33%, 90.64%, and 84.34%, respectively. This indicated that the peak strength of sandstone can be improved by increasing the confining pressure. This is because at a high confining pressure, the lateral deformation of sandstone is constrained, the rock particles are continuously compressed and compacted in the axial direction, and the compacted particles are subjected to axial compression until the rock is sheared.

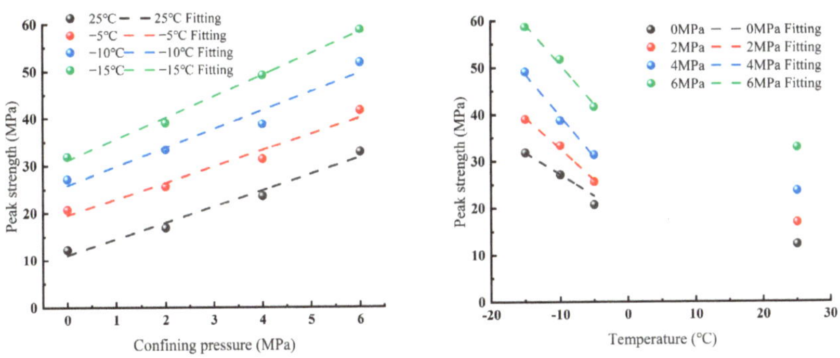

(a) (b)

Figure 6. Sandstone peak intensity and the relationship between confining pressure and temperature. (**a**) Relationship between peak strength and confining pressure. (**b**) Relationship between peak strength and temperature.

According to Figure 6b, as the confining pressure was constant, the peak strength linearly increased with the decrease in the negative temperature, but the peak strength at room temperature was different from that at negative temperature. This was because under the condition of negative temperature, the bearing capacity of the rock not only depends on the rock particles, but also on the ice–water mixture generated by the temperature reduction in the rock. The water in the rock can be divided into unfrozen water and ice. The ice not only has a certain strength, but also contributes to the cementation between rock particles, greatly improving the bearing capacity of the rock. As depicted in Figure 6b, as the temperature dropped from room temperature to −5 °C, the peak strengths at the confining pressures of 0, 2, 4, and 6 MPa increased by 40.83%, 34.16%, 24.51%, and 26.84%, respectively. As the confining pressure was 0 MPa, and the temperature dropped from −5 to −15 °C, the growth rates of rock peak strength were 40.83%, 31.09%, and 17.29%, respectively. When the confining pressure was 6 MPa, the corresponding values were 26.84%, 24.35%, and 13.41%, respectively. This shows that under triaxial compression, the strength of rock decreased with the decrease in the negative temperature. This was because with the decrease in the temperature, the content of unfrozen water in the rock decreased gradually, but the water content in the rock was fixed, and the water that can be frozen in the rock gradually decreased, leading to a decrease in the growth rate of rock peak strength.

3.3.2. Relationship between the Freezing Temperature/Confining Pressure and the Elastic Modulus of Sandstone

Elastic modulus reflects the ability of rock to withstand elastic deformation. The elastic modulus is essentially the bonding strength between micro atoms or molecules. Any change in the temperature and confining pressure obviously affects the elastic modulus. The variations in the elastic modulus of sandstone as a function of the temperature and confining pressure are depicted in Figure 7. The elastic modulus of rock was linearly proportional to the confining pressure. As the confining pressure of rock increased from 0 to 6 MPa, the elastic modulus measured at 25, −5, −10, and −15 °C increased by 123.04%, 108.19%, 64.89%, and 45.88, respectively. This was because as the confining pressure increased, the lateral constraint increased, and the lateral strain of rock decreased, which undoubtedly had a restraining effect on the destruction of rock and improved its resistance to deformation.

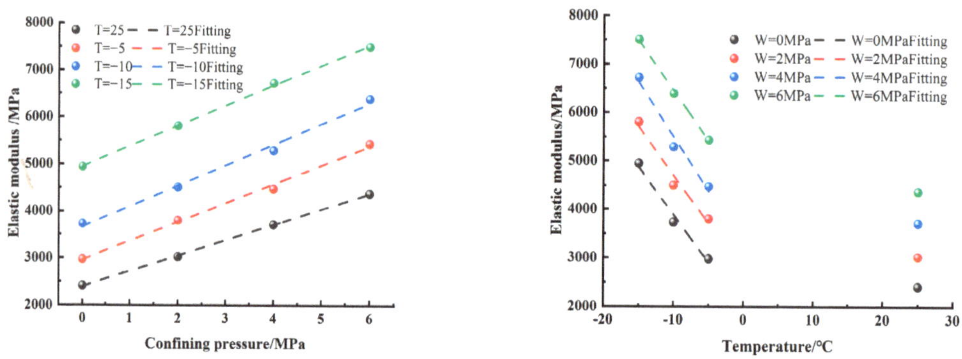

(a) (b)

Figure 7. Sandstone elastic modulus and the relationship between confining pressure and temperature. (**a**) Relationship between elastic modulus and confining pressure. (**b**) Relationship between elastic modulus and temperature.

It can be seen in Figure 7a that with the decrease in the temperature, the variation trend of the elastic modulus was similar to that of the peak strength, i.e., the elastic modulus of rock increased as the temperature decreased. As the test temperature of rock decreased from room temperature to negative temperature, the elastic modulus corresponding to the confining pressures of 0, 2, 4, and 6 MPa increased by 23.51%, 25.64%, 10.88%, and 15.28%, respectively. When the test temperature dropped from −5 to −15 °C, the elastic modulus of rock under different confining pressures increased to a higher extent, and the corresponding growth rates from low to high confining pressures were 73.09%, 52.75%, 72.21%, and 39.82%, respectively. The above results are mainly attributed to the fact that a part of pore water in the rock under negative temperature froze into ice, which had a certain strength, and its cementation improved the ability of rock mass to resist deformation. Therefore, the elastic modulus of rock was not only determined by the rock skeleton but also by the ice in the pores.

3.3.3. Relationship between the Freezing Temperature/Confining Pressure and Poisson's Ratio of Sandstone

The variations in the Poisson's ratio of sandstone as a function of the confining pressure and temperature are depicted in Figure 8. It is clear that the Poisson's ratio of rock quadratically decreased with the rise in the confining pressure. This was because the lateral deformation was restrained with the rise in the confining pressure. According to the formula of Poisson's ratio, the lateral strain decreases, and it is more influenced by the confining pressure than the axial strain. Therefore, the Poisson's ratio decreases in a quadratic manner with the in the confining pressure [25,26]. As displayed in Figure 8b, the Poisson's ratio of rock decreased as the temperature was reduced at the same confining pressure, which was quite different from the relation between temperature and the rock's elastic modulus.

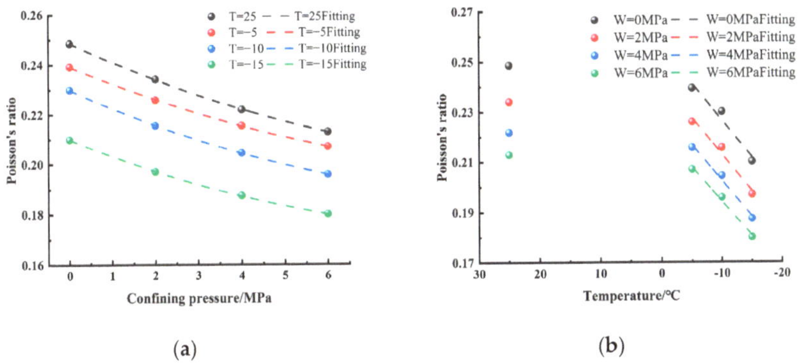

Figure 8. Poisson's ratio of sandstone and the relationship between confining pressure and temperature. (**a**) Relationship between Poisson's ratio and confining pressure. (**b**) Relationship between Poisson's ratio and temperature.

4. Constitutive Relationship

4.1. Constitutive Relationship before Yield Point

According to the above tests, the room-temperature triaxial compression test was selected, and the stress–strain curve before the yield point was obtained, as depicted in Figure 9. It can be seen that the pores and cracks in the red sandstone were compacted at the onset of compression, and the strain vs. stress curve had a concave parabolic shape. The compaction of pores and cracks led to a large degree of deformation of the rock. Then, the internal defects were compacted, and the rock particles bore the 3D stress. The rock skeleton was able to bear a small deformation, and the stress–strain curve in this phase showed a linear growth.

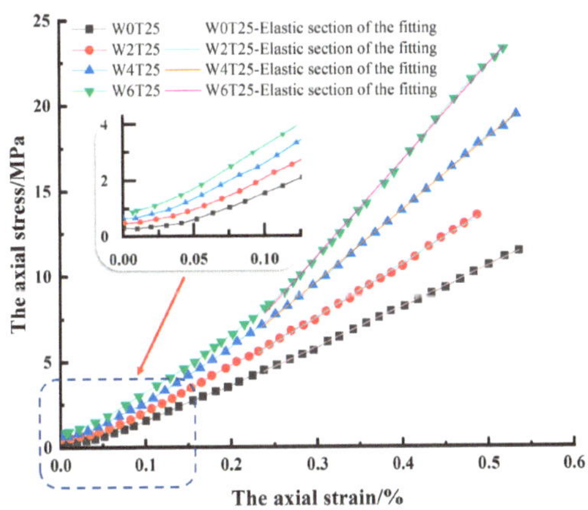

Figure 9. Stress–strain curve before yield point.

For accurately describing the constitutive relationship in the two stages, the TPHM [27,28] was used. Hooke's law based on natural strain was used to describe the crack compaction section, while Hooke's law based on engineering strain was used to describe the elastic stage.

$$\left.\begin{array}{l}\varepsilon_1 = \frac{3-\gamma_h}{3E_r}[\sigma_1 - \mu(\sigma_2+\sigma_3)] + \frac{\gamma_h}{3}[1-\exp(-\frac{\sigma_1}{E_h})] \\ \varepsilon_2 = \frac{3-\gamma_h}{3E_r}[\sigma_2 - \mu(\sigma_1+\sigma_3)] + \frac{\gamma_h}{3}[1-\exp(-\frac{\sigma_2}{E_h})] \\ \varepsilon_3 = \frac{3-\gamma_h}{3E_r}[\sigma_3 - \mu(\sigma_1+\sigma_2)] + \frac{\gamma_h}{3}[1-\exp(-\frac{\sigma_3}{E_h})]\end{array}\right\} \quad (1)$$

where ε_1, ε_2, and ε_3 are the strains measured in the triaxial test of rock. σ_1, σ_2, and σ_3 are the corresponding nominal principal stresses. E_r and E_h represent the modulus of elasticity of the hard and soft components, respectively (soft components refers to the pores and fractures in the rock, and hard components corresponds to the rock particle skeleton, except the soft components), and γ_h is the proportion of soft components in the rock, and μ represents the Poisson's ratio of the hard components.

Under the triaxial stress $\sigma_2 = \sigma_3$, $\varepsilon_2 = \varepsilon_3 = 0$, the principal strain of rock under triaxial confining pressure can be obtained as follows:

$$\varepsilon_1 = \frac{3-\gamma_h}{3E_r}[\sigma_1 - \mu(\sigma_2+\sigma_3)] + \frac{\gamma_h}{3}[1-\exp(-\frac{\sigma_1}{E_h})] \quad (2)$$

Equation (2) is the constitutive equation of sandstone before the yield point. The axial strain and stress can be measured by the test. The crucial step is the determination of three parameters: γ_h, E_r, and E_h. To determine each parameter, the ratio $\gamma_r{}'$ of the hard components to the whole rock mass and the ratio $\gamma_h{}'$ of the soft components to the whole rock mass are introduced. The sum of the two is 1, and the proportional relationship between $\gamma_r{}'$ and $\gamma_h{}'$ can be expressed as

$$\gamma_h{}' = \frac{\gamma_h}{3} \quad (3)$$

On the basis of the stress–strain curve acquired from this test, the value of $\frac{3E_r}{3-\gamma_h}$ in Equation (2) can be determined from the slope of the straight line in the elastic stage of the rock. The intersection point of the extension of the straight line in the elastic section and the strain axis is $\gamma_h{}'$. After determining the parameters γ_h and E_r, the value of E_h can be obtained according to the low stress stage, i.e., the stress and strain data points of the compaction section are substituted in Equation (2). The stress–strain curve and the parametric solution of the TPHM model are depicted in Figure 10.

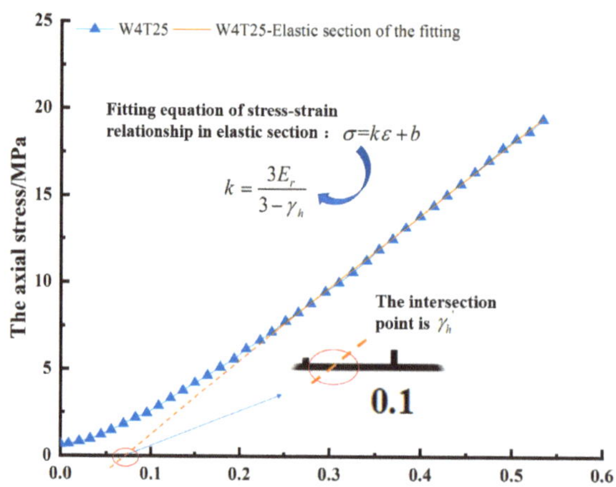

Figure 10. Relationship between model parameter solution and stress–strain curve.

It can be seen in Figure 8 that the slope of the elastic stage of sandstone obviously increased with the increase in the confining pressure. Further, the value of $\gamma_h{}'$ decreased. Therefore, it can be concluded that the parameters of TPHM model were related to the stress state of sandstone. The constitutive equation established by using the above single test curve is not universal. Therefore, to reduce the error of the constitutive equation, the triaxial stress–strain curve under different temperatures was obtained on the basis of the parametric solution of the TPHM model. The variations in the γ_h, E_r, and E_h as a function of the confining pressure are presented in Figure 11.

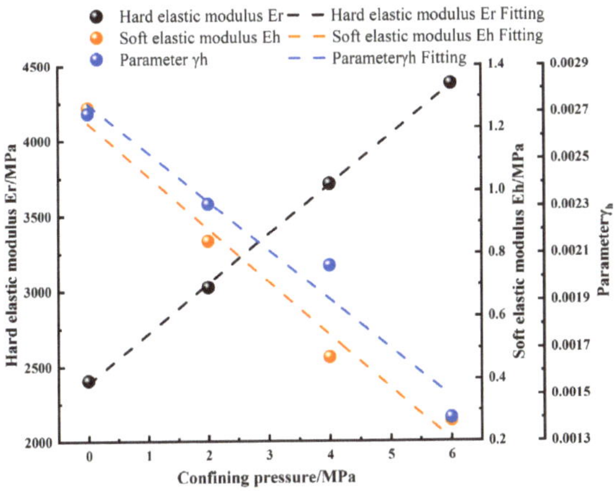

Figure 11. Relationship between model parameters and confining pressure.

Further, the corresponding fitting equations are obtained by linear fitting:

$$E_r = 329.78\sigma_3 + 2389.26 \tag{4}$$

$$E_h = -0.16849\sigma_3 + 1.21486 \tag{5}$$

$$\gamma_h = -0.000207\sigma_3 + 0.00273 \tag{6}$$

By substituting the fitting equations into Equation (2), a double-strain Hooke constitutive model of red sandstone with only three basic mechanical parameters can be obtained as follows:

$$\begin{aligned} \varepsilon_1 = {} & \tfrac{0.000207\sigma_3+2.99727}{989.34\sigma_3+7167.78}(\sigma_1 - 2\mu\sigma_3) + \\ & \tfrac{-0.000207\sigma_3+0.00273}{3}[1 - \exp(-\tfrac{\sigma_1}{-0.16849\sigma_3+1.21486})] \end{aligned} \tag{7}$$

The stress–strain curves under the confining pressures of 0, 2, 4, and 6 MPa were obtained by using the above constitutive equations, as shown in Figure 12. Compared with the other models, the optimized TPHM was simpler and allowed for higher-level simulation. It was able to reflect the crack compaction and elastic deformation phases before the yield point more effectively. However, it is a macro-scale approximation, and it is based on Hooke's laws of natural and engineering strains. Moreover, it cannot reflect the stress–strain relationship beyond the yield point.

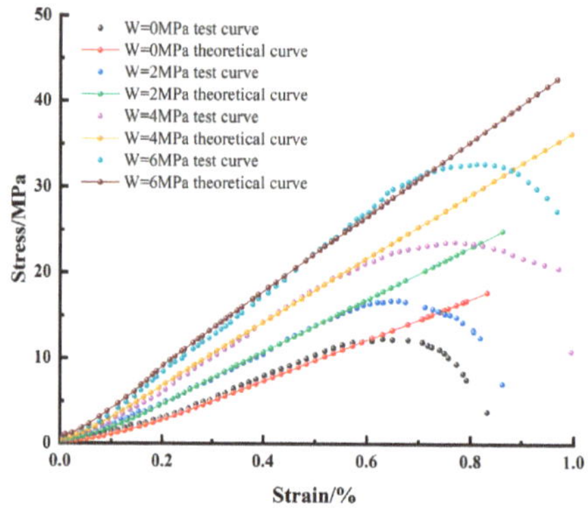

Figure 12. Comparison between experimental data and model data.

4.2. Constitutive Relationship after the Yield Point

On the basis of the Lemaitre strain equivalence hypothesis [29–34], the rock damage constitutive equation can be established as follows:

$$[\sigma]= [E][1-D][\varepsilon] = [1-D][\sigma^*] \tag{8}$$

where $[\sigma]$ and $[\sigma^*]$ are the stress matrix and the effective stress matrix, respectively. $[E]$ is the elastic matrix, $[\varepsilon]$ is the axial strain matrix, $(1-D)$ is the relative area of the rock that can effectively bear the internal force during compression, and D is the rock damage parameter.

Assuming that the number of damaged micro-elements during rock compression is N_h, the damage parameter D is defined as the ratio of N_h to the total number of micro-elements N [35–37], i.e.,

$$D = \frac{N_h}{N} \tag{9}$$

According to Figure 1, the red sandstone contains many internal pores and fissures, and there are weak layers with different micro-element strengths in the rock. Considering the continuous damage of rock during triaxial compression, it is assumed that the micro-element strength and micro-element force are isotropic and obey Hooke's Law. Before rock failure, the distribution variable F of the micro-element intensity follows Weibull distribution, and the probability density function is defined as follows:

$$P(F) = \frac{m}{F_0}\left(\frac{F}{F_0}\right)^{m-1}\exp\left[-\left(\frac{F}{F_0}\right)^{m}\right] \tag{10}$$

where m and F_0 represent the *Weibull* distribution parameters.

Hence, the number of destroyed micro-elements in the interval $[F, F + \mathrm{d}F]$ can be expressed as $NP(y)dy$. When the rock is loaded to some degree F, the number of damaged micro-elements can be obtained as follows:

$$N_h(F) = \int_0^F NP(y)dy = N\left\{1-\exp\left[-\left(\frac{F}{F_0}\right)\right]\right\} \tag{11}$$

Substituting the above equation into Equation (9), the rock damage evolution equation can be derived as

$$D = \frac{N_h}{N} = 1-\exp\left[-\left(\frac{F}{F_0}\right)^{m}\right] \tag{12}$$

According to Equation (12), the micro-element strength F must be calculated to determine the damage parameter D. The traditional method is to use uniaxial strain as the distribution variable, which cannot describe the influence of rock stress state on the micro-element strength. Therefore, the distribution variable of the micro-element intensity is used to establish the failure criterion of rock. On the basis of the simple parametric form of Drucker–Prager (D-P) criterion, the infinitesimal strength can be obtained.

$$F = f(\sigma) = \alpha I_1 + \sqrt{J_2} \tag{13}$$

where α is the cohesion, and φ is the internal friction angle of rock. Further, the first and second invariants of the stress tensor are represented by I_1 and J_2, respectively.

$$\left.\begin{array}{l} \alpha = \frac{\sqrt{3}\sin\varphi}{3\sqrt{3+\sin^2\varphi}} \\ I_1 = \sigma^*{}_x + \sigma^*{}_y + \sigma^*{}_z = \sigma^*{}_1 + \sigma^*{}_2 + \sigma^*{}_3 \\ J_2 = \frac{1}{6}[(\sigma^*{}_1 - \sigma^*{}_2)^2 + (\sigma^*{}_2 - \sigma^*{}_3)^2 + (\sigma^*{}_3 - \sigma^*{}_1)^2] \end{array}\right\} \tag{14}$$

where $\sigma_1{}^*$, $\sigma_2{}^*$, and $\sigma_3{}^*$ are the effective stress values of rock under compression, and the corresponding nominal stress values are σ_1, σ_2, and σ_3. The nominal stress of rock can be measured by triaxial test, and $\sigma_1{}^* = \sigma_2{}^*$, $\sigma_1 = \sigma_2$. According to the Hooke's Law and the concept of stress tensor,

$$\left.\begin{array}{l} I_1 = \frac{E\varepsilon_1(\sigma_1+2\sigma_3)}{\sigma_1-2\mu\sigma_3} \\ \sqrt{J_2} = \frac{E\varepsilon_1(\sigma_1-\sigma_3)}{\sqrt{3}(\sigma_1-2\sigma_3)} \end{array}\right\} \tag{15}$$

Combining Equations (8), (12), and (13) and Hooke's law, and using the D-P strength criterion, the damage constitutive equation of rock is derived as follows:

$$\sigma_1 = E\varepsilon_1(1-D) + 2\mu\sigma_3 = E\varepsilon_1 \cdot \exp[-(\frac{\alpha I_1 + \sqrt{J_2}}{F_0})^m] + 2\mu\sigma_3 \tag{16}$$

It can be seen from the established damage constitutive relationship (16) that the Weibull distribution parameters F_0 and m need to be determined in advance, and the peak point of the stress–strain curve under triaxial compression can be used to obtain the parameter m, while the function F_0 is related to F [38–42], and the specific expression is

$$m = \frac{1}{\ln \frac{E\varepsilon_1{}^0}{(\sigma_1)_{\max} - \mu(\sigma_2+\sigma_3)}} \tag{17}$$

$$F_0 = \left(\frac{F_1{}^m}{\ln \frac{E\varepsilon_1{}^0}{(\sigma_1)_{\max} - \mu(\sigma_2+\sigma_3)}} \right)^{\frac{1}{m}} \tag{18}$$

where $(\sigma_1)_{\max}$ is the peak stress of the stress–strain curve, $\varepsilon_1{}^0$ is the strain value corresponding to the peak stress, and $F_1{}^0$ is the micro-element strength at the peak point.

To prove the validity of the constitutive model, the internal friction angle of sand is considered to be 28.56° according to the triaxial freezing test. Using Equations (17) and (18), the values of model parameters m and F_0 can be determined. The constitutive model equation of soft rock damage is derived by substituting m and F_0 into Equation (16). The constitutive model curve is compared with the test curve in Figure 13.

The following inferences can be drawn by comparing the test results with the modeling results.

1. On the basis of the D-P strength criterion, the damage-softening constitutive model of rock is more suitable for describing the stress–strain curve of rock, especially beyond the yield point, with a high fitting degree.
2. The stress before the yield point described by the constitutive relation is generally high, and it cannot be used to accurately describe the crack compaction and elastic deformation stages of Cretaceous red sandstone under 3D stress.

3. Compared with the test curve under high confining pressure, the model curve is more consistent with the test stress–strain curve obtained under low confining pressure, and under the negative temperature state, the fitting degree of the model curve rises with the decrease in the temperature, which may be ascribed to the increase in the elastic modulus of rock as the temperature decreases.

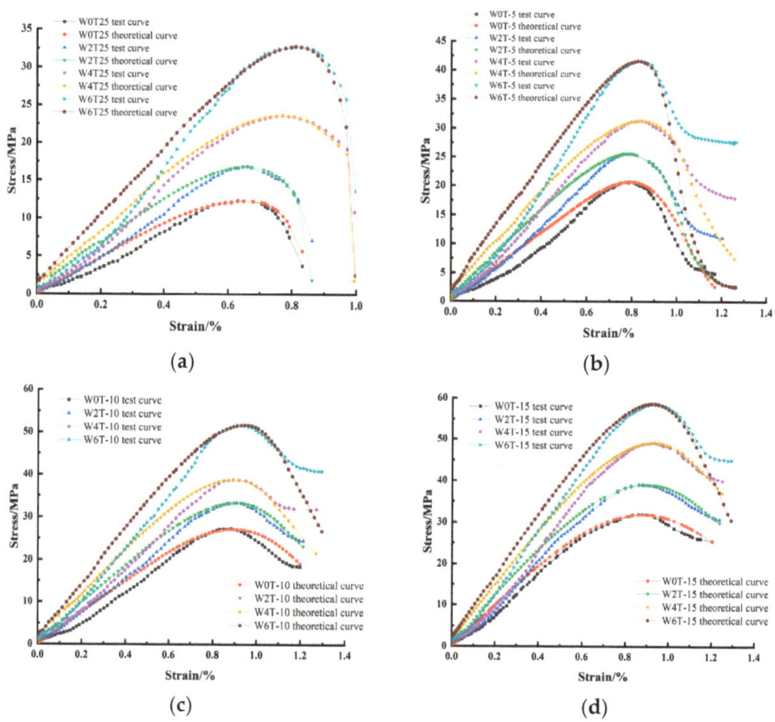

Figure 13. Comparison between experimental data and model data. (**a**) T = 25 °C; (**b**) T = −5 °C; (**c**) T = −10 °C; (**d**) T = −15 °C.

4.3. Model Validation

Cretaceous rock is different from ordinary rocks. It contains a large number of pore structures. The stress–strain curve under triaxial stress had an obvious crack compression section, and a strain softening section was observed under negative temperature. The double strain Hooke constitutive model can effectively describe the stress and strain state before the yield point, but it cannot reveal these characteristics after the yield point. However, the stress–strain curve obtained by the damage constitutive model on the basis of the D-P criterion was in good agreement with the test stress-strain curve after the rock yield point. Therefore, the constitutive relationship of red sandstone can be more accurately described by establishing the constitutive model in sections. However, there is no unified standard for determining the rock yield point. The peak point of the stress–strain curve, namely, the peak stress, is generally regarded as the yield stress of rock in classical rock theory, which can not only lead to the underestimation of rock plastic deformation, but also affect the calculation of hardening parameters [43–49].

In fact, the rock is damaged only when it reaches a certain stress state under the 3D compression process. According to previous studies, the yield stress value of sandstone under various conditions is 73–78% of the peak stress value before the peak point. For simplifying the calculations, 78% of the peak stress value before the peak point was selected as the yield point in this study. On the basis of the experimental results under different confining pressures at room temperature, the established constitutive relationship before and after the yield point was compared with the test results, as shown in Figure 14.

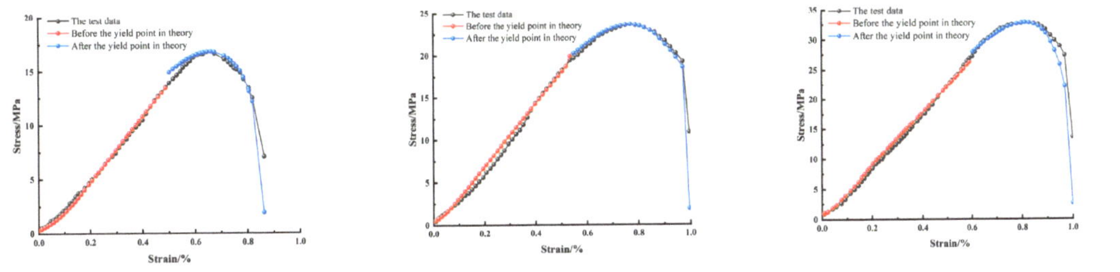

(**a**) (**b**) (**c**)

Figure 14. Comparison between experimental data and model data. (**a**) W = 2 MPa; (**b**) W = 4 MPa; (**c**) W = 6 MPa.

It is clear from Figure 14 that the established model can competently fit the stress–strain curve under different confining pressures. However, the model did not predict the sudden drop of stress at the point of rock failure, and there was no good contact between the joints in the segmented model, especially the fitting curve when the confining pressure was 2 MPa. The difference between the theoretical values of the two models at the joint was 1.4 MPa, which was strongly related to the failure of the established damage constitutive model to better fit the elastic stage. To quantify the fitting degree of the model, the model deviation was analyzed, and the specific expressions are as follows:

$$\delta = \frac{\sum_{i=1}^{n} |\sigma_i{}^* - \sigma_i|}{n} \tag{19}$$

$$\overline{\delta} = \frac{\sum_{i=1}^{n} |\sigma_i{}^* - \sigma_i|}{\sum_{i=1}^{n} \sigma_i} \times 100 \tag{20}$$

where δ and $\overline{\delta}$ are the absolute deviation and relative deviation between the modeling and experimental results, respectively, and σ_i and $\sigma_i{}^*$ are the corresponding test stress and theoretical stress values, respectively (i = 1, 2, 3 …). n represents the number of data points. The model parameters obtained from the above equation are listed in Table 4.

Table 4. Results of model deviation.

Confining Pressure/MPa	Absolute Deviation δ /MPa	Relative Deviation $\overline{\delta}$ /%
2	0.312	3.28
4	0.526	3.82
6	0.706	3.68

According to the above results, the absolute deviation of the established rock constitutive model did not exceed 1 MPa, and the relative deviation was less than 5%, which indicates that the model was able to accurately describe the stress–strain behavior of Cretaceous red sandstone under triaxial stress.

5. Conclusions

In this study, the mechanical properties and constitutive relationship of Cretaceous water-rich sandstone at different freezing temperatures and confining pressures were examined under the effect of surrounding rocks. The main results of the study are summarized as follows:

1. Triaxial compression tests were conducted on the frozen samples of weakly cemented red sandstone in the Luohe Formation of Lower Cretaceous in Lijiagou air-return shaft of Wenjiapo Mine under different confining pressures. The stress–strain curves of rock subjected to triaxial compression exhibited four phases: crack compaction, elastic deformation, yielding phase, and failure phase. Different from the deformation characteristics under confining pressure, there was no strain-softening stage behind the peak of rock under uniaxial compression, which manifested as a sudden stress drop and rock failure.
2. The peak strength and the elastic modulus of rock increased with the decrease in the temperature when the confining pressure was constant and showed a linear growth trend under negative temperature conditions, but the rate of growth gradually decreased. When the test temperature was constant, as the confining pressure increased, the peak strength and the elastic modulus increased, while the Poisson's ratio decreased.
3. The TPHM was able to effectively describe the constitutive relationship of rock before the yield point. On the basis of the Lemaitre strain equivalence hypothesis, the micro-elements in the sandstone were assumed to obey Weibull distribution. According to the D-P criterion, the damage constitutive relationship of rock was established to describe the stress–strain relationship after the yield point.
4. The results based on the constitutive relationship were compared with the test results. The absolute error between the two was not more than 1 MPa, and the relative error was less than 5%. This indicated that the established constitutive model was not only better able to describe the rock fracture compression and elastic stages before the yield point, but also revealed the post-peak strain softening section, verifying the applicability and rationality of the model.

Author Contributions: Conceptualization, Z.Y. and S.S.; methodology, S.S. and Z.Y.; software, S.S., Y.X., C.W. and K.H.; validation, S.S., Y.X., and K.H.; formal analysis, Y.X.; investigation, Y.X., C.W. and S.S.; resources, Z.Y.; data curation, S.S.; writing—original draft preparation, S.S.; writing—review and editing, Y.X., C.W. and Z.Y.; visualization, K.H. and C.W.; supervision, Z.Y. and C.W.; project administration, Z.Y. and C.W.; funding acquisition, Z.Y. All authors have read and agreed to the published version of the manuscript.

Funding: This research was funded by the Anhui Provincial Fund: Temperature Crack Control in Winter Prefabrication Construction of Drilling Shaft Lining in Alpine Region (YJS20210410).

Institutional Review Board Statement: Not applicable.

Informed Consent Statement: Not applicable.

Data Availability Statement: Not applicable.

Conflicts of Interest: The authors declare no conflict of interest.

References

1. Li, Y.L.; Lu, B.; Yang, R.S.; Lin, H.; Wang, S.S.; Liu, C.H. Cemented backfilling mining technology with continuous mining and continuous backfilling method for underground coal mine and typical engineering cases. *J. China Coal Soc.* **2022**, *47*, 1055–1071. (In Chinese)
2. Chen, Y.L.; Cui, H.D.; Li, M.; Pu, H.; Liu, F.M.; Zhang, L.Y. Dynamic mechanical properties and failure mechanism of saturated coal-measure sandstone in open pit mine with damage under real-time low-temperature conditions. *J. China Coal Soc.* **2022**, *47*, 1168–1179. (In Chinese)
3. Yao, Z.S.; Cheng, H.; Rong, C.X. Shaft structural design and optimization of deep freezing bedrock shaft in west area. *J. China Coal Soc.* **2010**, *35*, 760–764. (In Chinese)
4. Yuan, L. Theory and technology considerations on high-quality development of coal main energy security in China. *Bull. Chin. Acad. Sci.* **2023**, *38*, 11–22. (In Chinese)

5. Dong, S.N.; Liu, H.X.; Wang, H. Reserch status and prospect on the mechanism of dynamic sand inrush at weakly cemented strata in working face with thick bedrock. *J. China Coal Soc.* **2022**, *47*, 274–285. (In Chinese)
6. Li, H.; Li, J.; Li, L.; Xu, H.; Wei, J. Prevention of water and sand inrush during mining of extremely thick coal seams under unconsolidated Cenozoic alluvium. *Bull. Eng. Geol. Environ.* **2020**, *79*, 3271–3283. [CrossRef]
7. Li, Q.H.; Li, S.X. Experimental study on mechanical properties of ultra-deep sandstone. *Chin. J. Rock Mech. Eng.* **2021**, *40*, 948–957.
8. Tan, W.H.; Wang, P.F.; Wang, J.L.; Liu, J.J.; Ma, X.W. Mechanical properties of granite under uniaxial compression based on CT technology. *J. Basic Sci. Eng.* **2020**, *28*, 1489–1498. (In Chinese)
9. Meng, Z.P.; Peng, S.P.; Zhang, S.H. Triaxial test on physical and mechanical properties of sandstone for different diagenesis degree. *Chin. J. Geotech. Eng.* **2003**, *25*, 140–143. (In Chinese)
10. Meng, Q.B.; Liu, J.F.; Xie, L.X.; Pu, H.; Yang, Y.G.; Huang, B.X.; Qian, W. Experimental mechanical strength and deformation characteristics of deep damaged–fractured rock. *Bull. Eng. Geol. Environ.* **2021**, *81*, 32. [CrossRef]
11. Zhou, Z.; Ma, W.; Zhang, S.; Mu, Y.; Li, G. Experimental investigation of the path-dependent strength and deformation behaviours of frozen loess. *Eng. Geol.* **2020**, *265*, 105449. [CrossRef]
12. Li, Q.L.; Wang, M.; Fu, Q.; Li, T.; Liu, D.; Hou, R.; Li, H.; Cui, S.; Ji, Y. Short-term influence of biochar on soil temperature, liquid moisture content and soybean growth in a seasonal frozen soil area. *J. Environ. Manag.* **2020**, *266*, 110609. [CrossRef]
13. Yoshikawa, T.; Noda, T. Triaxial test on water absorption compression of unsaturated soil and its soil-water-air-coupled elastoplastic finite deformation analysis. *Soils Found.* **2020**, *60*, 1151–1170. [CrossRef]
14. Zhang, C.; Yang, C.Q.; Bai, Y. Investigation of damage evolution and its model of rock-like brittle materials. *Rock Soil Mech.* **2021**, *42*, 2344–2354. (In Chinese)
15. Li, W.T.; Li, S.C.; Feng, X.D.; Li, S.C.; Yuan, C. Study of post-peak strain softening mechanical properties of rock based on mohr-coulomb criterion. *Chin. J. Rock Mech. Eng.* **2011**, *30*, 1460–1466. (In Chinese)
16. Zhang, C.H.; Zhao, Q.S.; Huang, L.; Ye, S.; Yu, Y.J. Post-peak strain softening mechanical model of rock considering confining pressure effect. *Rock Soil Mech.* **2010**, *31*, 193–197. (In Chinese)
17. Lu, Y.H.; Chen, M.; Yuan, J.B.; Jin, Y.; Teng, X.Q. Borehole instability mechanism of a deviated well in anisotropic formations. *Acta Pet. Sin.* **2013**, *34*, 563–568. (In Chinese)
18. Cao, W.G.; Zhao, M.H.; Tang, X.J. Study on simulation of statistical damage in the full process of rock failure. *Chin. J. Geotech. Eng.* **2003**, *34*, 184–187. (In Chinese)
19. Cao, W.G.; Fang, Z.L.; Tang, X.J. A study of statistical constitutive model for soft and damage rocks. *Chin. J. Rock Mech. Eng.* **1998**, *17*, 628–633. (In Chinese)
20. Huang, H.F.; Ju, N.P.; Li, L.; Xiao, J.W.; Li, M.; Bai, J.; Lv, X.T. Improved Harris function based statistical damage softening model for rocks. *J. Eng. Geol.* **2018**, *26*, 520–527. (In Chinese)
21. Zhang, W.Z.; Chen, C.X.; Yu, M.Y. Study of mechanical properties and constitutive relation of weathered sandstone. *Rock Soil Mech.* **2009**, *30*, 33–36. (In Chinese)
22. Muralha, J.; Grasselli, G.; Tatone, B.; Blümel, M.; Chryssanthakis, P.; Jiang, Y.J. ISRM suggested method for laboratory determination of the shear strength of rock joints: Revised version. *Rock Mech. Rock Eng.* **2014**, *47*, 291–302. [CrossRef]
23. Wang, J.B.; Zhang, Q.; Song, Z.P.; Zhang, Y.W. Creep properties and damage constitutive model of salt rock under uniaxial compression. *Int. J. Damage Mech.* **2020**, *29*, 902–922. [CrossRef]
24. Yan, Z.Q.; Li, Z.; Tan, Y.Z.; Ma, L.J.; Yu, L.Y.; Li, H.Y. Coupling effects of strain rate and low temperature on the dynamic mechanical properties of frozen water-saturated sandstone. *Water* **2022**, *14*, 3513. [CrossRef]
25. Yu, J.; Fu, G.F.; Chen, X.; Guo, X.Y. Experimental study on mechanical properties of sandstone after freezing-thawing cycles under triaxial confining pressure unloading. *Chin. J. Rock Mech. Eng.* **2015**, *34*, 2001–2009. (In Chinese)
26. Zhang, H.M.; Meng, X.Z.; Liu, X.Y. Establishment of constitutive model and analysis of damage characteristics of frozen-thawed rock under load. *Arab. J. Geosci.* **2021**, *14*, 1277. [CrossRef]
27. Li, L.C.; Zhao, Y. Investigation on nonlinear elastic behaviour of rocks based on a two-part hooke's model. *Chin. J. Rock Mech. Eng.* **2012**, *31*, 2119–2126. (In Chinese)
28. Zhao, Y.; Liu, H.H. An elastic stress–strain relationship for porous rock under anisotropic stress conditions. *Rock Mech. Rock Eng.* **2012**, *45*, 389–399. [CrossRef]
29. Azam, S., Hojjatollah, R.; Ali, N. An anisotropic continuum damage mechanics model for the simulation of yield surface evolution and ratcheting. *Int. J. Appl. Mech.* **2022**, *14*, 2250044.
30. Peng, Z.X.; Zeng, Y.W. Microcrack propagation-based damage mechanics model of rock. *J. Northeast. Univ. (Nat. Sci.)* **2022**, *43*, 1784–1791. (In Chinese)
31. Hou, J.L.; Lv, J.N.; Ricoeur, A.; Hu, Y.F.; Zuo, H.; Chen, Y.H.; Li, Q. The *M*-integral in fracture and damage mechanics: A review of developments and applications. *Eng. Fract. Mech.* **2022**, *273*, 108741. [CrossRef]
32. Zhao, S.L.; Yang, Z.J.; Fu, X.D.; Fang, Z. Shear damage mechanism of coarse-grained materials considering strain localization. *Rock Soil Mech.* **2023**, *44*, 31–42. (In Chinese)
33. Morro, A. A phase-field approach to continuum damage mechanics. *Materials* **2022**, *15*, 7671. [CrossRef] [PubMed]
34. Lemaitre, J. How to use damage mechanics. *Nucl. Eng. Des.* **1984**, *80*, 233–245. [CrossRef]
35. Krajcinovic, D.; Silva, M.A.G. Statistical aspects of the continuous damage theory. *Int. J. Solids Struct.* **1982**, *18*, 551–562. [CrossRef]

36. Zhao, G.J.; Chen, C.; Yan, H. A thermal damage constitutive model for oil shale based on weibull statistical theory. *Math. Prob. Eng.* **2019**, *2019*, 4962586. [CrossRef]
37. Bu, P.; Li, Y.L.; Li, Y.; Wen, L.F.; Wang, J.; Zhang, X. Creep damage coupling model of concrete based on the statistical damage theory. *J. Build. Eng.* **2023**, *63*, 105437. [CrossRef]
38. Liu, C.X.; Yang, L.D.; Cao, W.G. A statistical damage softening constitutive model for rock and back analysis of parameters. *Chin. J. Undergr. Space Eng.* **2007**, *3*, 453–457. (In Chinese)
39. Huang, H.F.; Ju, N.P.; Lan, K.W.; Li, M.; Kong, W.; Guo, L.Y. Statistical damage softening model for rock and back analysis of its parameters. *J. Yangtze River Sci. Res. Inst.* **2018**, *35*, 102–106. (In Chinese)
40. Pan, J.L.; Guo, Q.F.; Tian, L.M.; Wu, X.; Wang, M. Study on rock statistical damage softening constitutive model and its parameters based on the unified strength theory. *Min. Res. Dev.* **2019**, *39*, 38–42. (In Chinese)
41. Walton, G.; Sinha, S. Challenges associated with numerical back analysis in rock mechanics. *J. Rock Mech. Geotech. Eng.* **2022**, *14*, 2058–2071. [CrossRef]
42. Zhao, H.B.; Chen, B.R.; Li, S.J.; Li, Z.; Zhu, C. Updating the models and uncertainty of mechanical parameters for rock tunnels using Bayesian inference. *Geosci. Front.* **2021**, *12*, 230–242. [CrossRef]
43. Liu, X.S.; Ning, J.G.; Tan, Y.L.; Gu, Q.H. Damage constitutive model based on energy dissipation for intact rock subjected to cyclic loading. *Int. J. Rock Mech. Min. Sci.* **2016**, *85*, 27–32. [CrossRef]
44. Pan, J.L.; Cai, M.F.; Li, P.; Guo, Q.F. A damage constitutive model of rock-like materials containing a single crack under the action of chemical corrosion and uniaxial compression. *J. Cent. South Univ.* **2022**, *29*, 486–498. [CrossRef]
45. Zhu, Z.N.; Tian, H.; Wang, R.; Jiang, G.S.; Dou, B.; Mei, G. Statistical thermal damage constitutive model of rocks based on Weibull distribution. *Arab. J. Geosci.* **2021**, *14*, 495. [CrossRef]
46. Fang, Y.; Yao, Z.S.; Huang, X.W.; Li, X.W.; Diao, N.H.; Hu, K.; Li, H. Permeability evolution characteristics and microanalysis of reactive powder concrete of drilling shaft lining under stress-seepage coupling. *Constr. Build. Mater.* **2022**, *331*, 127336. [CrossRef]
47. Li, H.C.; Zhang, S. A constitutive damage model of rock based on the assumption of modified Lemaitre strain equivalence hypothesis. *Rock Soil Mech.* **2017**, *38*, 1321–1326, 1334. (In Chinese)
48. Shi, Z.M.; Li, J.T.; Wang, J.; Chen, J.; Lin, H.; Cao, P. Experimental and numerical study on fracture characteristics and constitutive model of sandstone under freeze-thaw-fatigue. *Int. J. Fatigue* **2023**, *166*, 107236. [CrossRef]
49. Fang, Y.; Yao, Z.S.; Huang, X.W.; Li, X.W.; Xu, Y.J.; Hu, K.; Shu, S.Y. Durability analysis and damage constitutive modeling of fiber-reinforced reactive powder concrete of drilling shaft in porous water-bearing rock stratum. *Case. Stud. Constr. Mater.* **2023**, *18*, e01911. [CrossRef]

applied sciences

MDPI

Article

Partitioning of Dissolved Organic Carbon, Major Elements, and Trace Metals during Laboratory Freezing of Organic Leachates from Permafrost Peatlands

Irina S. Ivanova [1], Liudmila S. Shirokova [1,2], Jean-Luc Rols [3] and Oleg S. Pokrovsky [4,5,*]

1. Tomsk Branch of the Trofimuk Institute of Petroleum Geology, Geophysics of Siberian Branch, Russian Academy of Sciences, 4 Akademichesky Pr., 634021 Tomsk, Russia; ivanovais_1986@mail.ru (I.S.I.); lshirocova@yandex.ru (L.S.S.)
2. N. Laverov Federal Center for Integrated Arctic Research, Ural Branch of the Russian Academy of Sciences, 23 Nab Severnoi Dviny, 163000 Arkhangelsk, Russia
3. Laboratoire Ecologie Fonctionnelle et Environnement, CNRS, Toulouse INP, Université Toulouse 3—Paul Sabatier (UPS), 31062 Toulouse, France; jean-luc.rols@univ-tlse3.fr
4. BIO-GEO-CLIM Laboratory, Tomsk State University, 35 Lenina Pr., 634510 Tomsk, Russia
5. Geosciences and Environment Toulouse, UMR 5563 CNRS, University of Toulouse, 14 Avenue Edouard Belin, 31400 Toulouse, France

* Correspondence: oleg.pokrovsky@get.omp.eu; Tel.: +33-561-332-625

Featured Application: Elaboration of unified experimental protocol for studying freezing and thawing of organic-rich (humic) natural waters under laboratory conditions.

Abstract: Climate change is likely to modify the freezing–thawing cycles in soils and surface waters of permafrost-affected and subarctic regions. However, the change of solution chemical composition during ice formation and the evolution of the remaining fluids remain very poorly known. Towards a better understanding of dissolved (<0.45 μm) organic carbon, as well as major and trace element behavior in permafrost peatland environments, here we performed laboratory freezing of peat leachates, from complete freezing to complete thawing, in order to quantify the partitioning of solutes between the aqueous solution and the remaining ice. Freezing experiments were conducted, with and without polyurethane insulation. Two main types of experiments involved (i) progressive freezing, when we started from liquid leachates (filtered <0.45 μm) and allowed them to freeze at −18 °C, and (ii) progressive thawing, where first, we froze solid a series of <0.45 μm filtered leachates and then monitored their thawing at room temperature, 20 °C. We hypothesized the existence of two main groups of solutes, behaving conservatively or non-conservatively during freezing, depending on their incorporation into the ice or their ability to coagulate in the form of insoluble minerals or amorphous materials in the fluid phase. An unexpected result of this work was that, despite a sizable degree of element concentration in the remaining fluid and possible coagulation of organic, organo-mineral, and inorganic compounds, the freezing and subsequent thawing produced final concentrations of most solutes which were not drastically different from the initial concentrations in the original leachates prior to freezing. This demonstrates the high stability of dissolved (<0.45 μm) organic carbon, iron, aluminum, and some trace metals to the repetitive freezing and thawing of surface waters in permafrost peatlands.

Keywords: peat; freezing; thawing; major; trace elements; organic carbon; experiments

1. Introduction

The climate warming in high latitudes leads to the change in frequency of the seasonal thawing-freezing regime [1,2] which will be mostly pronounced during autumn and spring, i.e., the so called "shoulder" seasons or transition periods [3]. It is known that spring and autumn are extremely important for the export of dissolved and particulate load by rivers and element cycling in the lakes of Arctic wetlands [4–6]. During these transition periods, the majority of organic carbon and inorganic solutes originate from leaching of supra-permafrost soil and vegetation [7,8]. In addition to the effect on the carbon and major elements cycling, these processes largely control the release of many trace elements from soils to the aquatic systems [5,9,10]. However, in contrast to relatively good empirical and modeling understanding of the physical aspects of soil freeze/thaw processes [1,11], the behavior of aqueous solutes (<0.45 μm) upon freezing and thawing of organic matter-rich surface and soil waters of permafrost peatlands remains poorly understood. It is known that both the peat porewaters [12,13] and surface waters [5,7] of these regions are strongly enriched in dissolved organic carbon (DOC) and some trace metals. These solutes are often present in the form of organic and organo-mineral colloids [10,14,15], which are likely to coagulate upon ice formation, for example, during the full freezing of surface depressions and shallow thermokarst lakes and ponds [4]. In contrast to the sizable number of works devoted to characterizing the impact of freezing and thawing on mineral soil properties [16–34], the freezing/thawing impact on the chemical composition of porewater and ice has not been well described, and most often, the researchers have to rely on modeling predictions [35].

Although laboratory experiments allow for the identification of governing factors during freezing/thawing effects [36–43], their applications to naturally relevant aquatic settings of high latitude permafrost peatlands remain very restricted. Previous works in permafrost peatlands reported translocation of microorganisms and changes in porewater chemistry (pH, UV absorbance, DOC, major and trace element concentrations) after the thawing and bidirectional freezing of peat cores [44,45]. Another recent study examined the impact of freezing/thawing cycles (FTC) on the surface waters from Northeast European permafrost peatland [46]. The latter authors reported only minor (<5–15%) changes of DOC and labile ion concentrations, whereas several trace elements (Fe, Al, P, Mn, As, and REE) exhibited a sizable decrease in their concentrations after FTC. These authors concluded an overwhelming impact of colloidal status of DOC and trace metal on this coagulation/dissolution processes during freezing and thawing. However, the quantitative assessments of these effects are still missing. In the present study, we aimed at (1) elaborating an optimal experimental procedure allowing for a discrete sampling of fluid during complete freezing progress; (2) quantifying the partitioning coefficients of elements between water and ice, and (3) testing the relationship between different elements, depending on their speciation and affinity to dissolved organic matter.

We hypothesize the existence of several group of contrasting solutes—(1) those incorporating into forming ice; (2) those remaining in the fluid phase via concentrating due to freezing front migration but not subjected to coagulation, and (3) less soluble complex organo-mineral amorphous compounds or certain minerals, subjected to reversible or irreversible coagulation. Testing these mechanistic hypotheses constituted the first objective of this study. The second objective was to use laboratory leachates of natural substrates (peat) as surrogates for the direct experimental modeling of chemical and physical processes occurring in shallow surface waters and topsoil horizons during the autumn—spring period. We anticipate that achieving these objectives should allow a uniform approach for the laboratory modeling of natural aquatic processes in permafrost peatlands and provide useful background for assessing the impact of climate change on chemical composition of surface waters in these environmentally-important regions.

2. Materials and Methods

2.1. Organic Substrates from Permafrost Peatlands Used for Aqueous Leachate Preparation

Peat core samples were collected at the end of the summer in the two largest permafrost peatlands of Northern Eurasia—the Bolshezemelskaya Tundra (NE Europe) and the Western Siberia Lowland (Figure 1). Two peat horizons, thawed (surface, 0 to 10 cm) and deep (frozen, between 40 and 50 cm) were collected from at the sampling sites is between 30 and 40 cm, and the permafrost is discontinuous. The dominant vegetation on the mound is mosses and lichens, with some dwarf shrubs. The detailed description of the environmental context of the sampling sites is provided elsewhere (for Bolshezemelskaya Tundra [47,48] and for Western Siberia [49,50]).

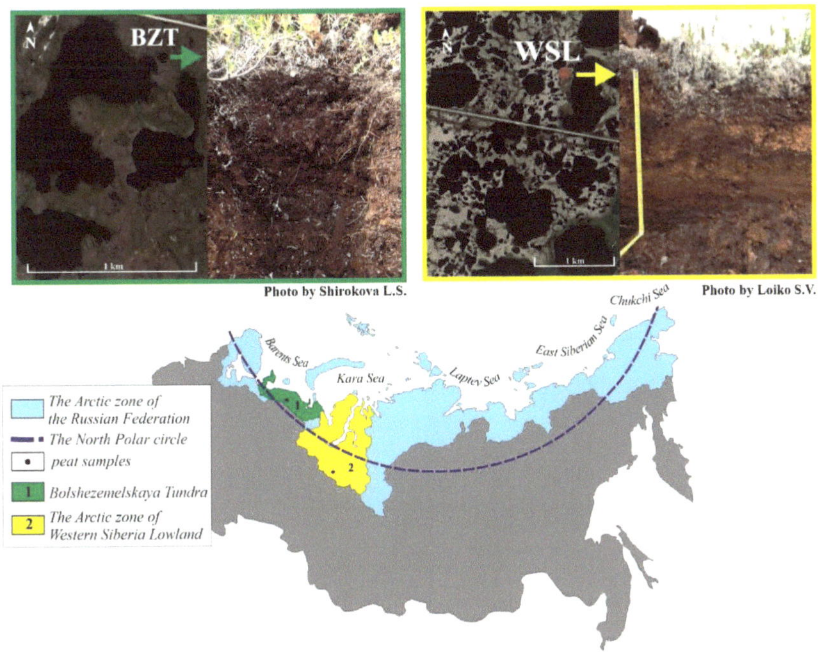

Figure 1. Location map of study area in the Bolshezemelskaya Tundra (1, BZT) and the Western Siberia Lowland (2, WSL), along with photos of sampled peat cores and their environmental context (flat-mound bogs with thermokarst lakes).

To prepare aqueous leachates, 10 g of dry sample (thawed and frozen peat) was reacted in a shaker at 25 °C with 1 L of Milli-Q water for 24 h under aerobic conditions, with periodic aeration. The resulting suspension was filtered through a sterile single-use Millipore filter unit and placed in broad mouth, PES jars of 150 mL volume.

2.2. Freezing Experiments of Aqueous Leachates

Freezing experiments were conducted in two main setups: with polyurethane insulation (approximately 10 cm around the jars and the bottom), and without insulation, as illustrated in Figure 2. Inside the plastic jar, we placed a flexible tube which was connected to a syringe and which allowed for the sampling of fluid in the course of the experiment. The freezing occurred from the borders to the center of the jar, in the case of no insulated reactors, and from the surface to the bottom, in the case of insulated reactors (Figure 2). As such, the sampling tube was located at the center of the jar, for the non-insulated reactors, and at the bottom of the jar, in case of the insulated reactors.

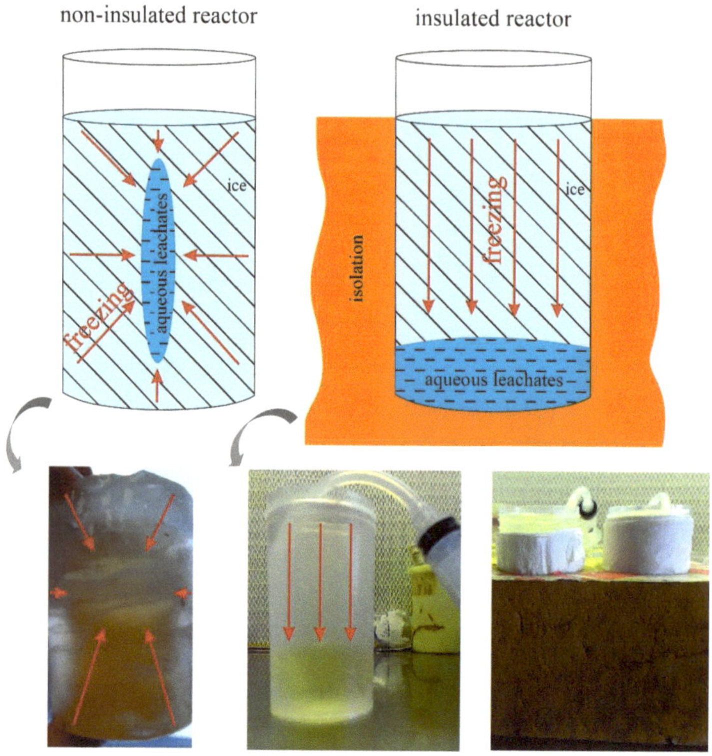

Figure 2. A scheme of experiments and photos of experimental reactors.

The two main type of experiments involved (i) progressive freezing, in which we started from liquid leachates (filtered <0.45 μm) and allowed them to freeze at −18 °C, and (ii) progressive thawing, in which we first froze solid a series of <0.45 μm filtered leachates, and then monitored their thawing at room temperature, 20 °C. In order to avoid the change in the fluid: ice ratio during sampling, we used a single-reactor sacrificial sampling technique. For this, instead of consecutive sampling from the same reactor, 10 identical reactors with filtered leachates were prepared simultaneously and placed in the freezer. Each sampling used the entire reactor, which was then removed from the freezer and discarded.

The sampling was performed regularly as follows. Samples of the aqueous leachate of peats with an insulant were taken after 6, 8, 11, and 15 h, whereas samples of the aqueous leachate of peat BZT were taken after 6, 11, 29, and 31 h. Samples of the aqueous leachate of peats without insulant were taken after 4, 5, and 6 h, and samples of aqueous leachate of peat BZT were taken after 4 and 5 h. During the thawing of the aqueous leachate of WSL peat, samples were collected after 2, 3, and 6 h, and samples of the aqueous leachate of peat BZT were collected after 1, 2, 3, 5, and 7 h. Typically, between 30 and 50% of the remaining or formed fluid was collected; the volume of sampled fluid did not exceed 15% of the initial volume. Immediately before sampling, the reactors were vigorously shaken to homogenize the precipitate that could have been formed in the remaining fluid. This guaranteed the avoidance of the dilution or concentration of solutes relative to precipitates formed during the freezing of peat leachates.

2.3. Chemical Analyses

A non-filtered subsample of the fluid which was extracted from freezing reactor was used to measure pH (uncertainty of ±0.01 pH units, WTW inoLab-pH7110) and specific conductivity (±0.1 μS cm^{-1}, Consort C830). The rest of the 15 mL was filtered through a 0.45 μm Minisart® (Fisher Scientific, Illkirch-Graffenstaden, France) syringe filter. In the <0.45 μm filtrates, the DOC and DIC were analyzed by high-temperature catalytic oxidation using TOC-VCSN, Shimadzu® (Genzo Shimadzu Sr., Kyoto, Japan), with an uncertainty of ± 2% and a detection limit of 0.1 mg L^{-1}. The DIC was measured after sample acidification with HCl, and the DOC was analyzed in the acidified samples after sparging with C-free air for 3 min at 100 mL min^{-1} as non-purgeable organic carbon (NPOC). The internationally certified water samples (MISSISSIPPI-03) were used to check the validity and reproducibility of the analysis. The UV-absorbance of the water samples was measured using a 10 mm quartz cuvette on a CARY-50 UV-Vis spectrophotometer (Varian, Belrose, Australia) to assess the aromaticity of pore fluids via specific UV absorbance ($SUVA_{254}$). Major cations, Si, P, and ~40 trace elements (TE) were measured in He and Ar operating modes, with a quadrupole ICP-MS Agilent 7500 ce (Agilent Technologies, Santa Clara, CA, USA). Indium and rhenium (approximately 3 μg L^{-1}) were used as internal standards to correct for instrumental drift and eventual matrix effects. The appropriate corrections for oxide and hydroxide isobaric interferences were applied for the Rare Earth Elements (REE). Three in-house standard solutions (1, 10, and 100 μg L^{-1} of each element in 2% HNO_3) were measured every 10 samples. The data tables present the results for the elements, exhibiting a good agreement (±10%) between the certified or recommended values and our measurements, or for cases in which we obtain a good reproducibility (the relative standard deviation of our various measurements of standards lower than 10%), even if no certified or recommended data are available. During ICP MS analysis, the SLRS-5 international standard [51,52] was measured at the beginning of the analytical session and after each 20 samples to assess the external accuracy and sensitivity of the instrument. All certified major (Ca, Mg, K, Na, Si) and trace elements (Al, As, B, Ba, Co, Cr, Cu, Fe, Ga, Li, Mn, Mo, Ni, Pb) and all naturally-occurring REEs (La, Ce, Pr, Nd, Sm, Eu, Gd, Dy, Ho, Er, Tm, Yb, Lu, Sb, Sr, Th, Ti, U, V, Zn) concentrations of the SLRS-5 standard and the measured concentrations agreed, with an uncertainty of 10–20%. The agreement for Cd, Cs, and Hf was between 30 and 50%. For all major and most trace elements, the concentrations in the blanks were below analytical detection limits (≤0.1–1 ng L^{-1} for Cd, Ba, Y, Zr, Nb, REE, Hf, Pb, Th, U; 1 ng L^{-1} for Ga, Ge, Rb, Sr, Sb; ≤10 ng L^{-1} for Ti, V, Cr, Mn, Fe, Co, Ni, Cu, Zn, As). Some rare elements, such as Sn, Nb, W, Tl, Ta, and Bi, which were not certified in the reference materials, were also measured, but their concentrations were presented only in the case when three independent subsamples provided a <20% agreement.

2.4. Data Interpretation

In order to identify the group of solutes depending on their partitioning between the forming ice and the remaining solution, we normalized the concentration of main non–conservative solutes (DOC, Fe, Al, P, Mn, Sr, trivalent and tetravalent hydrolysates) to that of major and inert components which were not subjected to coagulation/mineral precipitation, such as K, Na, or Cl. Note that we could not use Mg, Ca, Si, and sulfate for such a normalization because these ions can form sparingly soluble salts (Mg, Ca carbonates, Ca sulfate, Mg hydrous silicate) upon progressive concentration of the remaining fluid during freezing. Assuming that there is no (or very little) incorporation of soluble labile ions into the ice structure, we calculated the degree of element accumulation in the remaining fluid at each time (C_t) relative to the initial concentration (C_0), i.e., the element concentration factor (F_{conc}), as follows:

$$F_{conc} = C_t/C_0. \quad (1)$$

To differentiate between conservative and non-conservative elements, the F_{conc} value of each element at each time t of sampling was normalized to that of Na and traced as a function of time for each consecutive sampling.

2.5. Statistical Analysis

The data were processed by means of mathematical statistics using the MS Excel and Statistica software (version 13.2, Microsoft) at a significance level value of 5%. Additionally, in order to better differentiate between groups of elements, we performed pairwise (Pearson) correlations between elements in the fluid phase, considering all substrates and all types of treatments together. Significance criterion was set at $p < 0.05$. Significance of the difference in element concentration between different experimental treatments (freezing and thawing; with and without isolation; surface and deep peat horizon; peat from NE European tundra and Western Siberia) was examined by a pairwise Mann–Whitney test at $p < 0.05$.

3. Results

3.1. Initial Leachate Composition and the Impact of Reactor Design and Freezing Mode on Element Concentration in the Remaining Fluid

There were significant differences in element concentrations in the initial leachate of frozen and thawed peat horizons, as well as between the peat of NE European tundra (BZT) and Western Siberia (WSL). Thus, the leachates of peat from the NE European Tundra demonstrated sizably lower pH than those of the WSL (4.74 and 5.8–6.2, respectively), with a 2.5 times higher DOC concentration (Table 1).

Table 1. Element concentrations in the initial leachates (<0.45 μm) of frozen peat from NE European tundra (BZT) and frozen and thawed peat horizons of Western Siberia (F WSL and WSL, respectively).

Element	Western Siberian Lowland		European Tundra	Element	Western Siberian Lowland		European Tundra
	WSL	F WSL	BZT		WSL	F WSL	BZT
S.C., μS cm^{-1}	10	20	15.5	Zr, μg/L	0.0192	0.0878	0.4806
UV_{245}, nm	0.386	0.167	1.037	Nb, μg/L	0.0015	0.0007	0.0095
pH	5.78	6.2	4.74	Mo, μg/L	0.1083	0.9193	0.034
DOC, mg/L	9.1	8.0	27.4	Cd, μg/L	0.0607	0.0579	0.0045
Li, μg/L	0.01	0.01	NA	Sb, μg/L	0.0144	0.0566	0.007
B, μg/L	59.2	54.8	72.3	Cs, μg/L	0.0004	0.0004	0.0015
Na, μg/L	915	686	1472	Ba, μg/L	163.8	10.56	245
Mg, μg/L	73	8	150.3	La, μg/L	0.0214	0.0137	0.0375
Al, μg/L	15.6	9.8	78.1	Ce, μg/L	0.032	0.0277	0.069
Si, μg/L	32	19	388	Pr, μg/L	0.0031	0.0024	0.0084
P, μg/L	9.2	15.0	68.3	Nd, μg/L	0.0141	0.0096	0.0362
K, μg/L	215	113.3	732.5	Sm, μg/L	0.0049	0.0017	0.0101
Ca, μg/L	289	257.3	571.7	Eu, μg/L	0.0119	0.0011	0.0192
Ti, μg/L	0.106	0.1155	0.439	Gd, μg/L	0.0054	0.0031	0.0147
V, μg/L	0.360	0.2248	0.176	Tb, μg/L	0.0005	0.0003	0.0017
Cr, μg/L	0.040	0.06	0.460	Dy, μg/L	0.004	0.0023	0.0095
Mn, μg/L	1.26	0.1153	0.0323	Ho, μg/L	0.0007	0.0003	0.0017
Fe, μg/L	22.9	12.3	56.0	Er, μg/L	0.0068	0.0047	0.0058
Co, μg/L	0.015	0.0104	0.0373	Tm, μg/L	0.0005	0.0001	0.0009
Ni, μg/L	0.17	0.18	0.26	Yb, μg/L	0.0024	0.0009	0.0059
Cu, μg/L	0.674	0.7167	25.4	Lu, μg/L	0.0003	0.0002	0.0008
Zn, μg/L	212	8.62	420	Hf, μg/L	0.0019	0.0089	0.0343
Ga, μg/L	0.0002	0.0006	0.0181	W, μg/L	0.0069	0.0558	0.005
As, μg/L	0.1754	0.1352	0.2781	Pb, μg/L	0.044	0.025	1.406
Rb, μg/L	0.08	0.05	0.26	Th, μg/L	0.0019	0.0017	0.0204
Sr, μg/L	4.96	2.417	5.048	U, μg/L	0.0013	0.003	0.0074
Y, μg/L	0.0248	0.0114	0.0514				

Note: S.C.—specific conductivity.

As a result of such high acidity of organic-rich waters, the leachates of BZT peat were sizably, by a factor of 2 to 3, richer in Fe, Al, and other trace metals compared to those from the WSL peat. The leachates from the active horizon of the WSL peat layer exhibited 0.4 unit lower pH values, 1.5 to 2.0 times higher concentrations of Si, K, Fe, Al, Sr, and 10–20 times higher Mn, Zn, and Ba levels than the leachates from the frozen peat horizons.

3.2. Evolution of pH, DOC, and Metal Concentration in the Fluid Phase during the Freezing and Thawing of Peat Leachates

Preliminary examination of the concentration evolution in the course of freezing and thawing (in both directions) demonstrated that there was no significant (at $p < 0.05$) difference between the two treatments in terms of pH, DOC, major, and trace element concentration evolution. Progressive freezing of the aqueous peat leachates led to a decrease in pH relative to the initial solutions (Figure 3), as well as a general increase in DOC and element concentrations, but this increase was different among peat samples. Thus, a sizable—a factor of 2 to 3—increase in the concentration of most elements in the remaining fluid occurred during the freezing of leachates from the active (upper) layer peat of both Western Siberian (WSL) and NE European Tundra (BZT), shown as brown and black symbols, respectively, in Figure 3.

This was not the case for the leachates of frozen Siberian peat: this sample did not exhibit any systematic change in the concentration of solutes during progressive freezing and thawing (shown by blue symbols in Figure 3). Such a drastic difference in the behavior of the leachates from frozen and thawed peat horizons is not linked to pH or DOC behavior, but is likely due to the intrinsic properties of leachates, which depended on the position of the peat sample at the core, with respect to active layer depth.

Progressive thawing of completely frozen leachates produced a decrease in concentration after each subsequent sampling (shown as crosses in Figure 3); the curve approximating the element concentration versus time followed approximately that of progressive freezing. This was observed for most analyzed elements, including DOC, Fe, Al, P, some divalent trace metals, and trivalent and tetravalent hydrolysates.

We found that the initial concentration of some elements in leachates, prior to freezing, were not always recovered after full freezing and thawing of the reactors (Figure 4). This result is consistent with previous works on the freezing/thawing cycles of filtrates from northern peatlands that demonstrated a sizable coagulation and the removal of low-soluble elements bound to organic colloids [46,53].

Considering the element recovery from entire freezing/thawing cycle, three groups of elements could be distinguished separately for each substrate (Table 2).

Table 2. The degree of element recovery during the entire reversible freezing and thawing cycle, separately for <0.45 µm of leachates of peat from BZT, and from frozen (F WSL) and thawed (WSL). The significance of group distinction is at $p < 0.05$ (Mann–Whitney test).

WSL	F WSL	BZT
	$C_0/C_{final} < 0.5$	
Li, Cu, Pb, Th, U	Mg, Mn, Zn, Cd, Ba	Mg, Al, Si, V, Mn, Fe, Co, Ni, Ga, Rb, Zr, Cd, Sb, Cs, Ce, Pb, Th
	C_0/C_{final} 0.5 1.5	
DOC, B, Na, Mg, Al, P, K, Ca, Ti, V, Cr, Mn, Fe, Co, Zn, Ga, As, Rb, Sr, Zr, Mo, Cd, Sb, Cs, Ba, REEs, Hf	Li, B, Na, Al, Ti, V, Fe, Co, Ni, Cu, Ga, Zr, Mo, Sb, Cs, Ba, REEs, Hf, Pb, U	DOC, B, Na, P, K, Ca, Ti, Cr, Cu, As, Sr, Mo, Ba, REEs, Hf, W
	C_0/C_{final} 1.5–3	
Si, Ni, Yb	DOC, K, Ca, As, Sr	U
	$C_0/C_{final} > 3$ P, Rb, Th	Zn

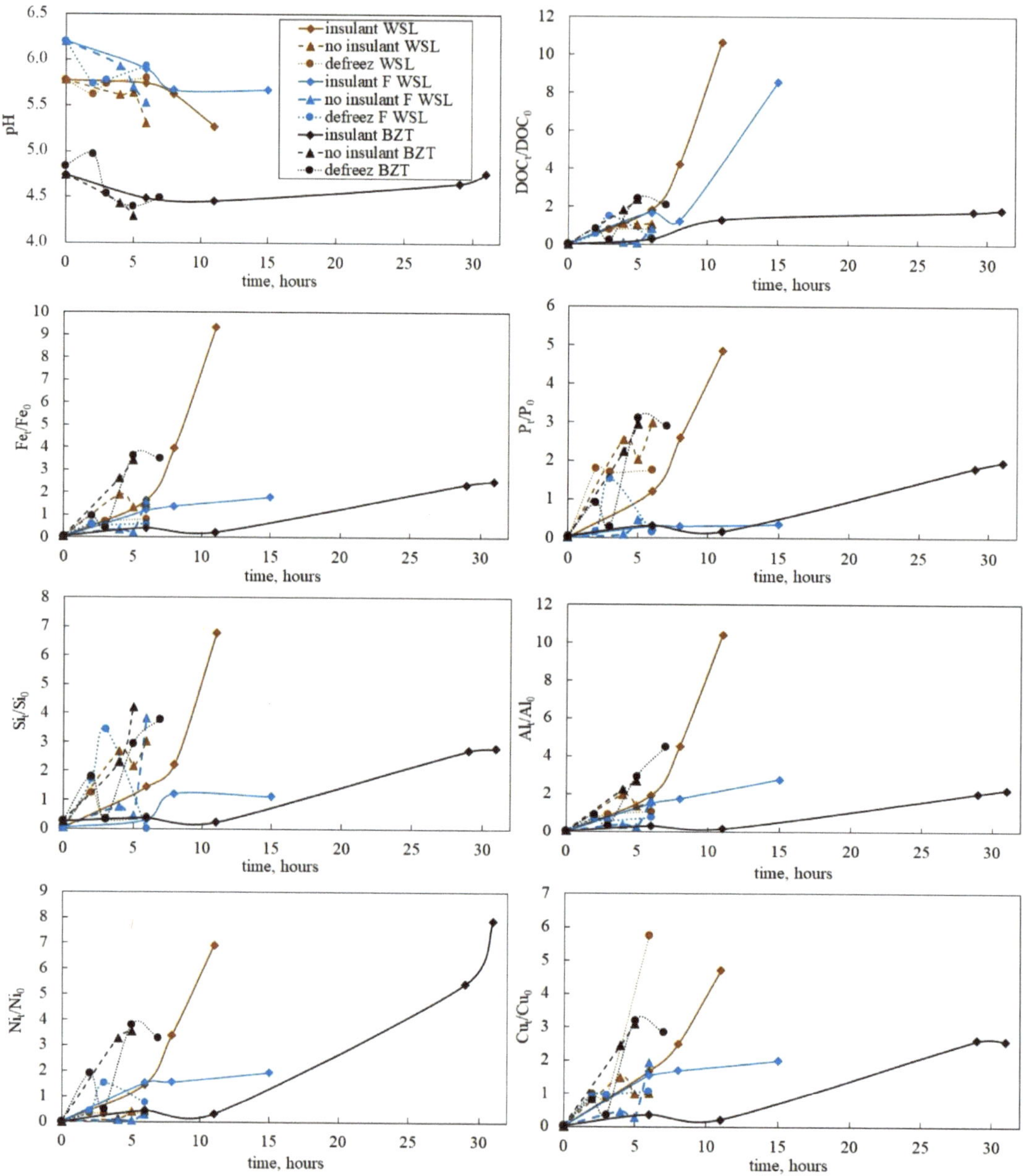

Figure 3. Temporal evolution of pH and solute (DOC, Fe, P, Si, Al, Ni, and Cu) concentrations in the remaining and forming fluid during progressive freezing and thawing, respectively, of peat leachates obtained from active (thawed) and permanently frozen (below active layer depth) peat horizons of permafrost peatlands in Northern Eurasia (at $p < 0.05$). Freezing is shown by diamonds and triangles, whereas thawing is shown by circles. The concentration of element at time t and initial (0) were normalized to that of sodium. Connecting lines are for guiding purposes. Analytical error bars are within the symbol size.

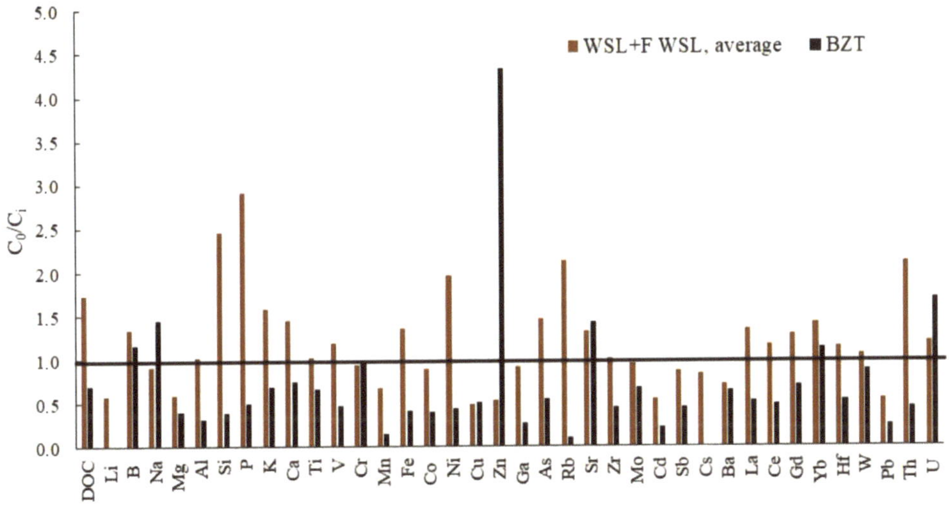

Figure 4. A histogram of ratios of the element concentration in the leachate prior to freezing (C_0) to the final concentration after the entire freezing/thawing cycle. The values above 1 indicate some removal of an element in the form of coagulates. The values below 1 likely represent experimental/analytical artifacts. A significant (at $p < 0.05$) difference from 1.0 is considered as a 30% deviation.

We discovered that the leachates from the frozen peat horizons (WSL) behaved differently from the leachates of the thawed horizons; namely, in the case of the leachate from frozen peat, the element concentration in the remaining fluid during progressive freezing did not increase significantly and remained rather constant over time. This contrasted with the temporal pattern of thawed peat leachates: the concentration of elements in the remaining fluid strongly increased upon freezing (Figure 3). This result could signify the efficient incorporation of solutes from frozen peat leachate into the forming ice, as reflected by the quite low partitioning coefficients of elements between the forming ice and the remaining fluid for this type of substrate. We have no explanation for the potential mechanism of the rather unexpected behavior of this particular leachate, but we can hypothesize that some volumes of the fluid, formed from the frozen leachate, partially preserved the peat dispersed ice structure and chemical composition (i.e., [9,10]) and could be directly incorporated into the forming ice, with a partitioning coefficient close to 1. Clearly, further freezing/thawing experiments, including the assessment of colloidal forms of aquatic leachates obtained from permanently frozen peat horizons (sampled below the active layer boundary) are needed to verify and develop this hypothesis.

3.3. Group of Elements Depending on Their Conservative and Non-Conservative Behavior during Freezing Revealed via Correlation Relationships

The correlations between element concentrations during freezing and thawing at different modes (with and without insulant, Figure 5 and Table 3) allowed for the distinguishing of two main groups of solutes: DOC, Fe, and Al as potential carriers of Si, P, K, Cr, Co, Ni, Cu, As, Pb, Nb, and trivalent and tetravalent hydrolysates, as well as Ca, Na, and K as major indicators of conservative solutes that did not exhibit any significant correlations with other solutes. It was found that while Fe, Al, and other trace metals, notably TE^{3+}, and TE^{4+}, exhibited strong correlations during freezing and thawing, the labile Na or Ca (as a major cation) did not correlate with Fe, Al, and DOC in any of the substrates. Note that the maximal number of elements correlated with three major components of the organo-mineral colloids (DOC, Fe, and Al) was observed in thawing rather than freezing experiments.

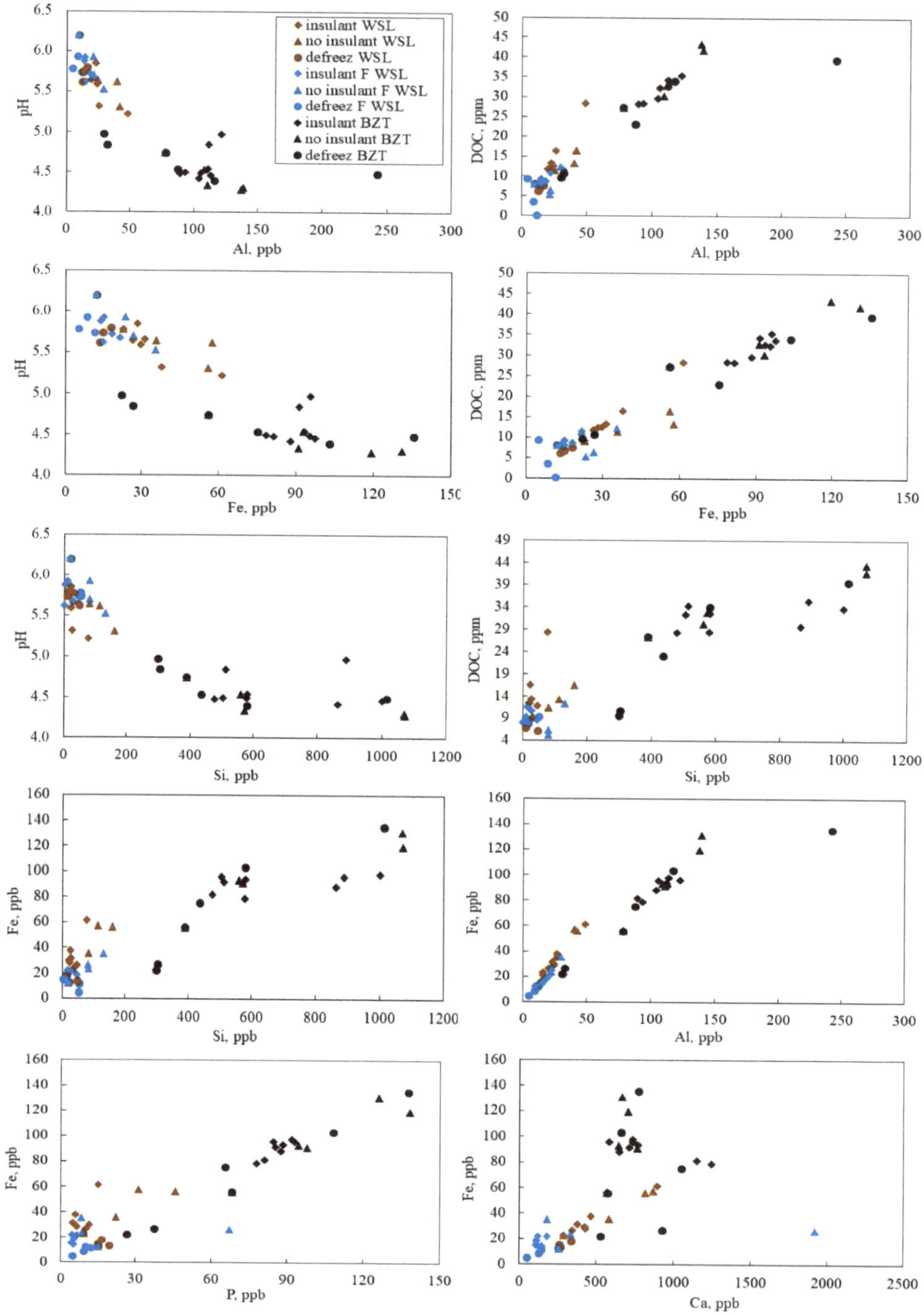

Figure 5. Examples of correlations between solute concentrations for all conducted experiments ($p < 0.05$). Analytical error bars are within the symbol size.

Table 3. Matrix correlation (Pearson, $p < 0.05$ significance level, labeled by asterisk) of element concentrations in aqueous solutions during freezing (left) and thawing (right) of peat leachates. Experiments, with and without insulant, are combined together.

	Freezing WSL + F WSL + BZT					Thawing WSL + F WSL + BZT			
	DOC	Na	Al	Fe		DOC	Na	Al	Fe
DOC	1	−0.0584	0.9752 *	0.9537 *	DOC	1	0.1815	0.9320 *	0.9695 *
Li	0.2841	0.6957	0.4028	0.4249	Li	0.9149 *	0.3597	0.9141 *	0.9396 *
B	0.0137	0.9399 *	0.0984	0.1099	B	0.2050	0.9889 *	0.1702	0.2581
Na	−0.0584	1	0.0585	0.0556	Na	−0.4262	1	0.1483	0.2255
Mg	0.2708	0.7249 *	0.3661	0.3465	Mg	0.6507	0.5771	0.6468	0.6935
Al	0.9752 *	0.0585	1	0.9714 *	Al	0.9320 *	0.1483	1	0.9684 *
Si	0.9433 *	0.0777	0.9784 *	0.9533 *	Si	0.9302 *	0.2209	0.9686 *	0.9516 *
P	0.8843 *	0.2769	0.9342 *	0.9110 *	P	0.9797 *	0.1920	0.9584 *	0.9802 *
K	0.8286 *	0.0625	0.8723 *	0.8240 *	K	0.8810 *	0.1027	0.9076 *	0.8650 *
Ca	0.2949	0.7511 *	0.3536	0.4123	Ca	0.6472	0.7605 *	0.6221	0.6901
Ti	0.9329 *	0.1386	0.9767 *	0.9448 *	Ti	0.9463 *	0.3545	0.8872 *	0.9503 *
V	−0.0965	−0.0566	−0.1673	0.0424	V	0.2062	−0.0739	0.2954	0.3317
Cr	0.7382 *	−0.0688	0.7430 *	0.6332	Cr	0.8028 *	0.5378	0.7283 *	0.7761 *
Mn	−0.3950	−0.2971	−0.4310	−0.3443	Mn	−0.4973	−0.2645	−0.3969	−0.4086
Fe	0.9537 *	0.0556	0.9714 *	1	Fe	0.9695 *	0.2254	0.9684 *	1
Co	0.5802	0.2920	0.6675	0.6908	Co	0.8884 *	0.1750	0.9240 *	0.9248 *
Ni	0.7402 *	−0.1376	0.7681 *	0.7031 *	Ni	0.9220 *	0.2549	0.9012 *	0.9375 *
Cu	0.9422 *	0.0668	0.9787 *	0.9200 *	Cu	0.9824 *	0.2509	0.9490 *	0.9864 *
Zn	0.3588	0.3950	0.3647	0.3455	Zn	0.2217	0.8124 *	0.1310	0.2148
Ga	0.9447 *	−0.0082	0.9617 *	0.8982 *	Ga	0.8970 *	0.1903	0.9884 *	0.9363 *
As	0.7844 *	−0.0093	0.7550 *	0.8698 *	As	0.9210 *	0.2233	0.9128 *	0.9598 *
Rb	0.7832 *	−0.1507	0.8052 *	0.7907 *	Rb	0.7769 *	−0.1060	0.9331 *	0.8367 *
Sr	−0.1877	0.5511	−0.1361	−0.0676	Sr	−0.0097	−0.1975	−0.0006	−0.0246
Y	0.8549 *	0.0521	0.8609 *	0.8934 *	Y	0.9522 *	0.3111	0.9301 *	0.9605 *
Zr	0.8053 *	0.0260	0.8602 *	0.7922 *	Zr	0.9661 *	0.2834	0.9565 *	0.9804 *
Nb	0.9546 *	0.1243	0.9925 *	0.9635 *	Nb	0.9577 *	0.3725	0.8911 *	0.9628 *
Mo	−0.5612	0.2892	−0.4798	−0.5046	Mo	−0.4110	−0.2568	−0.3710	−0.4251
Cd	−0.3985	−0.1202	−0.3732	−0.3835	Cd	−0.5188	−0.3213	−0.4304	−0.4827
Sb	−0.5286	0.1542	−0.4595	−0.4747	Sb	−0.3971	−0.2852	−0.3401	−0.3979
Cs	0.6519	−0.1790	0.6329	0.6351	Cs	0.6829	−0.0942	0.8846 *	0.7569 *
Ba	0.4225	−0.1811	0.4279	0.4552	Ba	0.7095 *	−0.0895	0.7918 *	0.7180 *
La	0.6586	−0.0468	0.6731	0.6531	La	0.9526 *	0.1168	0.9755 *	0.9722 *
Ce	0.6706	−0.0138	0.6699	0.6963	Ce	0.9631 *	0.1622	0.9812 *	0.9906 *
Nd	0.9474 *	0.0619	0.9602 *	0.9655 *	Nd	0.9643 *	0.2661	0.9659 *	0.9818 *
Hf	0.7619 *	0.0501	0.8238 *	0.7434 *	Hf	0.9608 *	0.3444	0.9306 *	0.9613 *
Pb	0.9462 *	0.0543	0.9799 *	0.9318 *	Pb	0.9271 *	0.1359	0.9964 *	0.9687 *
Th	0.9580 *	0.0672	0.9842 *	0.9479 *	Th	0.9781 *	0.1818	0.9742 *	0.9873 *
U	0.3306	0.4467	0.3885	0.3045	U	0.6095	0.5018	0.4967	0.5296

Further insights regarding the group of elemental patterns during freezing and thawing were obtained via consideration of correlations between element distribution factors (K_d solution/ice), as listed in Table 4. These correlations generally confirmed the group of elements described above; i.e., non-conservative, such as DOC, Fe, Al, trace metals, and notably trivalent and tetravalent hydrolysates; and conservative major cations (Na, Ca, Mg) and some anions (B). Overall, the number of significant correlations for K_d values was lower than that for concentration values, and the maximal number of inter-correlated elements was observed in the thawing experiments.

Table 4. Matrix correlation (Pearson, $p < 0.05$ significance level, labeled by asterisk) of element distribution coefficient (*Kd*) between the aqueous solution and ice during the freezing (left) and thawing (right) of peat leachates. Experiments, with and without insulant, are combined together.

Freezing WSL + F WSL + BZT (Insulant + No Insulant)						Thawing WSL + F WSL + BZT					
	DOC	**Na**	**Al**	**Ca**	**Fe**		**DOC**	**Na**	**Al**	**Ca**	**Fe**
DOC	1	−0.4156	0.5072	0.0843	0.5100	DOC	1	−0.3016	0.5712	0.1688	0.6304
Li	−0.1530	0.7967 *	0.4767	0.7761 *	0.4696	Li	0.2761	−0.0676	0.4645	0.3155	0.3766
B	−0.0739	0.6382	0.0415	0.6214	0.1158	B	−0.0679	0.9436 *	−0.0395	0.7411 *	0.1240
Na	−0.4156	1	0.1041	0.7425 *	0.1953	Na	−0.3016	1	−0.1697	0.5711	−0.0609
Mg	−0.2679	0.8074 *	0.1838	0.7995 *	0.1953	Mg	−0.2314	−0.0126	−0.1526	−0.7061	−0.2045
Al	0.5072	0.1041	1	0.2329	0.7560 *	Al	0.5712	−0.1697	1	0.2742	0.9317 *
Si	0.1082	0.4472	0.5676	0.3854	0.8156 *	Si	0.2021	−0.1342	0.4378	−0.4496	0.3752
P	0.1482	0.4320	0.2525	0.7756 *	0.5143	P	0.7269 *	−0.3039	0.4626	0.2122	0.4140
K	−0.0056	0.3047	0.6187	0.2150	0.4870	K	−0.2479	−0.0247	0.3868	−0.2146	0.2404
Ca	0.0843	0.7425 *	0.2329	1	0.3776	Ca	0.1688	0.5711	0.2742	1	0.4163
Ti	0.5325	0.1927	0.7393 *	0.3701	0.8923 *	Ti	0.4915	−0.0464	0.7842 *	0.4037	0.8648 *
V	0.7352 *	−0.1233	0.4677	0.3208	0.7364 *	V	0.5676	0.0204	0.8124 *	0.6671	0.9248 *
Cr	0.3214	−0.2685	0.3539	−0.1761	−0.1111	Cr	0.1342	0.0799	0.0982	0.2944	0.0721
Mn	−0.0527	−0.1320	0.4663	−0.3944	0.1574	Mn	−0.2300	0.4363	0.4854	0.2417	0.5073
Fe	0.5100	0.1953	0.7560 *	0.3776	1	Fe	0.6304	−0.0609	0.9317 *	0.4163	1
Co	0.0272	0.4662	0.7655 *	0.2426	0.5674	Co	−0.1855	0.0481	0.5476	−0.0876	0.4929
Ni	0.0342	−0.3260	−0.2851	−0.2410	−0.0968	Ni	0.6660	−0.0068	0.6953	0.3143	0.8425 *
Cu	−0.1651	0.3935	0.5024	0.1072	0.6297	Cu	0.2010	−0.1413	0.2787	0.1113	0.1933
Zn	−0.1390	0.0970	0.2811	−0.2016	0.1828	Zn	−0.4330	0.1063	−0.1833	−0.6541	−0.2768
Ga	0.5550	−0.2729	0.4173	0.0082	0.0492	Ga	0.5962	−0.2789	0.5572	−0.0347	0.4060
As	0.7066 *	−0.0685	0.3519	0.3874	0.6971	As	0.5922	−0.0274	0.8507 *	0.5036	0.8904 *
Rb	0.2731	−0.1322	0.6126	−0.0054	0.3259	Rb	0.5634	−0.3121	0.9508 *	0.1782	0.8402 *
Sr	−0.1735	0.7653 *	0.2376	0.8512 *	0.2824	Sr	−0.2950	−0.0756	0.1700	−0.3811	−0.0417
Y	0.1877	0.1517	0.5959	0.1345	0.6976	Y	0.5229	0.0566	0.9368 *	0.3604	0.9358 *
Zr	0.3338	−0.0874	0.4281	0.0913	0.3792	Zr	0.6300	−0.0906	0.9480 *	0.4148	0.9705 *
Nb	0.0318	0.5789	0.7029 *	0.4167	0.7073 *	Nb	0.0933	0.1119	0.6551	0.0888	0.7199 *
Mo	0.5238	0.2488	0.9165 *	0.4235	0.8317 *	Mo	0.0401	0.1440	0.6823	0.0976	0.6416
Cd	−0.0607	−0.1219	0.4597	−0.3656	0.2252	Cd	0.2183	−0.1742	0.8165 *	0.0156	0.8099 *
Sb	0.3598	−0.0489	0.4976	0.1477	0.5904	Sb	0.1665	−0.1549	0.5969	−0.0309	0.4949
Cs	0.5773	−0.2419	0.5398	0.0382	0.3247	Cs	0.5899	−0.2591	0.9123 *	0.2353	0.7915 *
Ba	−0.2300	0.0473	0.1627	−0.2749	0.0073	Ba	−0.3542	−0.0635	0.0718	−0.6617	−0.0606
La	−0.0399	−0.1035	−0.2501	−0.1054	−0.1870	La	0.5587	−0.1475	0.9793 *	0.2546	0.9591 *
Ce	0.1158	−0.0381	0.0226	0.0197	−0.0057	Ce	0.6383	−0.1569	0.9644 *	0.3963	0.9670 *
Pr	0.2602	−0.0751	0.1179	0.0857	0.1968	Pr	0.6663	−0.0715	0.9459 *	0.4638	0.9769 *
Nd	0.3362	0.0534	0.2615	0.2513	0.4887	Nd	0.6476	−0.0423	0.9222 *	0.5276	0.9306 *
Sm	0.4565	−0.1096	0.7472 *	−0.0797	0.6653	Sm	0.5272	0.0892	0.8231 *	0.3027	0.8126 *
Gd	0.2701	−0.0438	0.4191	0.0736	0.4928	Gd	0.5253	−0.2441	0.9277 *	0.2001	0.8383 *
Dy	0.3761	0.0351	0.4851	0.1995	0.7265 *	Dy	0.4651	0.0003	0.8763 *	0.2797	0.9322 *
Yb	0.2059	0.1526	0.5809	0.1827	0.5640	Yb	0.2427	0.2509	0.6487	0.2005	0.6722
Hf	0.2560	−0.0663	0.4668	0.0583	0.2919	Hf	0.5711	0.0393	0.8259 *	0.3117	0.8759 *
W	−0.0986	0.4389	−0.0700	0.4516	0.0428	W	−0.3747	0.7092 *	0.1301	0.3248	0.1942
Pb	−0.1740	0.1610	0.4887	−0.1852	0.2997	Pb	0.4775	−0.2282	0.8482 *	0.1417	0.7263 *
Th	0.6065	0.1494	0.5774	0.5807	0.7874 *	Th	0.5951	−0.1594	0.9138 *	0.4679	0.9234 *
U	0.2473	0.2469	0.6766	0.3174	0.5078	U	−0.2065	0.3369	0.1920	−0.0993	0.0739

4. Discussion

During freezing of natural porewaters that present in the permafrost peatlands, the solutes are known to be excluded downwards during the autumn freeze-up, when the freezing front propagates from the surface to the bottom of the peat profile [54,55]. During this process, the solutes can be accumulated by a factor of ten to hundreds of times in the remaining unfrozen part of the porewaters in mineral and organic soils [54,56,57]. The basic mechanism of DOC and inorganic solute accumulation in the residual fluid during ice formation in the reactor is simply a physically induced increase in their concentration at the freezing front propagation [58,59]. Numerous experiments on homogeneous and heterogeneous aqueous–solid systems demonstrated that the impurities (present in the form of soluble salts or suspended materials) are separated from aqueous solution by the so-called dynamic freezing front [60,61]. Other experiments on the freezing of natural surface waters of Arctic peatlands, similar to the peat porewater of the current study, demonstrated a progressive freezing of water from the reactor edges to the center, accompanied by the relative accumulation of DOM in the remaining liquid [46,53].

The temporal concentration patterns and elemental correlations demonstrated two contrasting group of solutes—labile elements and insoluble elements. The labile elements were present in essentially ionic forms (alkalis, alkaline–earth metals, oxyanions, and neutral molecules) which did not correlate with DOC, Fe, and Al, and exhibited rather similar concentrations in the fluid prior to freezing and after the entire freeze–thaw sequence. Contrasting to these soluble and labile solutes were divalent metals and insoluble TE^{3+} and TE^{4+} trace elements, which are known to be carried in the form of organic and organo-mineral (Fe, Al) colloids in the peat porewater (i.e., [15]) and dispersed ice of the peat cores [10]. Therefore, in agreement with previous studies of experimental bi-directional freezing of the entire peat cores (i.e., [44,45]), we hypothesize that, during freezing front propagation, large size organic-Fe-Al colloids are subjected to preferential exclusion relative to truly dissolved (inorganic) solutes from the ice. As a result, they could form precipitates either in the bulk of the remaining fluid or at the actual interface between the forming ice and the remaining aqueous solution.

It is known that high DOM concentrations diminish the proportion of DOC incorporated into the ice phase [20,62,63]. In the present study, we observed a positive correlation ($R = 0.94$, $p < 0.01$) between the initial DOC concentration in the leachates and the K_d value of DOC. Note however, that a rather high concentration of initial DOM and a fast freezing rate (at −20 °C) could drive some parts of ionic solutes and even organic colloids to be occluded in the bulk of developing ice in the form of liquid pockets, as is known for sea ice formation [64], and has been recently demonstrated for mineral soils [65,66].

Some decrease in DOM between the initial leachate and final fluid, after full freezing and thawing, is most likely linked to coagulation, precipitation, and removal (on filters) in the form of amorphous organic-rich solid phases. The detailed molecular mechanisms of this freeze-induced coagulation of ionic solutes and colloids are not known [35,45,46]. The chemical nature of coagulated material has not been investigated in the present study, but most likely includes, in addition to organic humic particles (e.g., [67]), organo-mineral composites of Fe and Al (hydr)oxides tightly linked to organic matter (OM). Such complex compounds are primary carriers of trace metal in soils and groundwater [68–70], as also evidenced by studies of colloidal matter in the surface waters of permafrost peatlands [10,14,15,71].

Despite these measurable, but rather minor, decreases in some component concentrations between the initial leachate and the final fluid obtained after the full freezing and thawing of the reactors, the present study generally evidenced a good recovery of DOC and most solutes over reversible freezing and thawing cycles. The results of the present study thus corroborate those reported earlier regarding the rather minor (i.e., <10–30%) effects of freeze–thaw cycles on DOC quantity and quality in acidic peat leachates [46]. Overall, this demonstrates the high stability of dissolved (<0.45 µm) OC, Fe, Al, and some trace metals to the repetitive freezing and thawing of surface waters in permafrost peatlands.

5. Conclusions

Experimental modeling of the reversible freezing and thawing of the aqueous leachates of peat originating from permafrost peatlands demonstrated a systematic evolution of the solute concentration of remaining/forming fluids in the course of progressive freezing/thawing, respectively. Element partitioning between the initial filtrate and the one formed after complete freezing and thawing varied, depending on the nature of the element and its relative affinity to organic matter, Fe/Al hydroxide colloids, or simple ionic complexes. In accord with the first hypothesis of this study, we revealed two groups of solutes: (1) those incorporating into forming ice; and (2) those remaining in the fluid phase via concentration due to freezing front migration, but not subjected to coagulation. In contrast, we did not note the existence of insoluble organo-mineral amorphous compounds or minerals subjected to irreversible coagulation. A likely reason for the latter observation is the rather acidic pH of peat leachates and high DOC concentrations, leading to the stabilization of typically insoluble solid phases known to be subjected to cryo-induced precipitation, such as Fe, Al hydroxide, or Ca carbonate minerals.

The leachates of frozen peat behaved drastically differently from the leachates of peat from the active layer, as the concentration of DOC, Fe and other solutes in the remaining fluid did not increase upon progressive freezing. We thus demonstrate the efficient incorporation of solutes from frozen peat leachates into the forming ice, as reflected by quite low partitioning coefficients of elements between forming ice and the remaining fluid for this type of substrate. It is thus possible that some volumes of the fluid, formed from frozen leachate, partially preserved the peat-dispersed ice structure and chemical composition and could be directly incorporated into forming ice, with a partitioning coefficient close to 1.

We also conclude that one can use a laboratory leachate of natural substrates, such as peat, as a surrogate for the direct experimental modeling of chemical and physical processes occurring in permafrost peatlands during the autumn—spring period. Overall, the present study provides a solid experimental basis for constructing a uniform approach for the laboratory modeling of soil fluid behavior in permafrost peatlands.

Author Contributions: Conceptualization, I.S.I. and O.S.P.; field sampling, I.S.I. and L.S.S.; analysis of collected samples, J.-L.R., I.S.I. and L.S.S.; experiments, I.S.I. and O.S.P.; visualization, I.S.I.; writing and editing O.S.P., I.S.I., L.S.S. and J.-L.R. All authors have read and agreed to the published version of the manuscript.

Funding: The study was funded by the Ministry of Education and Science of the Russian Federation (Agreement No. 075-15-2022-241).

Institutional Review Board Statement: Not applicable.

Informed Consent Statement: Not applicable.

Data Availability Statement: Not applicable.

Conflicts of Interest: The authors declare no conflict of interest.

References

1. Henry, H.A.L. Climate change and soil freezing dynamics: Historical trends and projected changes. *Clim. Change* **2008**, *87*, 421–434. [CrossRef]
2. Hayashi, M. The cold vadose zone: Hydrological and ecological significance of frozen-soil processes. *Vadoze Zone J.* **2013**, *12*, 1–8. [CrossRef]
3. Shogren, A.J.; Zarnetske, J.P.; Abbott, B.W.; Iannucci, F.; Bowden, W.B. We cannot shrug off the shoulder seasons: Addressing knowledge and data gaps in an Arctic headwater. *Environ. Res. Lett.* **2020**, *15*, 104027. [CrossRef]
4. Manasypov, R.M.; Vorobyev, S.N.; Loiko, S.V.; Kritzkov, I.V.; Shirokova, L.S.; Shevchenko, V.P.; Kirpotin, S.N.; Kulizhsky, S.P.; Kolesnichenko, L.G.; Zemtzov, V.A.; et al. Seasonal dynamics of organic carbon and metals in thermokarst lakes from the discontinuous permafrost zone of western Siberia. *Biogeosciences* **2015**, *12*, 3009–3028. [CrossRef]
5. Pokrovsky, O.S.; Manasypov, R.M.; Kopysov, S.G.; Krickov, I.V.; Shirokova, L.S.; Loiko, S.V.; Lim, A.G.; Kolesnichenko, L.G.; Vorobyev, S.N.; Kirpotin, S.N. Impact of permafrost thaw and climate warming on riverine export fluxes of carbon, nutrients and metals in Western Siberia. *Water* **2020**, *12*, 1817. [CrossRef]
6. Krickov, I.V.; Lim, A.G.; Shevchenko, V.P.; Starodymova, D.P.; Dara, O.M.; Kolesnichenko, Y.; Zinchenko, D.O.; Vorobyev, S.N.; Pokrovsky, O.S. Seasonal Variations of Mineralogical and Chemical Composition of Particulate Matter in a Large Boreal River and Its Tributaries. *Water* **2023**, *15*, 633. [CrossRef]
7. Ma, Q.; Jin, H.; Yu, C.; Bense, V.F. Dissolved organic carbon in permafrost regions: A review. *Sci. China Earth Sci.* **2019**, *62*, 349–364. [CrossRef]
8. Vonk, J.E.; Tank, S.E.; Bowden, W.B.; Laurion, I.; Vincent, W.F.; Alekseychik, P.; Amyot, M.; Billet, M.F.; Canario, J.; Cory, R.M.; et al. Reviews and syntheses: Effects of permafrost thaw on Arctic aquatic ecosystems. *Biogeosciences* **2015**, *12*, 7129–7167. [CrossRef]

9. Lim, A.G.; Loiko, S.V.; Kuzmina, D.M.; Krickov, I.V.; Shirokova, L.S.; Kulizhsky, S.P.; Vorobyev, S.N.; Pokrovsky, O.S. Dispersed ground ice of permafrost peatlands: A non-accounted for source of C, nutrients and metals. *Chemosphere* **2021**, *226*, 128953. [CrossRef]
10. Lim, A.G.; Loiko, S.V.; Kuzmina, D.M.; Shirokova, L.S.; Pokrovsky, O.S. Organic carbon, and major and trace elements reside in labile low-molecular form in the ground ice of permafrost peatlands: A case study of colloids in peat ice of Western Siberia. *Environ. Sci. Process. Impacts* **2022**, *24*, 1443–1459. [CrossRef]
11. Fu, Z.; Wu, Q.; Zhang, W.; He, H.; Wang, L. Water migration and segregated ice formation in frozen ground: Current advances and future perspectives. *Front. Earth Sci.* **2022**, *10*, 826961. [CrossRef]
12. Reeve, A.S.; Siegel, D.I.; Glaser, P.H. Geochemical controls on peatland pore water from the Hudson Bay Lowland: A multivariate statistical approach. *J. Hydrol.* **1996**, *181*, 285–304. [CrossRef]
13. Raudina, T.V.; Loiko, S.V.; Lim, A.G.; Krickov, I.V.; Shirokova, L.S.; Istigechev, G.I.; Kuzmina, D.M.; Kulizhsky, S.P.; Vorobyev, S.N.; Pokrovsky, O.S. Dissolved organic carbon and major and trace elements in peat porewater of sporadic, discontinuous, and continuous permafrost zones of western Siberia. *Biogeosciences* **2017**, *14*, 3561–3584. [CrossRef]
14. Pokrovsky, O.S.; Manasypov, R.M.; Loiko, S.V.; Shirokova, L.S. Organic and organo-mineral colloids in discontinuous permafrost zone. *Geochim. Cosmochim. Acta* **2016**, *188*, 1–20. [CrossRef]
15. Raudina, T.V.; Loiko, S.V.; Kuzmina, D.M.; Shirokova, L.S.; Kulizhsky, S.P.; Golovatskaya, E.A.; Pokrovsky, O.S. Colloidal organic carbon and trace elements in peat porewaters across a permafrost gradient in Western Siberia. *Geoderma* **2021**, *390*, 114971. [CrossRef]
16. DeLuca, T.H.; Keeney, D.R.; McCarty, G.W. Effect of freeze-thaw events on mineralization of soil nitrogen. *Biol. Fertil. Soils* **1992**, *14*, 116–120. [CrossRef]
17. Dietzel, M. Impact of cycling freezing on precipitation of silica in Me-SiO_2-H_2O systems and geochemical implications for cryosoils and sediments. *Chem. Geol.* **2005**, *216*, 79–88. [CrossRef]
18. Du, L.; Dyck, M.; Shotyk, W.; He, H.; Lv, J.; Cuss, C.; Bie, J. Lead immobilization processes in soils subjected to freeze-thaw cycles. *Ecotoxicol. Environ. Saf.* **2020**, *192*, 110288. [CrossRef]
19. Fitzhugh, R.D.; Driscoll, C.T.; Groffman, P.M.; Tierney, G.L.; Fahey, T.J.; Hardy, J.P. Soil freezing and the acid-base chemistry of soil solutions in a northern hardwood forest. *Soil Sci. Soc. Am. J.* **2003**, *67*, 1897–1908. [CrossRef]
20. Hentschel, K.; Borken, W.; Matzner, E. Repeated freeze–thaw events affect leaching losses of nitrogen and dissolved organic matter in a forest soil. *J. Plant. Nutr. Soil Sci.* **2008**, *171*, 699–706. [CrossRef]
21. Kim, E.-A.; Lee, H.K.; Choi, J.H. Effects of a controlled freeze-thaw event on dissolved and colloidal soil organic matter. *Environ. Sci. Pollut. Res.* **2017**, *24*, 1338–1346. [CrossRef] [PubMed]
22. Larsen, K.S.; Jonasson, S.; Michelsen, A. Repeated freeze–thaw cycles and their effects on biological processes in two arctic ecosystem types. *Appl. Soil Ecol.* **2002**, *21*, 187–195. [CrossRef]
23. Leuther, F.; Schlüter, S. Impact of freeze-thaw cycles on soil structure and soil hydraulic properties. *Soil* **2021**, *7*, 179–191. [CrossRef]
24. Mohanty, S.K.; Saiers, J.E.; Ryan, J.N. Colloid-facilitated mobilization of metals by freeze-thaw cycles. *Environ. Sci. Technol.* **2014**, *48*, 977–984. [CrossRef]
25. Mohanty, S.K.; Saiers, J.E.; Ryan, J.N. Colloid mobilization in a fractured soil during dry-wet cycles: Role of drying duration and flow path permeability. *Environ. Sci. Technol.* **2015**, *49*, 9100–9106. [CrossRef]
26. Ren, J.; Vanapalli, S.K. Effect of freeze–thaw cycling on the soil-freezing characteristic curve of five Canadian soils. *Vadose Zone J.* **2020**, *19*, e20039. [CrossRef]
27. Semenov, V.M.; Kogut, B.M.; Lukin, S.M. Effect of repeated drying-wetting-freezing-thawing cycles on the active soil organic carbon pool. *Eurasian Soil Sci.* **2014**, *47*, 276–286. [CrossRef]
28. Schimel, J.P.; Clein, J.S. Microbial response to freeze-thaw cycles in tundra and taiga soils. *Soil Biol. Biochem.* **1996**, *28*, 1061–1066. [CrossRef]
29. Vestgarden, L.S.; Austnes, K. Effects of freeze–thaw on C and N release from soils below different vegetation in a montane system: A laboratory experiment. *Glob. Change Biol.* **2009**, *15*, 876–887. [CrossRef]
30. Xiao, L.; Zhang, Y.; Li, P.; Xu, G.; Shi, P.; Zhang, Y. Effects of freeze-thaw cycles on aggregate-associated organic carbon and glomalin-related soil protein in natural-succession grassland and Chinese pine forest on the Loess Plateau. *Geoderma* **2019**, *334*, 1–8. [CrossRef]
31. Zhang, Z.; Ma, W.; Feng, W.; Xiao, D.; Hou, X. Reconstruction of soil particle composition during freeze-thaw cycling: A review. *Pedosphere* **2016**, *26*, 167–179. [CrossRef]
32. Nagare, R.M.; Schincariol, R.A.; Quinton, W.L.; Hayashi, M. Effects of freezing on soil temperature, freezing front propagation and moisture redistribution in peat: Laboratory investigations. *Hydrol. Earth Syst. Sci.* **2012**, *16*, 501–515. [CrossRef]
33. Nagare, R.M.; Schincariol, R.A.; Quinton, W.L.; Hayashi, M. Moving the field into the lab: Simulation of water and heat transport in Subarctic Peat. *Permafr. Periglac. Proc.* **2012**, *23*, 237–243. [CrossRef]
34. Smerdon, B.D.; Mendoza, C.A. Hysteretic freezing characteristics of riparian peatlands in the Western Boreal Forest of Canada. *Hydrol. Process.* **2010**, *24*, 1027–1038. [CrossRef]

35. McCarter, C.P.R.; Rezanezhad, F.; Quinton, W.L.; Gharedaghloo, B.; Lennartz, B.; Price, J.; Connon, R.; Van Cappellen, P. Pore-scale controls on hydrological and geochemical processes in peat: Implications on interacting processes. *Earth-Sci. Rev.* **2020**, *207*, 103227. [CrossRef]
36. Schwamborn, G.; Schirrmeister, L.; Frütsch, F.; Diekmann, B. Quartz weathering in freeze–thaw cycles: Experiment and application to the El'gygytgyn Crater Lake record for tracing Siberian permafrost history. *Geogr. Ann. Ser. A Phys. Geogr.* **2012**, *94*, 481–499. [CrossRef]
37. Wang, J.Y.; Song, C.C.; Hou, A.X.; Miao, Y.Q.; Yang, G.S.; Zhang, J. Effects of freezing thawing cycle on peatland active organic carbon fractions and enzyme activities in the Da Xing'anling Mountains, Northeast China. *Environ. Earth Sci.* **2014**, *72*, 1853–1860. [CrossRef]
38. Chen, J.; Xue, S.; Lin, Y.; Wang, C.; Wang, Q.; Han, Q. Effect of freezing–thawing on dissolved organic matter in water. *Desalination Water Treat.* **2016**, *57*, 17230–17240. [CrossRef]
39. Fellman, J.B.; D'Amore, D.V.; Hood, E. An evaluation of freezing as a preservation technique for analyzing dissolved organic C, N and P in surface water samples. *Sci. Total. Environ.* **2008**, *392*, 305–312. [CrossRef]
40. Jiang, N.; Juan, Y.; Tian, L.; Chen, X.; Sun, W.; Chen, L. Modification of the composition of dissolved nitrogen forms, nitrogen transformation processes, and diversity of bacterial communities by freeze–thaw events in temperate soils. *Pedobiologia* **2018**, *71*, 41–49. [CrossRef]
41. Kim, E.-A.; Choi, J.H. Changes in the mineral element compositions of soil colloidal matter caused by a controlled freeze-thaw event. *Geoderma* **2018**, *318*, 160–166. [CrossRef]
42. Chung, H.Y.; Jung, J.; Lee, D.H.; Kim, S.; Lee, M.K.; Lee, J.I.; Yoo, K.-C.; Lee, Y.I.; Kim, K. Chemical weathering of granite in ice and its implication for weathering in polar regions. *Minerals* **2020**, *10*, 185. [CrossRef]
43. Savenko, A.V.; Savenko, V.S.; Pokrovsky, O.S. Phase fractionation of chemical elements during the formation of ice in fresh surface waters. *Doklady Acad. Sci. Ser. Earth Sci. Geogr.* **2020**, *492*, 48–54. [CrossRef]
44. Morgalev, S.Y.; Morgaleva, T.G.; Morgalev, Y.N.; Loiko, S.V.; Manasypov, R.M.; Lim, A.G.; Pokrovsky, O.S. Experimental modeling of the bacterial community translocation during freezing and thawing of peat permafrost soils of Western Siberia. *IOP Conf. Ser. Earth Environ. Sci.* **2019**, *400*, 012017. [CrossRef]
45. Morgalev, S.Y.; Lim, A.G.; Morgaleva, T.G.; Morgalev, Y.N.; Manasypov, R.M.; Kuzmina, D.M.; Shirokova, L.S.; Orgogozo, L.; Loiko, S.V.; Pokrovsky, O.S. Fractionation of organic C, nutrients, metals and bacteria in peat porewater and ice after freezing and thawing. *Environ. Sci. Pollut. Res.* **2023**, *30*, 823–836. [CrossRef]
46. Payandi-Rolland, D.; Shirokova, L.S.; Labonne, F.; Bénézeth, P.; Pokrovsky, O.S. Low impact of freeze-thaw cycles on organic carbon and metals in waters of permafrost peatlands. *Chemosphere* **2021**, *279*, 130510. [CrossRef]
47. Shirokova, L.S.; Chupakov, A.V.; Zabelina, S.A.; Neverova, N.V.; Payandi-Rolland, D.; Causseraund, C.; Karlsson, J.; Pokrovsky, O.S. Humic surface waters of frozen peat bogs (permafrost zone) are highly resistant to bio- and photodegradation. *Biogeosciences* **2019**, *16*, 2511–2526. [CrossRef]
48. Shirokova, L.S.; Chupakov, A.V.; Ivanova, I.S.; Moreva, O.Y.; Zabelina, S.A.; Shutskiy, N.A.; Loiko, S.V.; Pokrovsky, O.S. Lichen, moss and peat control of C, nutrient and trace metal regime in lakes of permafrost peatlands. *Sci. Total Environ.* **2021**, *782*, 146737. [CrossRef]
49. Morgalev, Y.N.; Lushchaeva, I.V.; Morgaleva, T.G.; Kolesnichenko, L.G.; Loiko, S.V.; Krickov, I.V.; Lim, A.G.; Raudina, T.V.; Volkova, I.I.; Shirokova, L.S.; et al. Bacteria primarily metabolize at the active layer/permafrost border in the peat core from a permafrost region in western Siberia. *Polar Biol.* **2017**, *40*, 1645–1659. [CrossRef]
50. Aksenov, A.S.; Shirokova, L.S.; Kisil, O.Y.; Kolesova, S.N.; Lim, A.G.; Kuzmina, D.M.; Pouille, S.; Alexis, M.A.; Castrec-Rouelle, M.; Loiko, S.V.; et al. Bacterial number and genetic diversity in a permafrost peatland (western Siberia): Testing a link with organic matter quality and elementary composition of a peat soil profile. *Diversity* **2021**, *13*, 328. [CrossRef]
51. Heimburger, A.; Tharaud, M.; Monna, F.; Losno, R.; Desboeufs, K.; Nguyen, E. SLRS-5 elemental concentrations deduced from SLRS-5/SLRS-4 ratios of thirty-three uncertified elements. *Geostand. Geoanalyt. Res.* **2013**, *37*, 77–85. [CrossRef]
52. Yeghicheyan, D.; Bossy, C.; Bouhnik Le Coz, M.; Douchet, C.; Granier, G.; Heimburger, A.; Lacan, F.; Lanzanova, A.; Rousseau, T.C.C.; Seidel, J.-L.; et al. A compilation of silicon, rare earth element and twenty-one other trace element concentrations in the natural river water Reference Material SLRS-5 (NRC-CNRC). *Geostand. Geoanal. Res.* **2013**, *37*, 449–467. [CrossRef]
53. Pokrovsky, O.S.; Karlsson, J.; Giesler, R. Freeze-thaw cycles of Arctic thaw ponds remove colloidal metals and generate low-molecular-weight organic matter. *Biogeochemistry* **2018**, *137*, 321–336. [CrossRef]
54. Kokelj, S.V.; Burn, C.R. Geochemistry of the active layer and near-surface permafrost, Mackenzie delta region, Northwest Territories, Canada. *Can. J. Earth Sci.* **2005**, *42*, 37–48. [CrossRef]
55. French, H.; Shur, Y. The principles of cryostratigraphy. *Earth-Sci. Rev.* **2010**, *101*, 190–206. [CrossRef]
56. Lamhonwah, D.; Lafrenière, M.J.; Lamoureux, S.F.; Wolfe, B.B. Multi-year impacts of permafrost disturbance and thermal perturbation on High Arctic stream chemistry. *Arct. Sci.* **2016**, *3*, 254–276. [CrossRef]
57. Lamhonwah, D.; Lafrenière, M.J.; Lamoureux, S.F.; Wolfe, B.B. Evaluating the hydrological and hydrochemical responses of a High Arctic catchment during an exceptionally warm summer. *Hydrol. Process.* **2017**, *31*, 2296–2313. [CrossRef]
58. Ewing, S.A.; O'Donnell, J.A.; Aiken, G.R.; Butler, K.; Butman, D.; Windham-Myers, L.; Kanevskiy, M.Z. Long-term anoxia and release of ancient, labile carbon upon thaw of Pleistocene permafrost. *Geophys. Res. Lett.* **2015**, *42*, 10730–10738. [CrossRef]

59. Ostroumov, V.; Hoover, R.; Ostroumova, N.; Van Vliet-Lanoë, B.; Siegert, C.; Sorokovikov, V. Redistribution of soluble components during ice segregation in freezing ground. *Cold Reg. Sci. Technol.* **2001**, *32*, 175–182. [CrossRef]
60. Shafique, U.; Anwar, J.; uz-Zaman, W.; Rehman, R.; Salman, M.; Dar, A.; Jamil, N. Forced migration of soluble and suspended materials by freezing front in aqueous systems. *J. Hydro-Environ. Res.* **2012**, *6*, 221–226. [CrossRef]
61. Takenaka, N.; Bandow, H. Chemical kinetics of reactions in the unfrozen solution of ice. *J. Phys. Chem. A* **2007**, *111*, 8780–8786. [CrossRef] [PubMed]
62. Elliott, A.C.; Henry, H.A.L. Freeze–thaw cycle amplitude and freezing rate effects on extractable nitrogen in a temperate old field soil. *Biol. Fertil. Soils.* **2009**, *45*, 469–476. [CrossRef]
63. Xue, S.; Wen, Y.; Hui, X.; Zhang, L.; Zhang, Z.; Wang, J.; Zhang, Y. The migration and transformation of dissolved organic matter during the freezing processes of water. *J. Environ. Sci.* **2015**, *27*, 168–178. [CrossRef] [PubMed]
64. Petrich, C.; Eicken, H. Growth, Structure and Properties of Sea Ice. In *Sea Ice*, 2nd ed.; Thomas, D.N., Dieckmann, G.S., Eds.; Wiley Blackwell: Oxford, UK, 2010; Volume 2, pp. 425–467. [CrossRef]
65. Chen, C.; Huang, H.; Mo, X.; Xue, H.; Liu, M.; Chen, H. Insights into the kinetic processes of solute migration by unidirectional freezing in porous media with micromodel visualization at the pore-scale. *Sci. Total Environ.* **2021**, *784*, 147178. [CrossRef]
66. Ju, Z.; Du, Z.; Guo, K.; Liu, X. Irrigation with freezing saline water for 6 years alters salt ion distribution within soil aggregates. *J. Soil Sediments* **2019**, *19*, 97–105. [CrossRef]
67. Giesy, J.P.; Briese, L.A. Particulate formation due to freezing humic waters. *Water Resour. Res.* **1978**, *14*, 542–544. [CrossRef]
68. McCarthy, J.F.; Zachara, J.M. Subsurface transport of contaminants. *Environ. Sci. Technol.* **1989**, *23*, 496–502. [CrossRef]
69. Murphy, E.M.; Zachara, J.M. The role of sorbed humic substances on the distribution of organic and inorganic contaminants in groundwater. *Geoderma* **1995**, *67*, 103–124. [CrossRef]
70. Otero Farina, A.; Peacock, C.L.; Fiol, S.; Antelo, J.; Carvin, B. A universal adsorption behaviour for Cu uptake by iron (hydr) oxide organo-mineral composites. *Chem. Geol.* **2018**, *479*, 22–35. [CrossRef]
71. Krickov, I.V.; Pokrovsky, O.S.; Manasypov, R.M.; Lim, A.G.; Shirokova, L.S.; Viers, J. Colloidal transport of carbon and metals by western Siberian rivers during different seasons across a permafrost gradient. *Geochim. Cosmochim. Acta* **2019**, *265*, 221–241. [CrossRef]

applied sciences

MDPI

Article

Research on the Elastoplastic Theory and Evolution Law of Plastic Zone Contours of Horizontal Frozen Walls under Nonuniform Loads

Shilong Peng [1,2], Yuhao Xu [1,2,*], Guangyong Cao [1,2] and Lei Pei [1,2]

1 Anhui Provincial Key Laboratory of Building Structure and Underground Engineering, Anhui Jianzhu University, Hefei 230601, China; psl@ahjzu.edu.cn (S.P.)
2 College of Civil Engineering, Anhui Jianzhu University, Hefei 230601, China
* Correspondence: xyh@stu.ahjzu.edu.cn

Abstract: The study of the changes in stress and deformation of frozen walls during excavation has always been a hot topic in underground freezing engineering, and the size of the plastic zone is an important basis for evaluating the stability of frozen walls. In response to the shortcomings in the current design of horizontal frozen walls, based on the internal excavation unloading conditions of the frozen wall in artificial ground freezing, an elastoplastic mechanical model for the interaction between a circular horizontal freezing wall and unfrozen soil mass is established under nonuniform loads to obtain the corresponding solutions for stress and displacement in the system. In this study, considering the shear stress of the plastic zone, the method for solving the traditional plastic zone contour equation is modified; consequently, the modified solution of the contour equation of the plastic zone for the frozen wall is obtained. Using this theoretical solution, the influence of the external load p_0 and the lateral pressure coefficient λ on the contour line of plastic zone and tensile stress zone are analyzed by combining the project case, the calculation results show that: the $\lambda = 0.485$ is the critical point where the inner edge of the frozen wall just happens to have tensile stress. When $\lambda < 0.485$, the tensile stress zone is inevitable in the inner edge of the frozen wall vertical direction, and its range is only related to λ and increases with the decrease of λ. The p_0 only affects the magnitude of tensile stress in the region, but does not affect its range. At this time, the frozen wall compression plastic zone contour evolves from crescent shaped to ear shaped with the increase of p_0. When $0.485 < \lambda < 0.61$, there will be no tensile stress zone, the frozen wall compression plastic zone contour also evolves from crescent shaped to ear shaped with the increase of p_0. When $\lambda > 0.61$, there will be also no tensile stress zone, with the increase of p_0, the compression plastic zone contour evolves from the crescent shaped in the horizontal direction to the elliptical shaped, and there is no ear-shaped plastic zone in the whole evolution process. Based on our results, we show that our method can be used to provide a theoretical basis for the stability evaluation and parameter (thickness) design calculation of horizontal frozen walls under nonuniform loads.

Keywords: horizontal freezing method construction; circular frozen wall; nonuniform load; unload model; elastoplastic analysis

1. Introduction

Artificial ground freezing [1–4] is a special construction technology often used in the construction of underground structures in weak, water-bearing strata, such as shaft engineering, tunnel construction, and foundation pit excavation. Research on the stress and deformation of the frozen wall created during the artificial ground freezing process while performing excavations has always been a topic of interest in the field of underground freezing engineering; herein, the size of the plastic zone of the frozen wall is an important measure to evaluate the stability of the frozen wall. It should be noted that, in this paper, plastic zones refer to compression plastic zones. In the literature, in the case of both vertical or horizontal frozen walls, research on their plastic zone radius and stress expression for the case wherein the lateral pressure coefficient $\lambda = 1$—i.e., on the elastic or elastic–plastic characteristics of frozen wall under uniform load—is most common [5–11]. However, numerous measured data indicate that the in situ stress includes a certain degree of inhomogeneity, i.e., the ratio of horizontal stress to vertical stress (which is referred to as the lateral pressure coefficient λ) is not equal to 1 [12,13]. This inhomogeneous pressure plays an important role in the failure of underground structures.

Considering this, worldwide, research has been conducted considering nonuniform formation pressure on the surrounding rock of roadways [14–19]. Indeed, analytical solutions of models of excavation of circular tunnels in elastic homogeneous materials subject to nonuniform initial stresses (i.e., the solution known as Kirsch's solution) [20,21] are discussed in textbooks of rock mechanics and excavation analysis due to their importance as basic teaching/learning tools. Nevertheless, the research on the elastoplastic mechanical properties of the frozen wall formed by the artificial ground freezing method under nonuniform load conditions are still lacking and not explicitly developed [22,23]. In particular, under nonuniform load, the initial stress field of the frozen stratum is considerably different from that of the surrounding rock of roadways (as shown in Figures 1 and 2). In particular, as is clear from Figure 2, it is important to consider the interaction between the frozen wall and unfrozen soil in the frozen stratum while studying the mechanical characteristics of the frozen wall before the installation of a lining. However, as shown in Figure 1, before this installation, the surrounding rock of the roadway is just a single medium in the stratum; therefore, the latter case only involves an inhomogeneous stress analysis of a single medium. Therefore, the existing research results of the surrounding rock of roadways under nonuniform load are not suitable for analyzing the mechanical characteristics of horizontal frozen walls.

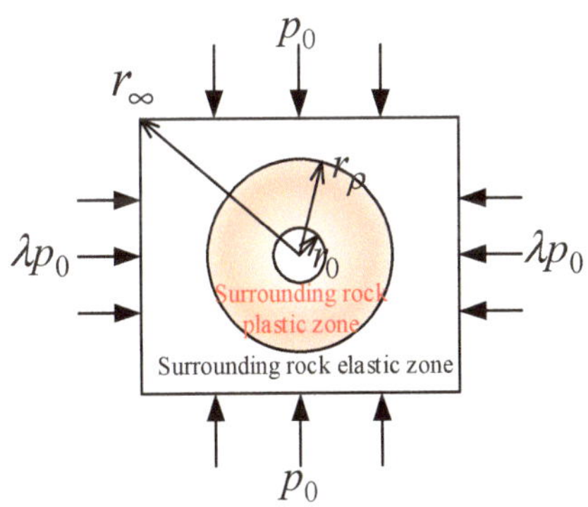

Figure 1. Initial stress field of the surrounding rock of the roadway.

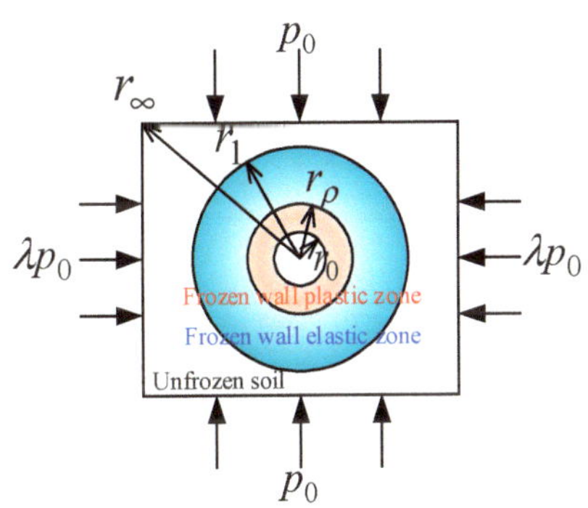

Figure 2. Initial stress field of the frozen stratum.

In the design theory of frozen walls, the vertical frozen wall theory has become relatively mature and has formed a relatively complete set of design theories [24]. However, the design of the horizontal frozen wall is not yet mature, mainly relying on the calculation formula for the vertical frozen wall of the neutral shaft in mining construction and combining the on-site experience method for design [25,26]. In recent years, scholars have focused on the design and research of horizontal frozen walls, which are still in the fields of numerical simulation, model testing, and on-site measurement. They mostly focus on the analysis of temperature and displacement fields [27–30]. Theoretical analysis only includes partial elastic design research, while elastic–plastic analysis has not yet seen relevant theoretical analysis research [6,22]. However, due to the different ground stresses in the vertical and horizontal directions, the external loads on the horizontal frozen wall are nonuniform, which fundamentally differs from the axisymmetric loads on the vertical frozen wall. Even the optimized design formula for the vertical frozen wall is theoretically not applicable to the horizontal frozen wall [23]. In order to gain a detailed understanding of the deformation and plastic zone contour characteristics of the horizontal frozen wall, and to design the horizontal frozen wall more reasonably, theoretical research on the elastic–plastic design of the horizontal frozen wall urgently needs to be addressed.

Traditionally, while studying the mechanical properties of frozen walls, the considered mechanical model was often assumed to be an infinitely long elastoplastic thick-walled cylinder under confining pressure loading conditions [6,31]. However, because the interaction between the frozen walls and surrounding unfrozen soil, as well as the radial unloading of the frozen wall, were often neglected, this led to an inaccurate analysis of the frozen walls, which consequently led to structural failures. However, in recent years, several scholars have used an unloading model to optimize their study of the mechanical properties of the frozen wall [8–10,32,33], which has led to suitable results. Nevertheless, analyses based on the elastoplastic theory for frozen walls using the excavation unloading model and considering the interaction between the frozen wall and unfrozen soil under nonuniform load conditions has not been conducted [34].

Therefore, in this study, considering the unloading effect of excavation, an elastoplastic mechanical model for the interaction between the horizontal frozen wall and surrounding unfrozen soil under nonuniform load conditions is established. In particular, the stress, deformation, and plastic zone contours in the elastic zone of the frozen wall are theoretically analyzed; in this case, the Mohr–Coulomb yield criterion is used as the plastic yielding criterion for the frozen wall. Based on the Mohr–Coulomb yield criterion and considering the shear stress in the plastic zone, the traditional method for calculating the plastic zone of the frozen wall is modified, and thus, a modified analytical solution of the contour equation for the plastic zone of the frozen wall is obtained. Finally, the distribution law of stress and displacement is analyzed for construction wherein the horizontal freezing method is used; based on this analysis, factors influencing the evolution of the elastoplastic zone are studied.

2. Basic Assumptions and Mechanical Models

2.1. Basic Assumptions

1. The horizontal frozen wall and surrounding unfrozen soil are regarded as infinite thick-walled cylinders; in addition, the plane strain model is used to analyze the mechanical characteristics of the horizontal frozen wall.
2. The frozen wall is homogeneous and composed of ideal elastoplastic material, while the unfrozen soil is also homogeneous, but composed of linear elastic material.
3. The influence of the force of the unfrozen soil and frozen wall's own weight is ignored in the model under study.
4. All the frozen soil inside the frozen wall is excavated in one instant, and there is no support on the inner edge of the frozen wall after the excavation; i.e., the radial load on the inner edge is 0.

5. There is complete contact between the frozen wall and unfrozen soil; i.e., the radial stress and shear stress on the contact surface are equal; in addition, the radial displacement and circumferential displacement are equal as well.
6. The stress field before and after freezing remains unchanged and is the same as the initial stress field. The initial stress field in the Cartesian coordinate system is expressed as follows:

$$\sigma_x^0 = \lambda p_0, \sigma_y^0 = p_0, \tau_{xy}^0 = 0 \tag{1}$$

Equation (1) can be represented in polar coordinates as follows:

$$\begin{cases} \sigma_r^0 = \lambda p_0 \cos^2\theta + p_0 \sin^2\theta = \frac{1+\lambda}{2} p_0 - \frac{1-\lambda}{2} p_0 \cos 2\theta \\ \sigma_\theta^0 = \lambda p_0 \sin^2\theta + p_0 \cos^2\theta = \frac{1+\lambda}{2} p_0 + \frac{1-\lambda}{2} p_0 \cos 2\theta \\ \tau_{r\theta}^0 = (1-\lambda) p_0 \sin\theta \cos\theta = \frac{1-\lambda}{2} p_0 \sin 2\theta \end{cases} \tag{2}$$

where σ^0 represents the initial stress, which is equivalent to the initial stress field given by σ_r^0, σ_θ^0, and $\tau_{r\theta}^0$, with compressive stress being considered positive; p_0 is the initial vertical stress value, $p_0 = \gamma h$; h is the buried depth; γ is the soil weight; and λ is the horizontal lateral pressure coefficient.

2.2. Mechanical Models

The mechanical model considered in our study is shown in Figures 2 and 3. In particular, Figure 2 shows the mechanical model for the initial stress field of the frozen stratum, while Figure 3 shows the mechanical model for the interaction between the frozen wall and surrounding rock after excavation and unloading. Although the contour shape of the elastoplastic zone of the frozen wall does not remain a regular circle under the action of a nonuniform stress field, for stress estimation, each contour can be approximately considered as a circle; then, the exact contour can be determined by solving for elastic stress of the frozen wall.

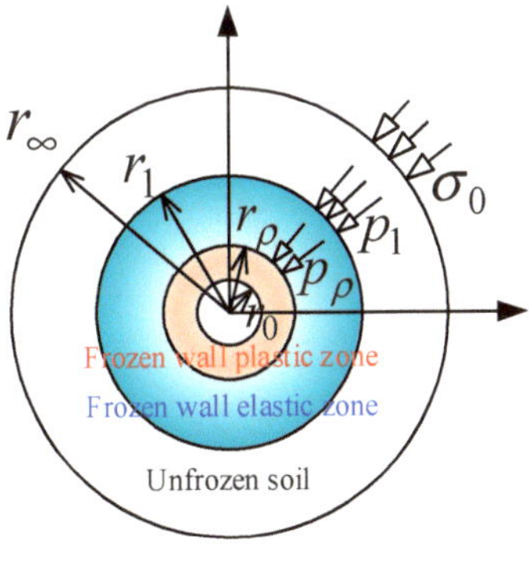

Figure 3. Mechanical model of the interaction between the frozen wall and surrounding rock after excavation and unloading.

Based on the abovementioned assumptions, the proposed mechanical model is divided into a frozen wall plastic zone, frozen wall elastic zone, and unfrozen soil elastic zone from the inside to the outside. In Figure 3, the stress boundary of the surrounding unfrozen soil at infinity is the initial stress field σ^0, load at the interface between the elastic zone of the unfrozen soil and outer surface of the elastic zone of the frozen wall is p_1, and load at the interface between the inner surface of the elastic zone of the frozen wall and outer surface of the plastic zone of the frozen wall is p_ρ.

2.3. Basic Mechanical Equations

The plane strain mechanical model shown in Figure 3 is used to calculate the stress and deformation of the frozen wall under nonuniform loading. The basic equations of elastic mechanics used in solving this model are well known. In particular, the basic mechanical equations in polar coordinates given in the classical literature [14,35] are discussed here.

First, the equilibrium differential equation for the plane strain mechanical model can be expressed as follows:

$$\begin{cases} \frac{\partial \sigma_r}{\partial r} + \frac{1}{r}\frac{\partial \tau_{r\theta}}{\partial \theta} + \frac{\sigma_r - \sigma_\theta}{r} = 0 \\ \frac{\partial \tau_{r\theta}}{\partial r} + \frac{1}{r}\frac{\partial \sigma_\theta}{\partial \theta} + \frac{2\tau_{r\theta}}{r} = 0 \end{cases} \tag{3}$$

where σ_r and σ_θ represent the normal stress components in the radial and hoop directions, respectively, while $\tau_{r\theta}$ represents the shear stress component.

Second, the geometric equation for the plane strain mechanical model can be expressed as follows:

$$\begin{cases} \varepsilon_r = \frac{\partial u}{\partial r} \\ \varepsilon_\theta = \frac{1}{r}\frac{\partial v}{\partial \theta} + \frac{u}{r} \\ \gamma_{r\theta} = \frac{1}{r}\frac{\partial u}{\partial \theta} + \frac{\partial v}{\partial r} - \frac{v}{r} \end{cases} \tag{4}$$

where u and v represent the displacement components in the radial and hoop directions, respectively. In addition, ε_r and ε_θ represent the normal strain components in the radial and hoop directions, respectively, while $\gamma_{r\theta}$ represents the shear strain component.

Third, the physical equation for the plane strain mechanical model can be expressed as follows:

$$\begin{cases} \varepsilon_r = \frac{1+\nu}{E}[(1-\nu)\sigma_r - \nu\sigma_\theta] \\ \varepsilon_\theta = \frac{1+\nu}{E}[(1-\nu)\sigma_\theta - \nu\sigma_r] \\ \gamma_{r\theta} = \frac{2(1+\nu)}{E}\tau_{r\theta} \end{cases} \tag{5}$$

where E and ν are the elastic modulus and Poisson's ratio of the stratigraphic material, respectively.

The above basic mechanical equations are applicable to both unfrozen soil as well as frozen wall.

3. Frozen Wall Elastic Problem

3.1. Solution of the Elastic Unloading Mechanical Model for the Frozen Wall

The elastic unloading mechanical model problem of the frozen wall can be obtained from the superposition of the equivalent initial stress field before excavation (Figure 4) and virtual excavation disturbance stress field (Figure 5) after the formation of the frozen wall.

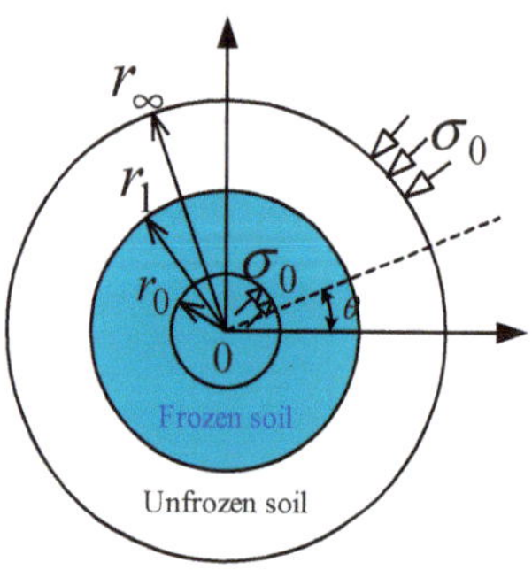

Figure 4. Equivalent initial stress field of the frozen wall before excavation.

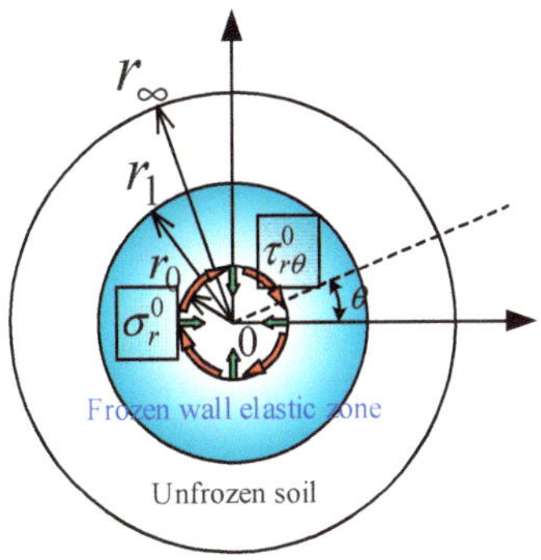

Figure 5. Virtual excavation disturbance stress field after the formation of the frozen wall.

Figure 4 shows that the initial stress σ^0 acts on both the inner edge of the frozen wall and outer edge of the unfrozen soil before excavation. In contrast, Figure 5 shows the virtual excavation disturbance stress field after excavation following the formation of the frozen wall, which is equivalent to the unloading force given by σ_r^0 and $\tau_{r\theta}^0$ at the inner edge of the frozen wall, where σ_r^0 is the equivalent radial load and $\tau_{r\theta}^0$ is the equivalent shear load. In the unloading model depicted in Figure 5, because the equivalent radial load and equivalent shear load of the inner edge of the frozen wall are trigonometrically related to the polar coordinate angle, it is assumed that the load and angle at the contact surface between the frozen wall and surrounding unfrozen soil are also related trigonometrically.

Depending on the characteristics of the model boundary conditions and plane strain problem, when the semi-inverse solution method is used to solve the abovementioned problems, the plane problem using the stress method can be generalized as solving a stress function that satisfies a compatible equation. Let Φ denote the stress function, then, the compatible equation in polar coordinates is given as follows:

$$\left(\frac{\partial^2}{\partial r^2}+\frac{1}{r}\frac{\partial}{\partial r}+\frac{1}{r^2}\frac{\partial^2}{\partial\theta^2}\right)\left(\frac{\partial^2\Phi}{\partial r^2}+\frac{1}{r}\frac{\partial\Phi}{\partial r}+\frac{1}{r^2}\frac{\partial^2\Phi}{\partial\theta^2}\right)=0 \quad (6)$$

If the stress function is known, the stress components can be obtained as follows:

$$\begin{cases}\sigma_r=\frac{1}{r}\frac{\partial\Phi}{\partial r}+\frac{1}{r^2}\frac{\partial^2\Phi}{\partial\theta^2}\\ \sigma_\theta=\frac{\partial^2\Phi}{\partial r^2}\\ \tau_{r\theta}=-\frac{1}{r}\frac{\partial^2\Phi}{\partial r\partial\theta}+\frac{1}{r^2}\frac{\partial\Phi}{\partial\theta}=-\frac{\partial}{\partial r}\left(\frac{1}{r}\frac{\partial\Phi}{\partial\theta}\right)\end{cases} \quad (7)$$

The stress functions of the unfrozen soil and frozen wall $\Phi_i(r,\theta)(i=1,2)$ are represented using separable forms (including $f_i(r)$ and $g_i(r)\text{con}2\theta$); these stress functions are similar and can be expressed as follows [36,37]:

$$\Phi_i(r,\theta)=f_i(r)+g_i(r)\cos 2\theta \quad (8)$$

Substituting the stress functions of Equation (8) into the compatible Equation (6) yields the following result:

$$\frac{d^4f_i(r)}{dr^4}+\frac{2}{r}\frac{d^3f_i(r)}{dr^3}-\frac{1}{r^2}\frac{d^2f_i(r)}{dr^2}+\frac{1}{r^3}\frac{df_i(r)}{dr}+\left[\frac{d^4g_i(r)}{dr^4}+\frac{2}{r}\frac{d^3g_i(r)}{dr^3}-\frac{9}{r^2}\frac{d^2g_i(r)}{dr^2}+\frac{9}{r^3}\frac{dg_i(r)}{dr}\right]\cos 2\theta=0 \quad (9)$$

Equation (9) is applicable to any arbitrary θ. Hence,

$$\begin{cases}\frac{d^4f_i(r)}{dr^4}+\frac{2}{r}\frac{d^3f_i(r)}{dr^3}-\frac{1}{r^2}\frac{d^2f_i(r)}{dr^2}+\frac{1}{r^3}\frac{df_i(r)}{dr}=0\\ \frac{d^4g_i(r)}{dr^4}+\frac{2}{r}\frac{d^3g_i(r)}{dr^3}-\frac{9}{r^2}\frac{d^2g_i(r)}{dr^2}+\frac{9}{r^3}\frac{dg_i(r)}{dr}=0\end{cases} \quad (10)$$

It is known that the constant term in the abovementioned stress function has no effect on the stresses and displacements of the unfrozen soil and frozen wall; therefore, we set the constant term in the stress function given by Equation (8) to 0. Hence, by solving the above differential equations, the appropriate stress function expression is obtained as follows:

$$\Phi_i(r,\theta) = a_i r^2 + b_i r^2 \ln r + c_i \ln r + (e_i + f_i r^4 + \frac{g_i}{r^2} + d_i r^2)\cos 2\theta \tag{11}$$

where a_i, b_i, c_i, d_i,e_i, f_i, and $g_i (i = 1,2)$ are undetermined parameters. When $i = 1$, the undetermined parameters in the stress function satisfying the semi-inverse solution of the frozen wall are expressed; in contrast, when $i = 2$, the undetermined parameters in the stress function satisfying the semi-inverse solution of the surrounding unfrozen soil are expressed. Substituting Equation (11) into the stress expression of Equation (7), the distribution function of the stress field for the frozen wall and surrounding unfrozen soil can be obtained as follows:

$$\begin{cases} \sigma_{ri} = -(2d_i + \frac{4e_i}{r^2} + \frac{6g_i}{r^4})\cos 2\theta + 2a_i + 2b_i \ln r + b_i + \frac{c_i}{r^2} \\ \sigma_{\theta i} = (2d_i + 12 f_i r^2 + \frac{6g_i}{r^4})\cos 2\theta + 2a_i + 2b_i \ln r + 3b_i - \frac{c_i}{r^2} \\ \tau_{r\theta i} = (6 f_i r^2 + 2d_i - \frac{2e_i}{r^2} - \frac{6g_i}{r^4})\sin 2\theta \end{cases} \tag{12}$$

Because the stresses of the surrounding unfrozen soil and frozen wall are solved for in the same manner, we only describe the solution for the stress field of the frozen wall below. The excavation of frozen soil inside the frozen wall can be considered equivalent to a stress relief field acting on the inner edge of the frozen wall, which induces stress redistribution and deformation of the frozen wall. After the excavation of frozen soil inside the frozen wall, the relief stress inside the frozen wall can be expressed as follows:

$$\begin{cases} \Delta\sigma_{r1} = \sigma_{r1} - \sigma_{r1}^0 = -\left[2d_1 + \frac{4e_1}{r^2} + \frac{6g_1}{r^4} - \frac{p_0(1-\lambda)}{2}\right]\cos 2\theta + 2a_1 + 2b_1 \ln r + b_1 + \frac{c_1}{r^2} - \frac{p_0(1+\lambda)}{2} \\ \Delta\sigma_{\theta 1} = \sigma_{\theta 1} - \sigma_{\theta 1}^0 = \left[2d_1 + 12 f_1 r^2 + \frac{6g_1}{r^4} - \frac{p_0(1-\lambda)}{2}\right]\cos 2\theta + 2a_1 + 2b_1 \ln r + 3b_1 - \frac{c_1}{r^2} - \frac{p_0(1+\lambda)}{2} \\ \Delta\tau_{r\theta 1} = \tau_{r\theta 1} - \tau_{r\theta 1}^0 = \left[6 f_1 r^2 + 2d_1 - \frac{2e_1}{r^2} - \frac{6g_1}{r^4} - \frac{p_0(1-\lambda)}{2}\right]\sin 2\theta \end{cases} \tag{13}$$

Furthermore, substituting Equation (13) into the physical as well as geometric equations and simplifying them, the elastic displacement of the frozen wall caused by the relief stress can be obtained as follows:

$$\begin{cases} \Delta u_1 = \frac{1+\mu_1}{E_1}\left\{-\left[2d_1 r - \frac{1}{1+\mu_1}\frac{4e_1}{r} - \frac{2g_1}{r^3} - \frac{p_0(1-\lambda)r}{2} + \frac{\mu_1}{1+\mu_1}4 f_1 r^3\right]\cos 2\theta \right. \\ \quad \left. + \frac{1-\mu_1}{1+\mu_1}2a_1 r + \frac{1-\mu_1}{1+\mu_1}2b_1 r \ln r - b_1 r - \frac{c_1}{r} - \frac{1-\mu_1}{1+\mu_1}\frac{p_0(1+\lambda)r}{2}\right\} \\ \Delta v_1 = \frac{1+\mu_1}{E_1}\left[2d_1 r + \frac{2 f_1 r^3(\mu_1+3)}{1+\mu_1} + \frac{2g_1}{r^3} - \frac{1-\mu_1}{1+\mu_1}\frac{2e_1}{r} - \frac{p_0(1-\lambda)r}{2}\right]\sin 2\theta + \frac{4b_1 r\theta}{E_1} \end{cases} \tag{14}$$

3.2. Determination of Undetermined Parameters

Based on the single value and continuous displacement conditions, we can say that $b_i = 0$, i.e.,

$$b_1 = b_2 = 0 \tag{15}$$

According to the Saint-Venant principle, the original rock stress field at a sufficiently far distance will be correspondingly less affected by the excavation and unloading effect inside the frozen wall.

In particular, when $r \to \infty$, the stress of the surrounding unfrozen soil after excavation and unloading of the frozen soil at the inner edge of the frozen wall can still be expressed by Equation (2). Comparing Equation (2) with Equation (12), we obtain:

$$\begin{cases} 2d_2 = p_0(1-\lambda)/2 \\ 2a_2 = p_0(1+\lambda)/2 \\ f_2 = 0 \end{cases} \tag{16}$$

Furthermore, as previously mentioned, because the radial stress and shear stress at the inner edge of the frozen wall are equal to zero after excavation and unloading, the following relationship can be obtained:

$$\begin{cases} (-2d_1 - \frac{4e_1}{r_0^2} - \frac{6g_1}{r_0^4})\cos 2\theta + 2a_1 + \frac{c_1}{r_0^2} = 0 \\ (6f_1 r_0^2 + 2d_1 - \frac{2e_1}{r_0^2} - \frac{6g_1}{r_0^4})\sin 2\theta = 0 \end{cases} \tag{17}$$

In a similar vein, because the radial stress, shear stress, radial displacement, and circumferential displacement at the contact surface between the frozen wall and surrounding unfrozen soil are equal, we obtain the following:

$$\begin{cases} 2d_1 + \frac{4e_1}{r_1^2} + \frac{6g_1}{r_1^4} = 2d_2 + \frac{4e_2}{r_1^2} + \frac{6g_2}{r_1^4} \\ 2a_1 + \frac{c_1}{r_1^2} = 2a_2 + \frac{c_2}{r_1^2} \\ 6f_1 r_1^2 + 2d_1 - \frac{2e_1}{r_1^2} - \frac{6g_1}{r_1^4} = 6f_2 r_1^2 + 2d_2 - \frac{2e_2}{r_1^2} - \frac{6g_2}{r_1^4} \\ \frac{1+\mu_1}{E_1}\left\{-\left[2d_1 r_1 - \frac{1}{1+\mu_1}\frac{4e_1}{r_1} - \frac{2g_1}{r_1^3} - \frac{p_0(1-\lambda)r_1}{2} + \frac{\mu_1}{1+\mu_1}4f_1 r_1^3\right]\right\} \\ \quad = \frac{1+\mu_2}{E_2}\left\{-\left[2d_2 r_1 - \frac{1}{1+\mu_2}\frac{4e_2}{r_1} - \frac{2g_2}{r_1^3} - \frac{p_0(1-\lambda)r_1}{2} + \frac{\mu_2}{1+\mu_2}4f_2 r_1^3\right]\right\} \\ \frac{1-\mu_1}{E_1}2a_1 r_1 - \frac{1-\mu_1}{E_1}\frac{p_0(1+\lambda)r_1}{2} - \frac{1+\mu_1}{E_1}\frac{c_1}{r_1} = \frac{1-\mu_2}{E_2}2a_2 r_1 - \frac{1-\mu_2}{E_2}\frac{p_0(1+\lambda)r_1}{2} - \frac{1+\mu_2}{E_2}\frac{c_1}{r_1} \\ \frac{1+\mu_1}{E_1}\left[2d_1 r_1 + \frac{2f_1 r_1^3(\mu_1+3)}{1+\mu_1} + \frac{2g_1}{r_1^3} - \frac{1-\mu_1}{1+\mu_1}\frac{2e_1}{r_1} - \frac{p_0(1-\lambda)r_1}{2}\right] \\ \quad = \frac{1+\mu_2}{E_2}\left[2d_2 r_1 + \frac{2f_2 r_1^3(\mu_2+3)}{1+\mu_2} + \frac{2g_2}{r_1^3} - \frac{1-\mu_2}{1+\mu_2}\frac{2e_2}{r_1} - \frac{p_0(1-\lambda)r_1}{2}\right] \end{cases} \tag{18}$$

Equations (8)–(11) are simultaneous equations that can be solved using MAPLE, which is a numerical software, to obtain the values of a_i, b_i, c_i, d_i, e_i, f_i, and g_i $(i = 1,2)$; then, the expressions for stress and displacement in the elastic zone of a horizontal frozen wall can be obtained by substituting the parameter values into Equations (5) and (7). The expressions for stress and displacement of unfrozen soil can be obtained in a similar manner.

4. Frozen Wall Elastic–Plastic Problem

4.1. Preliminary Determination of the Plastic Zone Radius

Yang et al. [9] derived an iterative equation to solve for the radius of the plastic zone of a frozen wall based on different yield criteria and excavation unloading effects in the case when the in situ geostress field is uniform. However, in most cases, the horizontal and vertical stresses are not equal; i.e., $\lambda \neq 1$. Researchers worldwide typically study the plastic zone of surrounding rocks under nonuniform stress field conditions based on the assumption that the surrounding rock after excavation is in an elastic state, subsequently calculating the stress in this elastic state according to the elasticity theory; finally, the stress of the elastic zone is substituted into the plastic yield condition to ascertain whether the surrounding rock has yielded [38]. At this time, the contour shape of the plastic zone of the surrounding rock is no longer a regular circle; instead it has sickle, cross, or other shape [17]. Although this abovementioned method is only an approximation, it is helpful for stress estimation. Furthermore, a similar method can be used for the elastoplastic analysis of a horizontal circular frozen wall under the action of a nonuniform stress field.

In the plane strain state, the relationship between the principal stresses and three polar stress components can be expressed as follows:

$$\left\{\begin{array}{l}\sigma_1 \\ \sigma_3\end{array} = \frac{\sigma_r+\sigma_\theta}{2} \pm \sqrt{\left(\frac{\sigma_r-\sigma_\theta}{2}\right)^2+\tau_{r\theta}^2}\right. \tag{19}$$

The Mohr–Coulomb yield criterion expressed in terms of principal stresses is as follows:

$$\sigma_1 = \frac{1+\sin\phi}{1-\sin\phi}\sigma_3 + \frac{2c\cos\phi}{1-\sin\phi} = A\sigma_3 + B \tag{20}$$

where $A = \frac{1+\sin\phi}{1-\sin\phi}$ and $B = \frac{2c\cos\phi}{1-\sin\phi}$.

Substituting Equation (19) into Equation (20), the expression of the yield condition can be obtained as follows:

$$\frac{\sigma_r+\sigma_\theta}{2} = -\frac{1+A}{1-A}\sqrt{\left(\frac{\sigma_r-\sigma_\theta}{2}\right)^2+\tau_{r\theta}^2} + \frac{B}{1-A} \tag{21}$$

Then, substituting Equation (12) into Equation (21), and simplifying it, we obtain

$$\begin{aligned} &12f_1\cos 2\theta\left(2a_1-\tfrac{B}{1-A}\right)r_p^2 - 4e_1\cos 2\theta\left(2a_1-\tfrac{B}{1-A}\right)\tfrac{1}{r_p^2} + \left(36f_1^2r^4+\tfrac{4e_1^2}{r_p^4}-24f_1e_1\right)\cos^2 2\theta \\ &= \left(\tfrac{1+A}{1-A}\right)^2\left[\begin{array}{c} 4d_1^2+36f_1^2r_p^4+24d_1f_1r_p^2+\tfrac{4e_1^2}{r_p^4}+\tfrac{36g_1^2}{r_p^8}+\tfrac{24e_1g_1}{r_p^6}+\tfrac{c_1^2}{r_p^4} \\ +\left(\tfrac{8d_1e_1}{r_p^2}+\tfrac{24d_1g_1}{r_p^4}+24f_1e_1+\tfrac{72f_1g_1}{r_p^2}\right)\cos 4\theta \\ -\left(\tfrac{4d_1c_1}{r_p^2}+12f_1c_1+\tfrac{4c_1e_1}{r_p^4}+\tfrac{12c_1g_1}{r_p^6}\right)\cos 2\theta \end{array}\right] - \left(2a_1-\tfrac{B}{1-A}\right)^2 \end{aligned} \tag{22}$$

Then, Equation (22) can be rewritten in the form of the plastic zone contour equation in the following manner:

$$f(r) = k_1r_p^4 + k_2r_p^2 + \frac{k_3}{r_p^2} + \frac{k_4}{r_p^4} + \frac{k_5}{r_p^6} + \frac{k_6}{r_p^8} + k_7 = 0 \tag{23}$$

where

$$k_1 = 36f_1^2\left(\sin^2 2\theta + \frac{4A}{1+A^2-2A}\right)$$

$$k_2 = 24d_1f_1\left(\frac{1+A}{1-A}\right)^2 - 12f_1\cos 2\theta\left(2a_1-\frac{B}{1-A}\right)$$

$$k_3 = (8d_1e_1+72f_1g_1)\left(\frac{1+A}{1-A}\right)^2\cos 4\theta - 4d_1c_1\left(\frac{1+A}{1-A}\right)^2\cos 2\theta + 4e_1\cos 2\theta\left(2a_1-\frac{B}{1-A}\right)$$

$$k_4 = (4e_1^2+24d_1g_1\cos 4\theta - 4c_1e_1\cos 2\theta + c_1^2)\left(\frac{1+A}{1-A}\right)^2 - 4e_1^2\cos^2 2\theta$$

$$k_5 = (24e_1g_1 - 12c_1g_1\cos 2\theta)\left(\frac{1+A}{1-A}\right)^2$$

$$k_6 = 36g_1^2\left(\frac{1+A}{1-A}\right)^2$$

$$k_7 = (4d_1^2+24f_1e_1\cos 4\theta - 12f_1c_1\cos 2\theta)\left(\frac{1+A}{1-A}\right)^2 + 24f_1e_1\cos^2 2\theta - \left(2a_1-\frac{B}{1-A}\right)^2$$

4.2. Solution of Plastic Zone Stress

In the plastic zone of the frozen wall, the radial stress σ_r^p is assumed to be the same as the stress distribution in its elastic zone [38]; in addition, the direction of the principal stress remains unchanged in the plastic and elastic zones. Based on these assumptions and considering Equation (21), the following relationship can be obtained:

$$\begin{cases} tg2\alpha = -\frac{2\tau_{r\theta}}{\sigma_r-\sigma_\theta} = \frac{2\tau_{r\theta}^p}{\sigma_\theta^p-\sigma_r^p} \\ \frac{\sigma_r^p+\sigma_\theta^p}{2} = -\frac{1+A}{1-A}\sqrt{\left(\frac{\sigma_r^p-\sigma_\theta^p}{2}\right)^2+\left(\tau_{r\theta}^p\right)^2}+\frac{B}{1-A} \end{cases} \tag{24}$$

where α is the direction of the principal stress of the frozen wall at any angle θ relative to the radius vector orientation of the stress (σ_r^p, σ_θ^p, $\tau_{r\theta}^p$) at the calculation point.

As indicated in Figure 6, the stress relationships can be represented in the form of a Mohr's stress circle; thus, the principal stress direction, radial stress, hoop stress, and shear stress distribution of the plastic zone of the frozen wall can be obtained as follows:

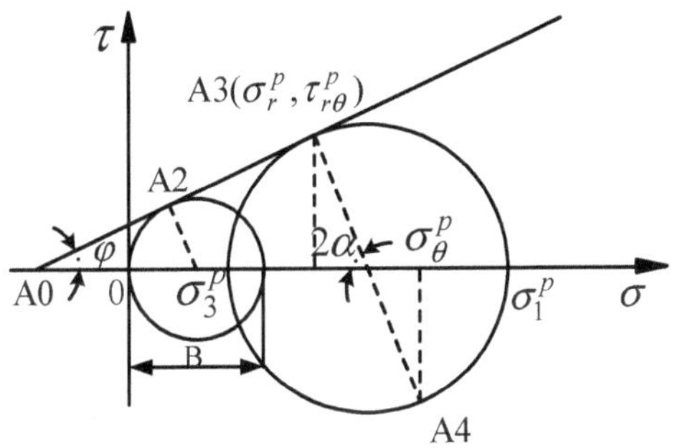

Figure 6. $\tau-\sigma$ curve of Mohr's stress circle.

$$\begin{cases} \sigma_r^p = -(2d_1+\frac{4e_1}{r^2}+\frac{6g_1}{r^4})\cos 2\theta+2a_1+2b_1\ln r+b_1+\frac{c_1}{r^2} \\ \sigma_\theta^p = \frac{(1+A)-(1-A)\cos 2\alpha}{(1+A)+(1-A)\cos 2\alpha}\sigma_r^p+\frac{2B\cos 2\alpha}{(1+A)+(1-A)\cos 2\alpha} \\ \tau_{r\theta}^p = \frac{(A-1)\sin 2\alpha}{(1+A)+(1-A)\cos 2\alpha}\sigma_r^p+\frac{B\sin 2\alpha}{(1+A)+(1-A)\cos 2\alpha} \end{cases} \quad \left(\sigma_\theta^p \geq \sigma_r^p\right) \tag{25}$$

where

$$\alpha = \frac{1}{2}\arctan\left[\frac{(6f_1r^6+2d_1r^4-2e_1r^2-6g_1)\sin 2\theta}{(6f_1r^6+2d_1r^4+2e_1r^2+6g_1)\cos 2\theta-c_1r^2}\right]$$

4.3. Modified Solution for the Plastic Zone Radius

The plastic zone contour equation given by Equation (22) does not consider the redistribution of stress in the plastic zone. In fact, when the frozen wall enters the plastic zone, its stress will be constantly adjusted and redistributed, resulting in continuous expansion of the plastic zone. In order to obtain a more accurate contour equation for the plastic zone than the previous one, a second approximation correction of the contour line of the plastic zone is performed considering stress redistribution of the plastic zone. In particular, for this second approximation, we use the Castner calculation method twice, wherein the initial plastic zone radius is calculated using the Castner method first, and then the stress values σ_r^p and $\tau_{r\theta}^p$ of the calculated points are obtained using the stress equation of the plastic zone; in this case, the contour line of the plastic zone is assumed to be a circle (the circle is made per the radius of the plastic zone for the calculated point). Finally, a new contour line equation for the plastic zone considering the action of σ_r^p and $\tau_{r\theta}^p$ is derived in a manner similar to Castner's method to obtain a more accurate plastic zone radius compared with that of the previous case.

After the approximation correction similar to Castner's method, the elastic stress field of the frozen wall with σ_r^p and $\tau_{r\theta}^p$ acting on the inner edge of the frozen wall is obtained as follows:

$$\begin{cases} \sigma_{r1}^e = -(2d_1 + \frac{4e_1}{r^2} + \frac{6g_1}{r^4})\cos 2\theta + 2a_1 + 2b_1 \ln r + b_1 + \frac{c_1}{r^2} + \sigma_r^p \frac{r_p^2}{r^2} \\ \sigma_{\theta 1}^e = (2d_1 + 12f_1 r^2 + \frac{6g_1}{r^4})\cos 2\theta + 2a_1 + 2b_1 \ln r + 3b_1 - \frac{c_1}{r^2} - \sigma_r^p \frac{r_p^2}{r^2} \\ \tau_{r\theta 1}^e = (6f_1 r^2 + 2d_1 - \frac{2e_1}{r^2} - \frac{6g_1}{r^4})\sin 2\theta \quad + \tau_{r\theta}^p \frac{r_p^2}{r^2} \end{cases} \tag{26}$$

By substituting Equation (26) into Equation (21), the modified implicit equation for the radius of the plastic zone can be obtained as follows:

$$\begin{aligned} & 12f_1 \cos 2\theta \left(2a_1 - \frac{B}{1-A}\right) r^2 - 4e_1 \cos 2\theta \left(2a_1 - \frac{B}{1-A}\right) \frac{1}{r^2} + \left(36f_1^2 r^4 + \frac{4e_1^2}{r^4} - 24f_1 e_1\right) \cos^2 2\theta \\ & = \left(\frac{1+A}{1-A}\right)^2 \left[\begin{array}{c} 4d_1^2 + 36f_1^2 r^4 + 24d_1 f_1 r^2 + \frac{4e_1^2}{r^4} + \frac{36g_1^2}{r^8} + \frac{24e_1 g_1}{r^6} + \left(\frac{c_1}{r^2} + \sigma_r^p \frac{r_0^2}{r^2}\right)^2 + \left(\tau_{r\theta}^p\right)^2 \frac{r_0^4}{r^4} \\ + \left(\frac{8d_1 e_1}{r^2} + \frac{24d_1 g_1}{r^4} + 24f_1 e_1 + \frac{72f_1 g_1}{r^2}\right) \cos 4\theta \\ - 4\left(d_1 + 3f_1 r^2 + \frac{e_1}{r^2} + \frac{3g_1}{r^4}\right)\left(\frac{c_1}{r^2} + \sigma_r^p \frac{r_0^2}{r^2}\right) \cos 2\theta \\ + \left(12f_1 \tau_{r\theta}^p r_0^2 + \frac{4d_1 \tau_{r\theta}^p r_0^2}{r^2} - \frac{4e_1 \tau_{r\theta}^p r_0^2}{r^4} - \frac{12g_1 \tau_{r\theta}^p r_0^2}{r^6}\right) \sin 2\theta \end{array} \right] - \left(2a_1 - \frac{B}{1-A}\right)^2 \end{aligned} \tag{27}$$

5. Engineering Examples and Analysis of the Evolution Law of the Elastoplastic Zone of the Frozen Wall

In the freezing design of the horizontal freezing method, the size of the plastic zone is an important basis for evaluating the stability of the frozen wall. In order to study the evolution law of the elastoplastic zone of the frozen wall under nonuniform load, we taking a horizontal frozen tunnel project as the background; in particular, the effects of the vertical external load p_0 and lateral pressure coefficient λ on the plastic zone contour and range of the tensile stress zone of the horizontal frozen wall are analyzed.

The tunnel is constructed using artificial freezing method, and the designed average temperature of the freezing wall is −10 °C. The buried depth of the tunnel is 187.0~248 m, with a longitudinal slope of $i = 1/1650$, and the initial pressure of the surrounding soil is between 0.3~0.9 MPa. The rock surrounding the tunnel is composed of N_2L^2 argillaceous siltstone, sandy mudstone, sandy gravel layer, and N_2L^{2s} loose sandstone. The elastic modulus and Poisson's ratio of frozen soil and unfrozen soil are respectively: $E_1 = 150$ MPa, $\mu_1 = 0.35$, $E_2 = 20$ MPa, $\mu_2 = 0.35$, and the excavation radius of frozen wall is 3.3 m. The horizontal freezing curtain and freezing hole layout of the tunnel are shown in Figures 7 and 8, respectively, and the relevant geological parameters are listed in Table 1.

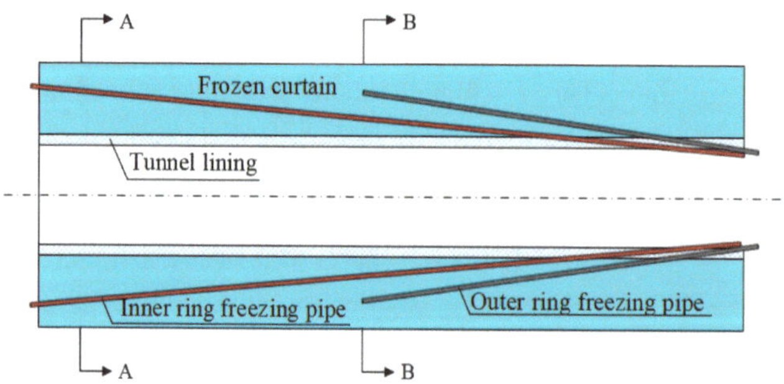

Figure 7. Schematic layout of the freezing pipes.

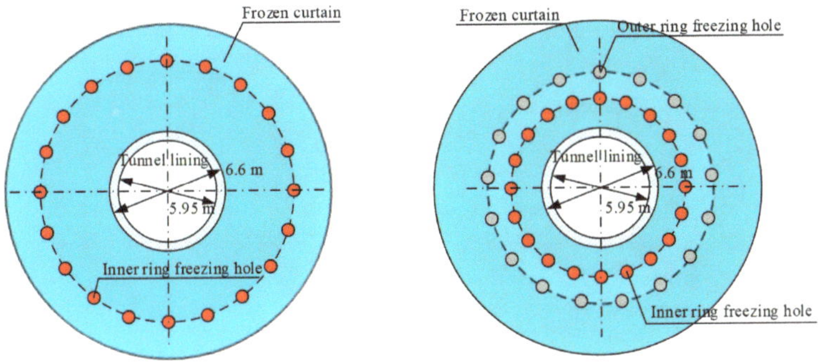

(**A**) A–A section (**B**) B–B section

Figure 8. Profile chart of the freezing holes.

Table 1. Physical and mechanical parameters of the frozen wall and unfrozen soil mass.

Parameters Material Types	Frozen Wall	Unfrozen Soil
Elastic modulus/MPa	150	20
Poisson ratio	0.35	0.35
Lateral pressure coefficient	0.8	0.8
Compressive strength/MPa	5.58	0.99
Cohesive strength/MPa	1.45	0.28
Internal friction angle/°	35	30
Bulk density/kN·m^3	19.2	22.0

Artificial ground freezing is an unstable thermal conductivity problem with complex boundary conditions, and the solution to its freezing temperature field requires consideration of factors such as phase transition, moving boundaries, and internal heat sources. Based on the basic theories of soil moisture migration, heat change, thermodynamics, and poromechanics during freezing, as well as Harlan's coupled model of water and heat, COMSOL multi-physics finite element software and MATLAB were used to build the numerical calculation models of the freezing temperature field of A–A and B–B sections of the horizontal freezing curtain of the tunnel [2,39,40], as shown in Figure 9. Via finite element calculation and analysis, it was found that the thickness of the frozen wall of the tunnel was 2.76~3.25 m; for the convenience of calculating the thickness of the frozen wall, it is taken as 3 m.

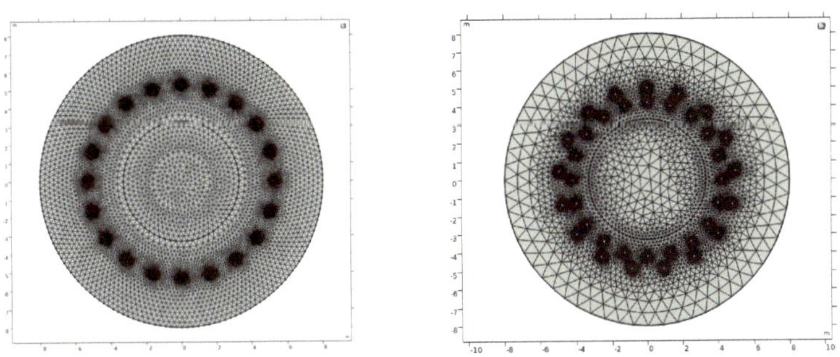

(**A**) Grid division of A–A section (**B**) Grid division of B–B section

Figure 9. *Cont.*

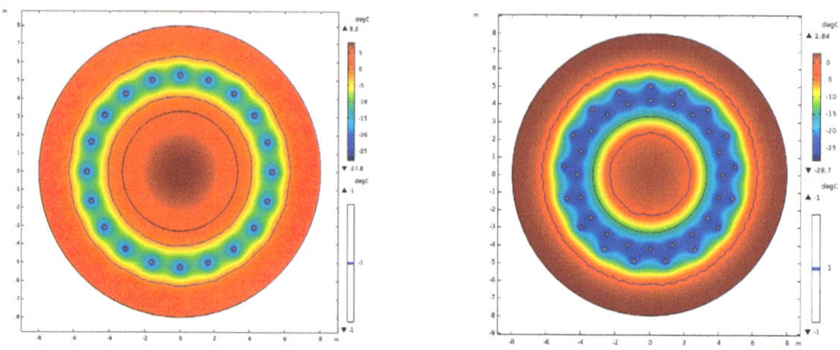

(C) Frozen temperature field distribution cloud map of A–A section

(D) Frozen temperature field distribution cloud map of B–B section

Figure 9. Grid division and frozen temperature field distribution cloud map.

5.1. Safety Analysis of Freezing Engineering

By substituting the relevant engineering and geological parameters into Equations (15)–(18) and solving the simultaneous equations for values of a_i, b_i, c_i, d_i, e_i, f_i, and g_i ($i = 1, 2$) using MAPLE, we obtain the values listed in the Table 2.

Table 2. Stress function parameters.

Parameters	Value	Parameters	Value
a_1	0.52	a_2	0.41
b_1	0.00	b_2	0.00
c_1	−11.36	c_2	−2.11
d_1	0.106	d_2	0.045
e_1	−2.06	e_2	−0.39
f_1	-6.60×10^{-4}	f_2	0.00
g_1	10.80	g_2	−1.72

The stress values of the frozen wall under the elastic state can be obtained by substituting the undetermined parameters into Equation (12); the corresponding values for principal stresses, principal stress direction, and yield state for $k\pi/12$ (k = 0, 1, 2, 3, 4, 5, 6) directions are listed in Table 3. Verification using Equation (21) indicates that plastic yield does not occur at the inner edge of the frozen wall of the tunnel; furthermore, the maximum circumferential stress at the inner edge of the frozen wall is 2.76 MPa, which meets the safety factor requirement of the compressive strength being twice the maximum circumferential stress. Under the actual working conditions for the tunnel considered in this study, for a frozen wall thickness of 3 m, it was calculated that, when the vertical external load increases to 1.83 MPa, the frozen wall reaches the elastic limit at the horizontal inner edge, and the vertical external load of the elastic limit is 2.03 times the design load, indicating that the tunnel freezing methodology leads to safe and reliable results.

Table 3. Stress and elastic–plastic state of the inner edge of the frozen wall.

Angle $\theta/°$	First Principal Stress σ_1/MPa	Second Principal Stress σ_2/MPa	Principal Stress Direction $\alpha+\theta/°$	Yield State
0°	2.76	0	0°	Not yielding
15°	2.67	0	15°	Not yielding
30°	2.42	0	30°	Not yielding
45°	2.09	0	45°	Not yielding
60°	1.75	0	60°	Not yielding
75°	1.51	0	75°	Not yielding
90°	1.42	0	90°	Not yielding

By substituting the parameters listed in Tables 1 and 2 in Equations (12) and (14), the stress and displacement distribution of the frozen wall in the No. 7 tunnel using the freezing method construction can be obtained as shown in Figures 10 and 11.

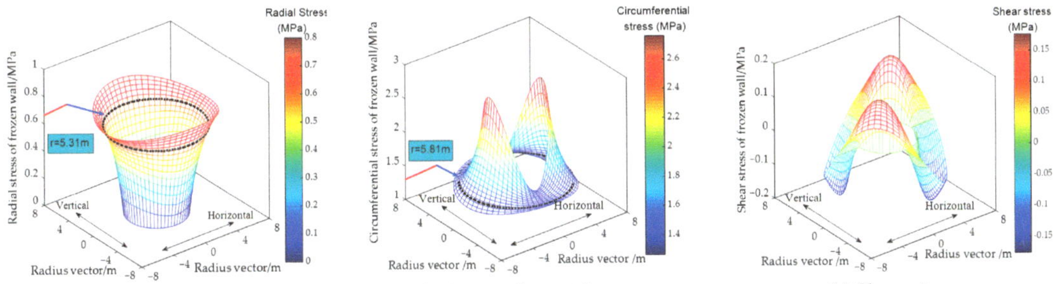

(a) Radial stress (b) Circumferential stress (c) Shear stress

Figure 10. Stress distribution for the frozen wall in the No. 7 Tunnel construction using the freezing method.

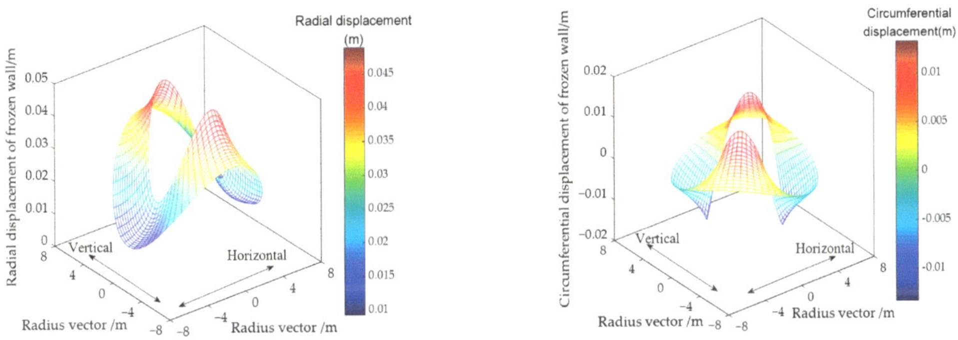

(a) Radial displacement (b) Circumferential displacement

Figure 11. Displacement distribution of the frozen wall.

As shown in Figure 10, because the frozen wall is subject to a nonuniform stress field, its radial stress increases at different gradients with the radius in each direction. In particular, when $\lambda < 1$, the radial stress in the horizontal direction increases faster than that in the vertical direction near the inner edge of the frozen wall; in addition, the magnitude of horizontal radial stress is larger than vertical radial stress. In contrast, near the outer edge, the radial stress in the vertical direction increases faster than that in the horizontal direction, and its value is also greater than that in the horizontal direction. However, for $r = 5.31m$, the radial stresses in all directions are equal to 0.64 MPa. Furthermore, the circumferential stress in the horizontal direction is significantly greater than that in the vertical direction. Moving along the vector radius from the inner edge to the outer edge, the horizontal circumferential stress decreases, while the vertical circumferential stress remains almost unchanged; in addition, the horizontal circumferential stress at the outer edge of the frozen wall is less than the vertical circumferential stress at the outer edge. As with radial stress distribution, the circumferential stresses in all directions are equal to 1.38 MPa at $r = 5.81m$. It should be noted that the shear stress is positive for angles in the range of 0–90° and 180–270°, whereas negative for angles in the range of 90–180° and 270–360°; the absolute value of shear stress is strictly symmetrical with respect to the horizontal, vertical, 45°, and 135° directions.

From Figure 11a, it can be seen that the radial displacement distribution of the elastic zone of the frozen wall shows that the vertical radial displacement of the frozen wall is greater than the horizontal radial displacement, and all of these displacements are positive values, i.e., indicating deformation to the excavation face; this can be attributed to the vertical outward load being is greater than the horizontal outward load. Furthermore, moving along a vector radius from the inner edge to outer edge, the radial displacement gradually decreases. From the circumferential displacement distribution shown in Figure 11b, we can see that the circumferential displacement is positive in the direction of 0–90° and 180–270°, negative in the direction of 90–180° and 270–360°, and the absolute value of circumferential displacement is strictly symmetrical with respect to the horizontal, vertical, 45°, and 135° directions.

5.2. Influence of External Load on the Plastic Zone Contour

Figure 12 shows the changes in the shape of the plastic zone contour of the frozen wall with an increase in p_0 for different λ.

If the nonuniformity of the two external loads is strong (as shown in Figure 12a with $\lambda = 0.3$), the tensile stress zone will clearly appear in the vertical direction at the inner edge of the frozen wall. In particular, when the value of p_0 is low, the frozen wall only appears as a crescent-shaped plastic zone within a certain range of the inner edge in the horizontal direction (e.g., contours of 2 MPa and 3 MPa in Figure 12a), and the radius of plastic zone in the direction of 0° and 180° is the largest. As p_0 increases, the plastic zone expands along the horizontal direction towards the outer edge of the freezing wall; in addition, it simultaneously expands along the inner edge of the freezing wall in the vertical direction. The contour of the plastic zone of the frozen wall gradually evolves from the crescent shape under low load to that of an ear-like shape (e.g., the contours of 5 MPa and 12 MPa in Figure 12a), and the plastic zone in the direction of 30–60° (the other three quadrants are similar in terms of the radius of the plastic zone) has the largest radius, which is the weakest position for plastic failure on the frozen wall. Furthermore, the horizontal inner edge of the freezing wall is in the plastic zone, while the outer edge is still in the elastic state; in contrast, for the vertical direction of the frozen wall, the inner edge lies in the tensile stress zone, whereas the outer edge lies in the elastic zone.

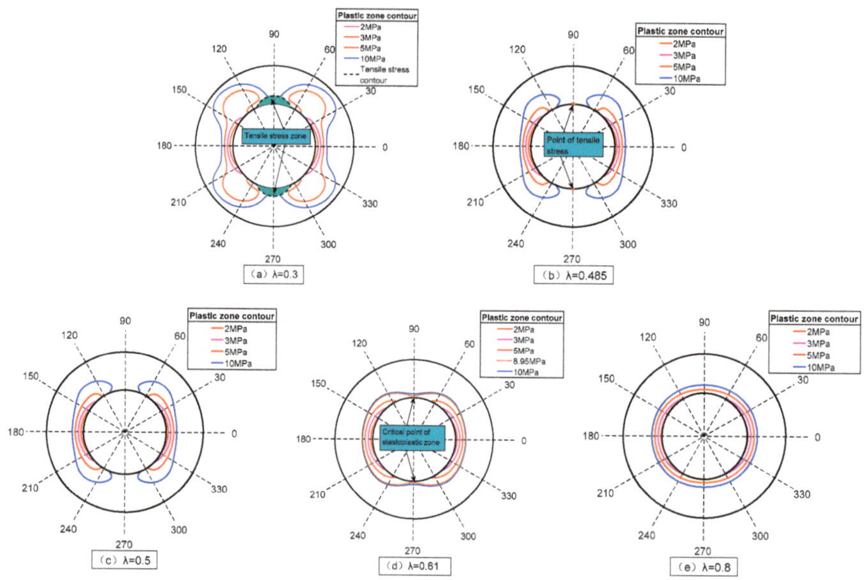

Figure 12. Relationship between the external load and plastic zone contour of the frozen wall.

If the nonuniformity of the two external loads is neither strong nor weak (as shown in Figure 12c with $\lambda = 0.5$), there is no tensile stress zone in the entire frozen wall. In particular, when the value of p_0 is low, the shape of the plastic zone of the frozen wall is the same as that of $\lambda = 0.3$. With an increase in p_0, the contour of the plastic zone of the frozen wall gradually evolves from a crescent shape to ear-like shape (e.g., contours of 5 MPa and 12 MPa in Figure 12c), and the plastic zone in the direction of 30–60° (the other three quadrants are similar in terms of radius of the plastic zone) has the largest radius, which is the weakest position of plastic failure of the frozen wall, but its evolution process is slower than the case of $\lambda = 0.3$. Furthermore, the vertical direction of the frozen wall is entirely in the elastic zone, whereas inner edge and outer edge of the horizontal frozen wall lie in the plastic and elastic zones, respectively.

If the nonuniformity of the two external loads is weak (as shown in Figure 12e with $\lambda = 0.8$), because the stress condition is similar to the stress state under uniform load ($\lambda = 1$), the contour of the plastic zone evolves from a crescent shape at the horizontal inner edge to an ellipse with an increase in the external load p_0 on the frozen wall (e.g., contour of 5 MPa in Figure 12e). Thus, in this case, no ear-shaped plastic zone is observed during the entire evolution process, and the radius of plastic zone in the direction of 0° and 180° is always the largest.

Based on our analysis, $\lambda = 0.485$ is the critical point for which the inner edge of the frozen wall just develops tensile stress under the abovementioned working conditions; this critical point for tensile stress is located on the vertical inner edge of the frozen wall. In contrast, when $\lambda<0.485$, the tensile stress zone is clearly observed on the inner edge of the vertical direction of the frozen wall, whereas, when $\lambda > 0.485$, there is no tensile stress zone on the frozen wall. Furthermore, $\lambda = 0.61$ is the critical point between the ear-shaped plastic zone and elliptical plastic zone of the frozen wall. Under the effect of high external load and when $\lambda < 0.610$, the shape of the plastic zone of the frozen wall is roughly that of an ear, whereas, for $\lambda > 0.610$, the shape of the plastic zone is elliptical. Thus, when $\lambda = 0.61$ and $p_0 = 8.95$ MPa, the inner edge of the frozen wall in the vertical direction is right at the critical point between the elastic and plastic states.

5.3. Influence of the Lateral Pressure Coefficient on the Plastic Zone Contour

The effect of the lateral pressure coefficient λ on the extent and shape of the plastic zone of frozen wall for different cases is shown in Figure 13.

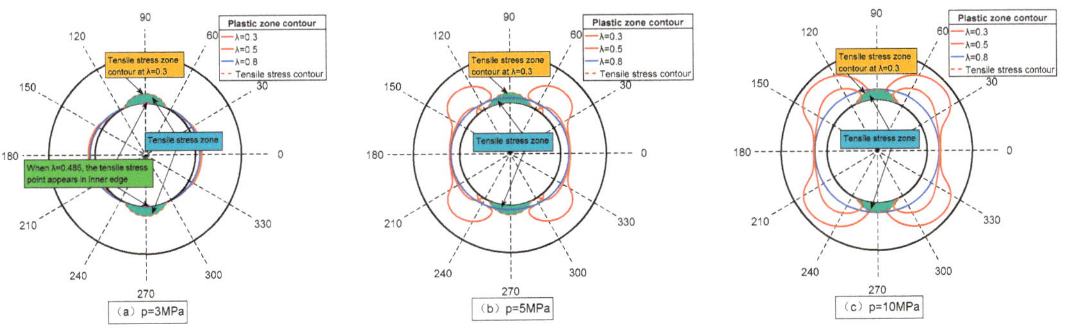

Figure 13. Relationship between the lateral pressure coefficient and plastic zone contour of the frozen wall.

In the case of $p_0 = 3$ MPa, when the value of λ is low, the contour of the plastic zone of the frozen wall is that of a short, thick crescent shape (e.g., contour of $\lambda = 0.3$ in Figure 13a), and the range of the horizontal plastic zone is the largest. In contrast, if λ is increased to 1, the crescent-shaped plastic zone gradually expands in the vertical direction along the inner edge of the frozen wall, and the horizontal plastic zone continuously reduces to form a long, thin, crescent-shaped contour, after which this contour becomes elliptical in shape ($\lambda = 0.8$). When $\lambda > 1$ and continues to increase, the horizontal plastic zone decreases gradually, while the vertical plastic zone increases and gradually separates in the horizontal direction, i.e., at 0°, finally forming two crescent shaped plastic zones in the vertical direction.

In the case of $p_0 = 5$ MPa or 10 MPa, for the same values of λ, the range of the tensile stress zone is the same as that of $p_0 =$ 3MPa. In particular, when λ = 0.3, the plastic zone contour of the frozen wall is ear-shaped, and the plastic zone in the direction of 30–60° (the other three quadrants are similar in terms of the radius of the plastic zone) has the largest radius, which is the weakest position for plastic failure on the frozen wall. When λ increases to 1, the plastic zone of the frozen wall gradually closes in the vertical direction along the inner edge, and the horizontal plastic zone continuously reduces until an ellipse is formed (λ = 0.8).

5.4. Discussion on Stability Evaluation of Horizontal Frozen Wall under Uneven Load

Based on the abovementioned analysis of the influence of the lateral pressure coefficient λ and vertical external load p_0 on the plastic zone contour and range of the tensile stress zone of the horizontal frozen wall, the following observations can be made:

(1) When λ<0.485, the tensile stress zone is clearly observable at the inner edge of the frozen wall in the vertical direction and its range is only related to λ; in particular, it increases with a decrease in λ, but is independent of the magnitude of p_0. Furthermore, p_0 only affects the magnitude of the tensile stress in the stress zone, but does not affect its range. Moreover, a reasonable vertical direction freezing reinforcement should be performed to increase the radius of the frozen wall and avoid tensile failure of the inner edge of the frozen wall.

(2) When $0.485 < \lambda < 0.610$, there is no tensile stress zone on the frozen wall. In such cases, the plastic failure of the horizontal inner edge under low load and inner edge in the direction 30–60° (the other three quadrants are similar in terms of the radius of the plastic zone) under high load should be avoided.

(3) When $\lambda > 0.610$, there is no tensile stress zone as in the previous case, and the plastic zone in the 0° and 180° directions has the largest radius, indicating the weakest positions of the plastic wall of the frozen wall. Thus, the compressive plastic failure of the inner edge of the frozen wall in the horizontal direction should be avoided.

For circular horizontal freezing engineering, in practice, λ should first be determined using an in situ stress test to judge whether tensile stress is present in the frozen wall, and the corresponding tensile failure of the tensile stress zone should be determined according to the tensile strength of the frozen soil obtained via the frozen soil mechanics test. By substituting the external load into the plastic contour equation of the frozen wall, the plastic zone can be determined; consequently, the maximum position of the plastic zone can be obtained. Finally, the freezing scheme can be improved based on the determined weak areas.

6. Conclusions

In this study, after deriving the analytical solution of the stress and displacement of the elastic–plastic zone of the frozen wall under nonuniform load, a specific engineering case is analyzed; in particular, the distribution law of stress and displacement of the frozen wall and corresponding influencing factors of the contour of its plastic zone are analyzed. Considering our observations, the following conclusions can be drawn:

The point where $\lambda = 0.485$ is the critical point at which the inner edge of the frozen wall just develops tensile stress. When $\lambda < 0.485$, the tensile stress zone is clearly observed at the inner edge of the frozen wall in the vertical direction, and its range is only related to λ; in particular, it increases with a decrease in λ. Furthermore, p_0 only affects the magnitude of the tensile stress in the region, but does not affect its range. When $\lambda > 0.485$, no tensile stress zone is observed.

When $\lambda < 0.61$, the frozen wall plastic zone contour evolves from a crescent shape to an ear shape as p_0 increases. In contrast, for $\lambda > 0.61$, the plastic zone contour evolves from a crescent shape in the horizontal direction to an elliptical shape with an increase in p_0, and there is no ear-shaped plastic zone observed during the entire evolution process.

When $p_0 = 3$ MPa, with an increase in λ tending towards 1, the contour of the plastic zone of the frozen wall evolves from a short, thick crescent shape to long, thin crescent shape until the intersection finally becomes an elliptical plastic zone; in contrast, when $p_0 = 5$ MPa or 10 MPa, for the same increase in λ, the contour of the plastic zone of the frozen wall gradually closes along the inner edge in the vertical direction evolving from an ear shape to ellipse.

Thus, this study provides a theoretical basis for the design and calculation of stress in horizontal frozen walls under nonuniform load conditions.

Author Contributions: Methodology, S.P.; Validation, Y.X.; Formal analysis, S.P. and G.C.; Resources, L.P.; Writing—original draft, S.P. All authors have read and agreed to the published version of the manuscript.

Funding: The work presented in this paper was financially supported by the National Natural Science Foundation of China (No. 52004003, 51878005, 51874005) and the Natural Science Foundation of Anhui (Grant No. 2208085QE142); The Open Project Program Foundation of Engineering Research Center of underground mine construction, Ministry of Education (No. JYBGCZX2021101, JYBGCZX2020102); Anhui Jianzhu University Teaching Research Project "Research and Application of Modular Teaching Mode for the Course of" Foundation Treatment "from the Perspective of" New Engineering "(2022jy01)".

Institutional Review Board Statement: Not applicable.

Informed Consent Statement: Not applicable.

Data Availability Statement: The datasets generated and analyzed during the current study are available from the corresponding author upon reasonable request.

Conflicts of Interest: The authors declare no conflict of interest.

References

1. Andersland, O.B.; Ladanyi, B. *An Introduction to Frozen Ground Engineering*; Springer Science & Business Media: New York, NY, USA, 2013.
2. Lin, J.; Cheng, H.; Cai, H.B.; Tang, B.; Cao, G.Y. Effect of Seepage Velocity on Formation of Shaft Frozen Wall in Loose Aquifer. *Adv. Mater. Sci. Eng.* **2018**, *2018*, 2307157. [CrossRef]
3. Li, K.Q.; Yin, Z.Y.; Liu, Y. Influences of spatial variability of hydrothermal properties on the freezing process in artificial ground freezing technique. *Comput. Geotech.* **2023**, *159*, 105448. [CrossRef]
4. Liu, Y.; Li, K.Q.; Li, D.Q.; Tang, X.S.; Gu, S.X. Coupled thermal–hydraulic modeling of artificial ground freezing with uncertainties in pipe inclination and thermal conductivity. *Acta Geotech.* **2022**, *17*, 257–274. [CrossRef]
5. Guan, H.D.; Zhou, X.M. Comparing study on elasto-plastic analysis of frozen wall based on interaction between frozen wall and surrounding rock. *Rock Soil Mech.* **2017**, *38*, 649–655.
6. Yang, R.S.; Wang, Q.X. Elastic analysis and design of circular horizontal frozen wall based on interaction between frozen wall and surrounding rock. *J. China Coal Soc.* **2016**, *41*, 1069–1077.
7. Cui, G.X. Mechanical analysis on cylindrical frozen soil wall and mine shaft lining in deep overburden. *Coal Sci. Technol.* **2008**, *36*, 17–21.
8. Yang, W.H.; Yang, Z.J.; Han, T.; Zhang, C.; Bo, D.-L. Elastic design theory of frozen soil wall based on interaction between frozen soil wall and surrounding rock. *Chin. J. Geotech. Eng.* **2012**, *34*, 516–520.
9. Yang, W.H.; Yang, Z.J.; Bo, D.L. Elastic-plastic design theory of frozen soil wall based on interaction between frozen soil wall and surrounding rock. *Chin. J. Geotech. Eng.* **2013**, *35*, 175–180.
10. Yang, W.H.; Du, Z.B.; Yang, Z.J.; Bo, D.-L. Plastic design theory of frozen soil wall based on interaction between frozen soil wall and surrounding rock. *Chin. J. Geotech. Eng.* **2013**, *35*, 1857–1862.
11. Li, K.Q.; Li, D.Q.; Liu, Y. Meso-scale investigations on the effective thermal conductivity of multi-phase materials using the finite element method. *Int. J. Heat Mass Transf.* **2020**, *151*, 119383. [CrossRef]
12. Amadei, B.; Stephansson, O. *Methods of In Situ, Stress Measurement, Rock Stress and Its Measurement*; Springer: Dordrecht, The Netherlands, 1997.
13. Ni, X.H. *In-Situ Stress Study and Applications*; China Coal Industry Publishing House: Beijing, China, 2007.
14. Jaeger, J.C.; Cook, N.G.W.; Zimmerman, R.W. *Fundamentals of Rock Mechanics*, 4th ed.; Blackwell Publishing Ltd.: Oxford, UK, 2007.

15. Hudson, J.A.; Harrison, J.P. *Engineering Rock Mechanics. An Introduction to the Principles*, 1st ed.; Elsevier Science, Ltd.: Oxford UK, 2000.
16. Sun, J.S.; Lu, W.B. Analytical elasto-plastic solutions to supporting rock masses of circular tunnels under asymmetric load. *Rock Soil Mech.* **2007**, *28*, 327–332.
17. Chen, L.W.; Peng, J.B.; Fan, W.; Sun, P. Analysis of surrounding rock mass plastic zone of round tunnel under non-uniform stress field based on the unified strength theory. *J. China Coal Soc.* **2007**, *32*, 20–23.
18. Pan, Y.; Zhao, G.M.; Meng, X.R. Elasto-plastic analysis on surrounding rock mass under non-uniform stress field. *J. China Coal Soc.* **2011**, *36*, 53–57.
19. Gao, Z.N.; Meng, X.R.; Wang, G.D. Analysis of the plastic zone of surrounding rock under seepage and a non-axisymmetric load *Mod. Tunn. Technol.* **2014**, *51*, 70–75.
20. Kirsch, G. Die Theorie der Elastizität und die Bedürfnisse der Festigkeitslehre. *Z. Ver. Dtsch. Ing.* **1898**, *42*, 797–807.
21. Pariseau, W.G. *Design Analysis in Rock Mechanics*; Taylor & Francis: Abingdon, UK, 2007.
22. Wang, Y.; Yang, W.H. Elastic design theory of circular horizontal frozen wall. *J. China Coal Soc.* **2015**, *40*, 2049–2056.
23. Simanjuntak, T.; Marence, M.; Mynett, A.E.; Schleiss, A.J. Pressure tunnels in non-uniform in situ stress conditions. *Tunn. Undergr. Space Technol.* **2014**, *42*, 227–236. [CrossRef]
24. Cui, G.; Yang, W.; Lü, H. *Frozen Wall and Shaft Lining in Deep Alluvium*; China University of Mining and Technology Press: Xuzhou, China, 1998.
25. Chen, Z.; Zhang, B. Calculation method for inclined shaftfrozen wall. *Mine Constr. Technol.* **2013**, *34*, 39–42.
26. Yanan, C. A Numerical Analysis of the Ice-Wall Thickness and Freezing Period for Guangzhou Metro Tunnel. Master's Thesis, Beijing Jiaotong University, Beijing, China, 2008.
27. Zhou, X.; Wang, M.; Tao, L.; Yang, S. Model test and prototype observation on artificial ground freezing and tunneling of Beijing subway. *Chin. J. Geotech. Eng.* **2003**, *25*, 676–679.
28. Geng, P.; Yan, Q.; He, C.; Bo, W. Numerical simulation of underground construction by horizontal ground freezing method. *Eng. Mech.* **2010**, *27*, 122–127.
29. Wu, Y.; Li, D.; Yang, M. Numerical simulation of tunnel with freezing method construction. *Chin. J. Rock Mech. Eng.* **2005**, *24* (Suppl. S2), 5851–5856.
30. Zhou, X.; Zheng, Y. Research on displacement field of tunnel freezing in high water clay based on simulation model test. *J. Tongji Univ. (Nat. Sci.)* **2000**, *28*, 472–476.
31. Zhou, T.J.; Zhou, G.Q. Calculation method of frozen soil wall thickness for inclined shaft in consideration of the shaft inclination. *J. China Univ. Min. Technol.* **2016**, *45*, 514–520.
32. Hu, X.D. A mechanical model of interaction of frozen soil wall and surrounding earth mass in unload state. *J. China Coal Soc.* **2001**, *26*, 507–511.
33. Hu, X.D. Determination of load on frozen soil wall in unload state. *J. Tongji Univ.* **2002**, *30*, 6–10.
34. Zhang, W.; Wang, B.S.; Wang, Y. Elastic Analysis of Nonhomogeneous Frozen Wall under Nonaxisymmetric Ground Stress Field and in State of Unloading. *Adv. Mater. Sci. Eng.* **2018**, *2018*, 2391431. [CrossRef]
35. Muskhelishvili, N.I. *Some Basic Problems of the Mathematical Theory of Elasticity*; P. Noordhoff Ltd.: Groningen, The Netherlands, 1953.
36. Lu, A.Z.; Zhang, N.; Xu, Y.Q.; Cui, P. Stress-displacement solution for a lined vertical borehole due to non-axisymmetric in situ stresses. *Int. J. Rock Mech. Min. Sci.* **2013**, *57*, 64–74. [CrossRef]
37. Carranza-Torres, C.; Rysdahl, B.; Kasim, M. On the elastic analysis of a circular lined tunnel considering the delayed installation of the support. *Int. J. Rock Mech. Min. Sci.* **2013**, *61*, 57–85. [CrossRef]
38. Yu, X.F.; Zheng, Y.R.; Liu, H.H.; Fang, Z.C. *Stability Analysis of Surrounding Rock of Underground Engineering*; China Coal Industry Publishing House: Beijing, China, 1983.
39. Cheng, H.; Lin, J.; Wang, B.; Rong, C.X. Mathematical model and test verification of seepage freezing in saturated sand layer. *Sci. Technol. Eng.* **2018**, *18*, 38–44.
40. Wang, B.; Rong, C.; Cheng, H.; Cai, H.; Dong, Y.; Yang, F. Temporal and spatial evolution of temperature field of single freezing pipe in large velocity infiltration configuration. *Cold Reg. Sci. Technol.* **2020**, *175*, 103080. [CrossRef]

applied sciences

MDPI

Article

Experimental and Theoretical Study of the Influence of Saline Soils on Frozen Wall Formation

Sergey Bublik [1,*], Mikhail Semin [1,*], Lev Levin [1], Andrey Brovka [2] and Ivan Dedyulya [2]

[1] Mining Institute of the Ural Branch of the Russian Academy of Sciences, 614007 Perm, Russia; aerolog_lev@mail.ru

[2] Institute of Nature Management, National Academy of Sciences of Belarus, 220072 Minsk, Belarus; andrew_brovka@mail.ru (A.B.); dedyulyaivan@yandex.by (I.D.)

* Correspondence: serega-bublik@mail.ru (S.B.); seminma@inbox.ru (M.S.)

Abstract: This paper examines the impact of salinity on the thermophysical properties of soils during artificial freezing. It focuses on analyzing heat and mass transfer in saline soils for constructing a frozen wall around a mineshaft at a potash salt deposit. The presence of salts in the groundwater near the contact point with water-protective strata is common in these deposits. Experimental studies were conducted on clay, chalk, and sand to understand the effect of salinity on the freezing temperature, unfrozen water content, specific heat capacity, and thermal conductivity of wet soil. These findings were used to simulate heat and mass transfer in saline soils using a one-dimensional model. The technique of circumferential averaging was introduced to account for the thermal impact of freeze pipes. The results indicate that higher soil salinity leads to a faster decrease in soil temperature under freezing conditions, although this dependence is weak for clay. This study also revealed that an increase in initial salinity results in a reduction in the thickness of the frozen wall. It was found that, for chalk and sand, there exists a range of initial salinity during which the frozen wall's thickness is almost independent of the initial salinity.

Keywords: artificial ground freezing; salinity of soil; mathematical modeling; thermophysical properties of soils

1. Introduction

The construction of mineshafts and subway tunnels under the condition of flooded soils is implemented using special methods. One of the most common of those methods is artificial ground freezing (AGF) [1,2]. The purpose of AGF is to build a protective structure, consisting of frozen soils, around the future mineshaft. This structure is most commonly referred to as a "frozen wall" (FW) [3]. The functions of an FW are to prevent the flooding of the shaft with groundwater and to strengthen the loose walls of the shaft before the construction of a permanent concrete lining [3]. The freezing of the soil is implemented with the help of freeze pipes mounted along the perimeter of the designed mineshaft (see Figure 1a).

The study and control of the formation of frozen walls are implemented using experimental and theoretical methods. Experimental methods involve monitoring the temperature distribution over the depth of several control boreholes drilled near the contour of the freeze pipes (see Figure 1b) [4]. The measured temperature distributions are used to adjust the parameters of the mathematical models of heat and mass transfer in the frozen media [5]. Further, the adjusted mathematical models are used to estimate the temperature distribution in the entire volume of the frozen soil to determine the actual thickness and continuity of the FW, with those being the main technical parameters of an FW.

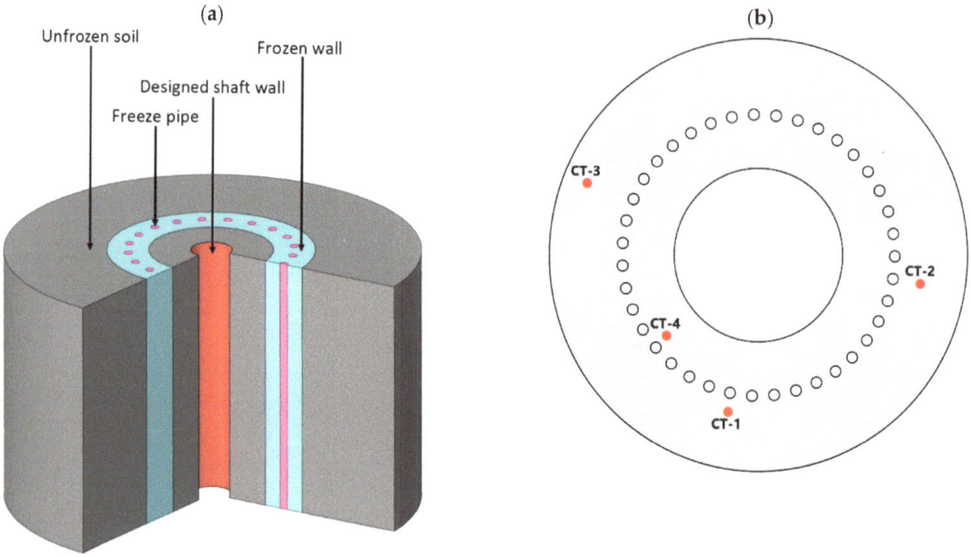

Figure 1. Scheme of the frozen wall (**a**) and an example of the location of control boreholes in a transverse horizontal section (**b**).

However, the mentioned methods for monitoring and predicting the state of an FW often do not consider the factor of soil salinity, which often occurs during the construction of mineshafts in potash salt deposits. The salinity of the soil leads to a change in its thermophysical and mechanical properties. In addition, an increase in the amount of dissolved salts in the porewater leads to a decrease in its freezing temperature [6,7], which can negatively affect the waterproofing and strength properties of an FW.

There are many studies devoted to the effect of salinity on the freezing of porewater. Tsytovich [6] presents extensive theoretical and experimental information about the general patterns of freezing of saline soils, as well as practical aspects of taking into account salinity in the construction of surface structures. Banin and Anderson [8] and Yong et al. [9] obtained experimental dependences of the thermophysical properties of several frozen soils on the concentration of salts in them. Lucas et al. [10] focused on the mathematical modeling of heat and mass transfer in frozen saline soils and the comparison of the obtained results with the experimental base. In it, Lucas et al. obtained equations for simultaneous heat and mass flows at one of the interfaces for a wide range of temperatures and concentrations. Liu et al. [11] investigated the electrical properties of frozen salted clay. Aleksyutina and Motenko [12] evaluated the effects of soil salinity and organic matter content on the thermal properties of frozen and thawed soils. Qiu et al. [13] proposed a thermal conductivity model of saline soil considering the heat dissipation due to heat radiation. Zhang et al. [14] proposed a coupled hydro-thermal-salt-mechanical model with phase change to simulate the one-side freezing process of saturated sulfate saline soil.

However, most studies are related to the natural freezing of the soil in permafrost regions, and they cannot be fully applied to the conditions involved with the artificial freezing of soils, because of the features of artificial freezing: higher temperature differences, faster changes in the amount of unfrozen groundwater over time, and higher values of rock and hydrostatic pressures due to greater depths.

The effects of salinity on the artificial freezing of soils are described in [7,15–17]. Lyu et al. [15] investigated the change in the freezing point with the increase in concentration of NaCl in the pore brine. A new theoretical equation was developed that relates thermal conductivity to water content and temperature. The works [16,17] present the results of laboratory studies on the effects of salinity on freezing temperatures in sand samples. In addition, the results of an experiment on the effects of water filtration and the artificial freezing of soil samples using two freeze pipes with different levels of salt content in the porewater are presented. However, these studies consider only the effect of salinity on the freezing temperature of porewater, while an analysis of the effect of salinity on the changes in the entire complex of basic thermophysical properties of soils was not implemented. The work [7] contains the results of laboratory studies on the effects of the salinity and initial water content on the freezing temperature of porewater, specific heat capacity, thermal conductivity, and unfrozen water content. However, as in the case of works [16,17], this study addresses only one type of soil—argillite-like clay—rather than sand. At the same time, the importance of studying a set of different soil types should be noted because the construction of underground structures often involves penetrating various types of soil layers. Dissolved salt can have different effects on the freezing processes of different types of soils.

Recently, there has been a tendency toward increased research using the numerical simulation of heat and mass transfer during AGF. This may be due to the growing need to design the construction of FWs in difficult hydrogeological conditions. For example, works [18,19] present the results of a study on the numerical modeling of AGF in saline soils, regarding the construction of underground tunnels. Based on the literature data on the experimentally measured thermal conductivity of frozen soils, Bi et al. [20] proposed a new neural network model for determining the dependence of thermal conductivity on temperature and other soil properties. Wan et al. [21] presents an analytical model for evaluating the thermal conductivity of sodium sulfate soils based on a generalized thermal conductivity theory of geomaterials.

At the same time, only a few works can be found that consider the presence of soils and their transfer via porewater during soil freezing. For example, in [22–24], complex, coupled thermo-hydro-mechanical models were developed with consideration given to changes in the soil salinity. However, the models presented in these works contain many equations, relationships, and unknown parameters. As a result, they are difficult to implement, and therefore, it is problematic to use them for the practical tasks of designing and controlling FW formation. Most often, when monitoring the state of an FW, carrying out a quick, near-real-time calculation of the temperature field in frozen soil is required. These calculations also usually require preliminary parametrization of the model according to experimental data from control boreholes. The process of parameterizing and adjusting the model also requires a series of trial calculations. For these reasons, in our work, to analyze the state of the FW, we tried to choose models that would be easy to implement, would be less demanding on computational resources, and would have fewer calibrated parameters. However, at the same time, these models should include the most significant physical processes during AGF and the formation of FWs.

Thus, based on all of the above-described considerations, it can be concluded that, at present, it is possible to find a lot of experimental studies on the effect of salinity on soil freezing in relation to permafrost problems, but only a small portion of the studies are directly related to AGF, and even fewer are directly related to the analysis of FW formation. In addition, the issue of the numerical simulation of heat and mass transfer processes in solving applied problems of monitoring and predicting FW formation in saline soils remains poorly covered. This is all further complicated by the fact that an extensive base of laboratory studies on the effect of salinity on the thermophysical properties of porewater is available only for sandy soils.

In view of this, this work aims to study the patterns of thermophysical processes in saline soils of various types in the AGF process. In particular, we want to find out how the soil salinity affects the FW growth rate and its continuity in relation to the construction of mine shafts in potash and rock salt deposits.

To achieve this goal, we solve two problems: the determination of the characteristic thermophysical parameters of porewater in various soils, depending on the concentration of the NaCl salt solution in the water; and the development of a coupled mathematical model of heat and mass transfer during AGF for the case of construction of a mineshaft at a potash salt deposit. The present work is a continuation of the work described in [7,25].

2. Experimental Study on the Effect of Salinity on Soil Thermophysical Properties

2.1. Sample Preparation

Here, we describe the experimental studies that were implemented to identify the dependence of the thermophysical properties of soils on negative temperatures and levels of salt concentration in porewater. They are important for the subsequent analysis of the characteristic features of the effects of salinity on heat transfer in soils during freezing. Porewater containing a solution of NaCl salt was considered. Argillite-like clay, chalk, and sand were studied. The choices of the soils and salt types were based on the fact that they are the most common for the above-salt strata of potash salt deposits.

This study was conducted according to a well-known technique [26].

To elucidate the patterns of salt influence on the phase composition of water and the thermophysical characteristics of soils, the samples' initial compositions needed to be as identical as possible. Consequently, investigations were carried out on samples with a compromised structure (pastes) at a water content and density that matched those of samples with a natural composition.

To achieve this, the chosen material was initially crushed using a laboratory mill and then sieved through a mesh with a diameter of 5 mm. Subsequently, it was meticulously mixed, placed within a sealed container, and left to age for a full calendar week. Preliminary experiments indicated that this duration was adequate for ensuring a consistent distribution of moisture throughout the sample's volume. Only after this period was the material's moisture content determined, which was regarded as its initial state.

Each type of soil was represented by nine samples. These samples were divided into three groups of three samples each. Each group of samples was associated with a certain soil water content. Additionally, the samples in each group were divided as follows: (1) samples with distilled water and the standard freezing temperature of the porewater; (2) samples with a NaCl content for which the freezing point of the porewater was 2 °C lower than the standard freezing temperature; (3) samples with a NaCl content for which the freezing temperature of the porewater was 4 °C lower than the standard freezing temperature. It is worth noting that, in the case of clay, studies were conducted for three values of initial water content. Chalk and sand studies were performed with only a single value for initial water content.

2.2. Methods of Measurement

All experiments were implemented in the Laboratory of Physicochemical Mechanics of Natural Dispersed Systems of the Institute of Nature Management of the National Academy of Sciences of Belarus. The unfrozen water content, the freezing temperature of the porewater, the specific heat capacity, and the thermal conductivity of the soil were the studied thermophysical parameters during the freezing process. A calorimeter was used to determine the unfrozen water content, the freezing temperature of the porewater, and the specific heat capacity. The principle underlying these measurements is the creation of adiabatic conditions, or conditions of controlled heat exchange, around a calorimetric cup filled with the sample under study. This allows for the determination of the thermophysical parameters under study (water content and specific heat capacity) at positive and negative temperatures, as well as the investigation of the changes in unfrozen water content during freezing and thawing cycles. The accuracy of the device is ±0.01 °C when measuring temperature, and the relative error in determining the heat of the phase transition is ±1%.

Experimentally measured masses of water and ice in the sample were used to calculate the unfrozen water content according to the following equation:

$$w = \frac{G_w - G_i}{G_S}, \quad (1)$$

where w is the unfrozen water content, G_w is the mass of water in the sample (kg), G_i is the mass of ice in the sample (kg), and G_S is the mass of the dry skeleton of the soil (kg).

The specific heat capacity was calculated for the dry skeleton of the soil using the following equation:

$$c_s = (1 + w_0)c_{meas} - wc_w - (w_0 - w)c_i - L\frac{dw}{dT}, \quad (2)$$

$$c_{meas} = \frac{\Delta Q_{meas} - C_{cal}\Delta T}{G_{\Sigma}\Delta T}, \quad (3)$$

where c_s is the specific heat capacity of the dry skeleton (J/(kg·°C)); c_{meas} is the measured effective specific heat capacity of the soil sample (taking into account the phase transition of the moisture) (J/(kg·°C)); c_w and c_i are the specific heat capacities of water and ice (J/(kg·°C)), respectively; C_{cal} is the heat capacity of an empty calorimeter (J/°C); w_0 is the initial soil moisture (kg/kg); L is the specific heat of the phase transition of the moisture (J/kg); ΔQ_{meas} is the amount of heat required to supply the calorimeter so that its temperature changes by ΔT (J); and G_{Σ} is the total mass of wet material (skeleton + water + ice) (kg).

The thermal conductivity of the soils was determined using a unique device based on the method of the stationary thermal state of the soil samples. The device features the use of original heat flow sensors mounted in metal heat-exchanger housings. This makes it possible to average the temperature field in the measurement planes and to protect the heat flow and temperature sensors from mechanical, physical, and chemical damage. The relative error in determining the thermal conductivity is 6%.

The calculation of the thermal conductivity coefficient was conducted in the stationary thermal state of the test sample according to the readings of the signals of the heat flow sensors with two heat exchangers. The first one is located on top of the sample, while the second one is located below it. Before the direct calculation of the thermal conductivity of the samples, the calibration parameters were determined based on the reference sample according to the following equations:

$$f = \frac{Q_2}{Q_1}, \quad (4)$$

$$K = \frac{\lambda_{ref} \cdot \Delta T}{(f \cdot q_1 + q_1) \cdot h}, \quad (5)$$

where f is the ratio of the heat fluxes of the upper and lower heat exchangers; Q_1 and Q_2 are the heat flows of the lower and upper heat exchangers in steady thermal mode, respectively (W/m^2); λ_{ref} is the thermal conductivity of the reference sample (W/(m·°C)); ΔT is the temperature difference of the heat exchangers (°C); and h is the sample height (m).

As a result, the thermal conductivity of the soil samples is determined using the following equation:

$$\lambda = K\frac{f \cdot q_1 + q_1}{\Delta T}h, \quad (6)$$

2.3. Results and Processing of Experimental Measurements

Table 1 shows the measured density of the wet soils and the specific heat capacity of the dry skeleton. Table 2 shows the dependence of the freezing temperature of the porewater with different levels of NaCl content in units of g/100 g of moisture. (Hereafter, "NaCl content" is replaced with "salinity".) In Tables 1 and 2, clay parameters are presented only for one value of water content: 0.26 kg/kg. These parameters appear to be the most interesting from the point of view of comparative analysis with the thermophysical properties of chalk, which has nearly the same water content: 0.25 kg/kg. The results of the experimental measurements of clay parameters with other water content values can be found in [7].

Table 1. Experimental measurements of soil density and specific heat capacity of dry skeleton.

Soil	Soil Density (kg/m^3)	Specific Heat Capacity of Dry Skeleton (J/(kg·°C))
Clay	2070	757
Chalk	1985	798
Sand	1730	692

Table 2. Experimental measurements of the dependence of the freezing temperature of porewater on the NaCl content.

Soil	NaCl (g/100 g)	Moisture (kg/kg)	Freezing Temperature (°C)
Clay	0.00	0.26	−0.23
	3.46		−2.55
	6.90		−5.00
Chalk	0.00	0.25	−0.30
	3.46		−2.31
	6.90		−4.51
Sand	0.00	0.11	−0.09
	5.20		−3.59

The obtained experimental dependencies for the freezing temperatures of the porewater were approximated by a linear dependence (see Figure 2). It should be noted that, in general, the freezing temperature of water may depend nonlinearly on salinity [27]. However, subsequently, the numerical simulation of the heat and mass transfer in the saline soils was implemented with salinity values in the range of 0.01 g/100 g to 13.8 g/100 g. For this range, the error of the linear approximation relative to the nonlinear one, according to the data from [27], was about 0.2 °C, which can be considered insignificant.

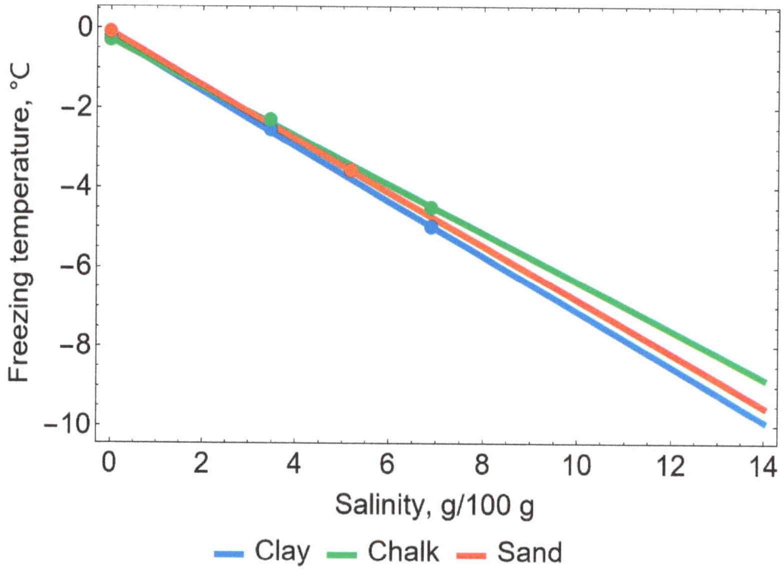

Figure 2. Dependence of the freezing temperature of porewater on salinity. (Experimental measurements are indicated by dots.)

Table 3 shows the results of the experimental measurements of the dependence of the unfrozen water content on temperature and salinity, and Table 4 shows the results for the measurements of thermal conductivity.

Table 3. Experimental measurements of the dependence of the unfrozen water content on temperature and salinity.

Temperature (°C)	Clay			Chalk			Sand	
	NaCl (g/100 g)							
	0.00	3.46	6.90	0.00	3.46	6.90	0.00	5.20
	Unfrozen Water Content (kg/kg)							
−0.5	0.981	-	-	0.488	-	-	0.168	-
−1.0	0.835	-	-	0.216	-	-	0.080	-
−2.0	0.650	-	-	0.132	-	-	0.053	-
−3.0	0.523	0.881	-	0.108	0.760	-	-	-
−4.0	0.446	0.758	-	-	-	-	-	0.779
−5.0	0.388	0.669	0.965	0.088	0.472	0.852	0.044	0.628
−6.0	0.346	0.596	0.827	0.084	0.412	0.728	-	0.531
−7.0	0.319	0.535	0.758	-	-	-	-	0.460
−8.0	0.296	0.485	0.696	-	-	-	-	0.407
−10.0	0.262	0.412	0.592	0.068	0.256	0.464	-	0.327
−15.0	0.223	0.323	0.446	0.052	0.176	0.332	0.035	0.239
−20.0	0.204	0.273	0.365	0.040	0.136	0.268	-	0.195
−25.0	0.185	0.242	0.327	-	-	-	0.035	0.177

Table 4. Experimental measurements of the dependence of the thermal conductivity on temperature and salinity.

Temperature (°C)	Clay			Chalk			Sand	
	NaCl (g/100 g)							
	0.00	3.46	6.90	0.00	3.46	6.90	0.00	5.20
	Thermal Conductivity (W/(m·°C))							
20	-	-	-	-	-	-	1.84	1.87
10	1.10	1.104	1.12	1.50	1.49	1.52	-	-
0	-	-	-	1.51	1.51	1.53	-	-
−1	1.44	1.104	1.12	1.86	1.51	1.53	2.18	1.87
−2	1.54	1.10	1.12	-	-	-	-	-
−3	1.61	1.13	1.12	2.08	1.59	1.56	2.38	1.87
−4	1.66	1.40	1.12	-	-	-	-	-
−5	1.69	1.47	1.12	2.15	1.83	1.77	-	-
−6	1.72	1.53	1.13	-	-	-	2.51	2.17
−7	1.73	1.57	1.38	-	-	-	-	-
−8	1.75	1.59	1.43	-	-	-	-	-
−10	1.77	1.65	1.51	2.19	2.00	1.92	2.57	2.30
−15	1.81	1.73	1.62	2.20	2.07	1.99	2.59	2.37
−20	1.85	1.78	1.69	2.21	2.10	2.01	2.60	2.42
−25	1.87	1.81	1.73	-	-	-	-	-

The functional dependence of the unfrozen water on temperature and salinity is given by a well-known empirical dependence [28,29]:

$$\gamma = \left[1 + \left(\frac{T_{liq} - T}{\omega}\right)^{\frac{1}{1-m}}\right]^{-m} \quad (7)$$

where γ is the unfrozen water content (kg/kg), T_{liq} is the freezing temperature of porewater (°C), T is the temperature (°C), ω is the characteristic cooling temperature associated with the most common pore radius (°C), and m is the index indicating the distribution of the pore radius relative to the average radius.

The unknown parameters of Function (7), ω and m, were determined via experimental measurements using the least squares method. The obtained values for various types of soils and salinity are presented in Table 5. Later, during the process of numerical simulation, the dependence of the value ω on salinity, according to Table 5, was approximated by a linear function, and the value of m was given as a constant due to its small change in the considered range of salinities. Its value is equal to the mean of m values for three soil samples with different salinities.

Table 5. Parameters of the approximate dependence of the unfrozen water content.

Soil	NaCl (g/100 g)							
	0.00	3.46	5.20	6.90	0.00	3.46	5.20	6.90
	ω (°C)				m			
Clay	1.00	1.23	-	1.39	0.36	0.32	-	0.29
Chalk	0.06	0.72	-	0.85	0.37	0.36	-	0.30
Sand	0.01	-	0.51	-	0.64	-	0.70	-

The functional dependence of the thermal conductivity was set by a known dependence on the unfrozen water content [30]:

$$\lambda = \lambda_s^{1-n}\lambda_w^{n\gamma}\lambda_i^{n(1-\gamma)} \tag{8}$$

where λ is the thermal conductivity of the soil (W/(m·°C)), n is the soil porosity, the "s" index corresponds to the dry skeleton, the "w" index corresponds to the water, and the "i" index corresponds to ice.

The porosity of soils can be calculated using the following equation:

$$n = w_0\frac{\rho}{\rho_w} \tag{9}$$

However, comparative analysis showed that it is not possible to achieve a good agreement with the experimental data when calculating the thermal conductivity according to Equation (8). It was assumed that the reasons for this are unaccounted for by physical factors. For example, these may include a change in the thermal conductivity of the solution due to a change in its salt concentration, the migration of moisture, or the limited applicability of Equation (8) for the types of soils under consideration. Another possible factor is the error (6%) in the experimental measurement of thermal conductivity, as noted earlier. In this regard, to calculate the coefficient of thermal conductivity, a correction factor, ξ, is introduced into Equation (8):

$$\lambda = \lambda_s^{1-\xi n}\lambda_w^{\xi n\gamma}\lambda_i^{\xi n(1-\gamma)} \tag{10}$$

Table 6 shows the calculated porosity and thermal conductivity of the dry skeleton for the studied soils (index "s" in Equation (10)) from the minimum mismatch with the experimental data on thermal conductivity. The mismatch was calculated as the root-mean-square error. It was assumed that the thermal conductivities of water and ice are 0.56/2.20 W/(m·°C), respectively.

Table 6. Porosity, specific heat capacity, and thermal conductivity of solid soil particles.

Soil	Porosity	Thermal Conductivity (W/(m·°C))
Clay	0.54	1.95
Chalk	0.50	2.27
Sand	0.20	2.75

Figures 3–5 show the dependencies of Equations (7) and (8) for each type of soil, with superimposed experimental data in the form of markers.

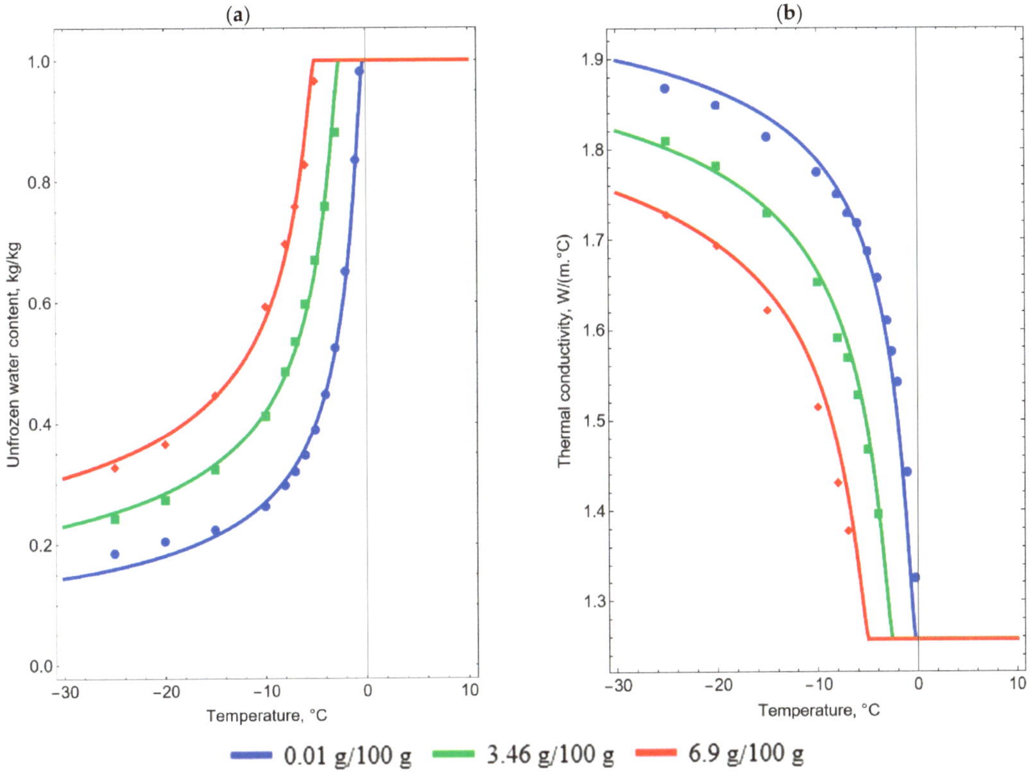

Figure 3. Dependence of the unfrozen water content (**a**) and the thermal conductivity (**b**) on temperature and salinity for clay.

Based on the measurements from Table 1 and measurements processed according to the unfrozen water content and thermal conductivity, the specific heat capacity of the soils was set according to a known dependence [30]:

$$\rho c_p = (\rho c_p)_s(1-n) + n\left[(\rho c_p)_l\gamma + (\rho c_p)_i(1-\gamma)\right], \tag{11}$$

where ρ is the soil density (kg/m^3), and c_p is the specific heat capacity of the soil at a constant pressure (J/(kg·°C)). Figure 6 shows the obtained dependencies. It was assumed that the density of water/ice is 1000/912 kg/m^3 and the specific heat capacity of water/ice is 4200/2100 J/(kg·°C).

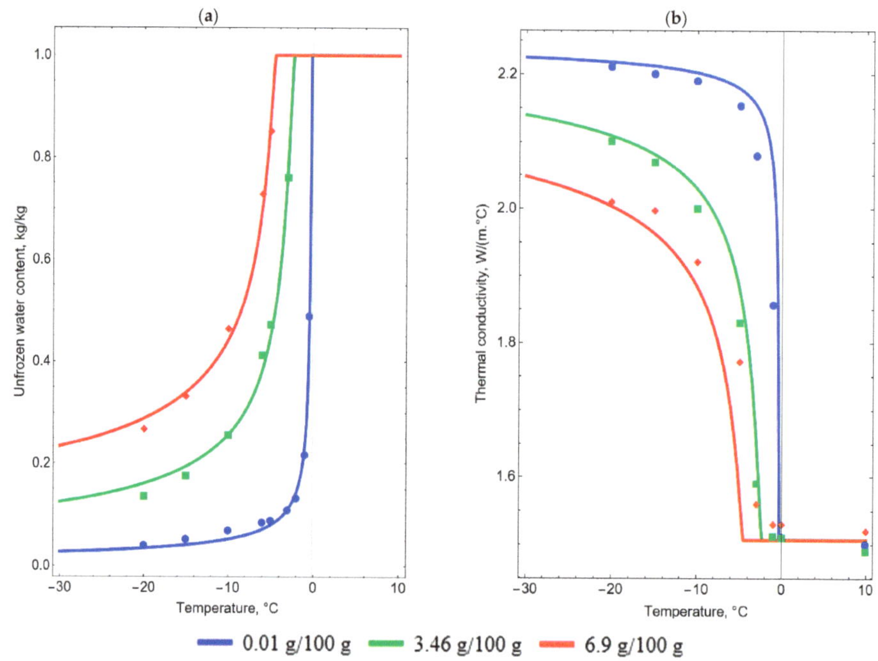

Figure 4. Dependence of the unfrozen water content (**a**) and the thermal conductivity (**b**) on temperature and salinity for chalk.

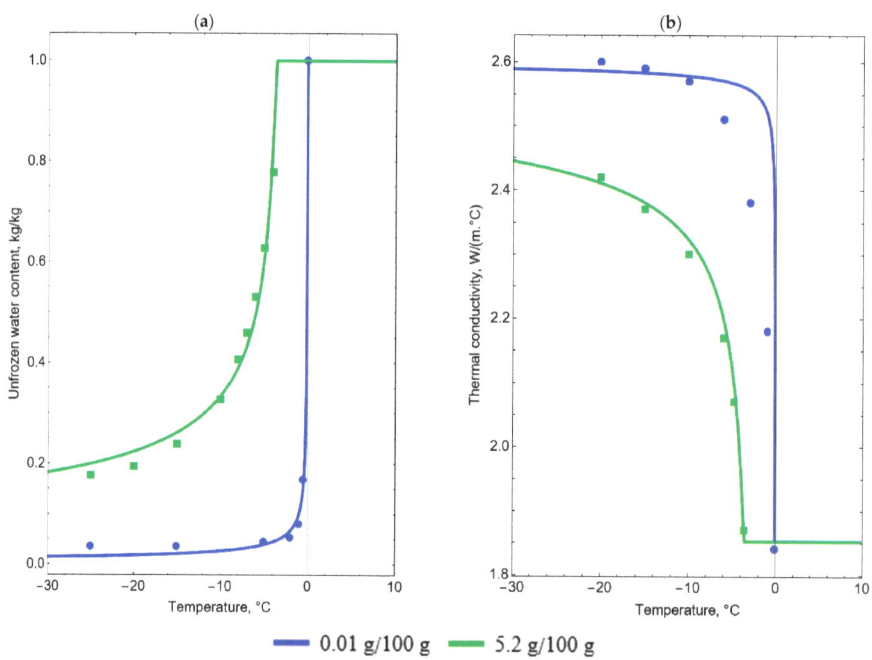

Figure 5. Dependence of the unfrozen water content (**a**) and the thermal conductivity (**b**) on temperature and salinity for sand.

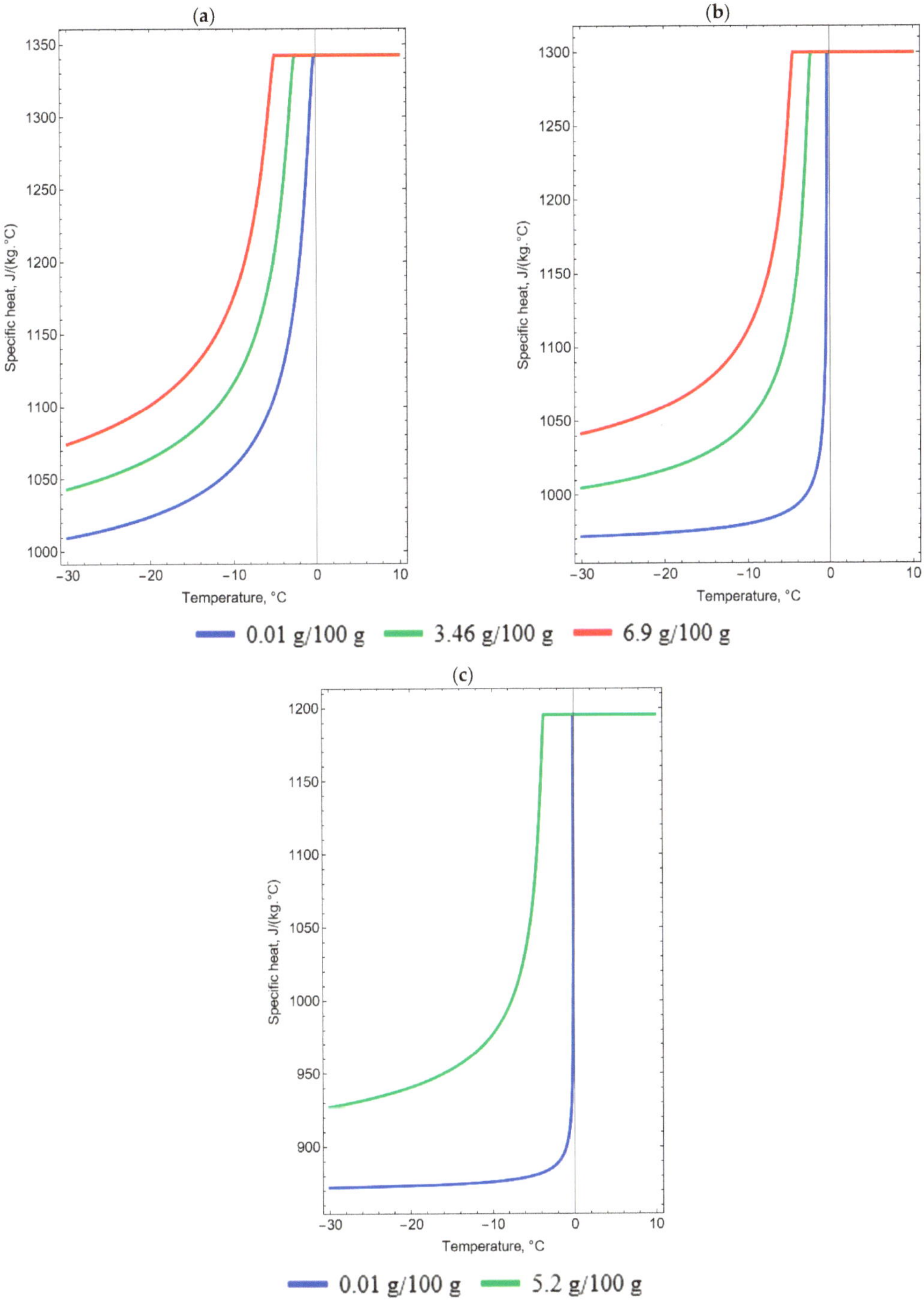

Figure 6. Dependencies of specific heat capacity on temperature and salinity for clay (**a**), chalk (**b**), and sand (**c**).

The comparison of thermophysical properties of soils in Figures 3–6 enabled us to observe the following similarities and differences among them:

1. As the salt concentration increases, all the considered soils exhibit an increase in the amount of unfrozen water, ceteris paribus. In all cases, clay contains the highest proportion of unfrozen water. Even when the salt concentration is nearly zero, clay retains a significant amount of unfrozen water. This fact is clear from the point of view that there is more bound water in clays than in sands and chalk [6,31]. In chalk and sand, a more pronounced reduction in the amount of unfrozen water can be observed with decreasing salt concentration. In a broader sense, it can be noted that freezing water in clay is considerably more challenging than in chalk and sand at any given salt concentration.
2. The impact of salt concentration on the thermal conductivity of all soils under consideration follows the same pattern—an increase in salt concentration leads to a decrease in thermal conductivity. Additionally, clay experiences a smoother decline in thermal conductivity as salt concentration rises, compared to chalk and sand. However, considering chalk, this observation holds true mainly for values close to zero, as the disparity in thermal conductivity between salt concentrations of 3.46 g/100 g and 6.9 g/100 g is noticeably smaller than the difference in thermal conductivity between salt concentrations of 0.01 g/100 g and 3.46 g/100 g.
3. The heat capacity of all soils increases with higher salt concentrations. Similar to the case of thermal conductivity, clay displays a more gradual change in heat capacity in response to changes in salt concentration compared to chalk and sand.

Figure 7 shows the temperature dependencies of the unfrozen water content, the coefficient of thermal conductivity (dashed lines), and the effective heat transfer coefficient (solid lines) for different types of soils at a salinity of 6.9 g/100 g. The difference between the curves makes it possible to understand the soils in which it is easier to freeze water and in which the temperature field changes faster. The effective heat transfer coefficient refers to the ratio of thermal conductivity to the derivative of the specific enthalpy of temperature:

$$a_{eff} = \lambda \left(\frac{\partial H}{\partial T} \right)^{-1}, \tag{12}$$

where a_{eff} is the effective heat transfer coefficient (m^2/c) and H is the specific heat of the soil (J/m^3). The specific enthalpy consists of the perceived heat and the latent heat of the phase transition [32]:

$$H = \int_{T_{ref}}^{T} \rho c_p dT + \rho \frac{w_0}{1 + w_0} L\gamma, \tag{13}$$

where T_{ref} is the reference temperature (°C) and L is the specific heat of the phase transition of water (J/kg). In accordance with [27], the dependence of the specific heat of the phase transition of water can be defined as follows:

$$L = L_0 + L'_0[T - (T_{norm} - 273.15)], \tag{14}$$

$$L_0 = 333427\left(1 - 0.42c - 15.47c^{4.145} + 0.529c^{1.477}\right), \tag{15}$$

$$L'_0 = 2458.25\left(1 - 0.4524c^{0.79} + 145.824c^{3.56} + 614412c^{10.17}\right), \tag{16}$$

$$T_{norm} = 273.1526\left(1 - 0.215c - 1.312c^{2.702} - 20.572c^{6.176}\right), \tag{17}$$

where c is the concentration of NaCl solution in the porewater, which is associated with salinity by the following ratio:

$$c = \frac{s\rho_w}{100\rho_{NaCl} + s\rho_w} \tag{18}$$

where s is the salinity (g/100 g) and ρ_{NaCl} is the NaCl density (kg/m^3). The NaCl density is multiplied by 100 to match the salinity unit.

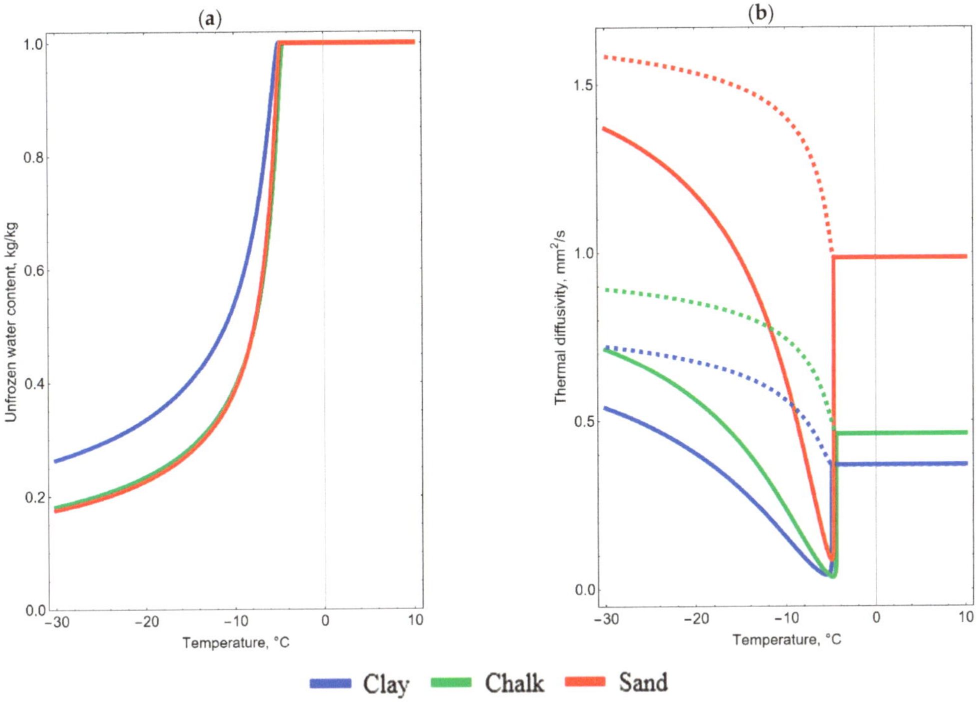

Figure 7. Dependencies of the unfrozen water content (**a**) and the effective heat transfer coefficient (**b**) of soils on the temperature for clay, chalk, and sand at a salinity of 6.9 g/100 g.

The analysis of the thermodynamic processes using the parameter of effective thermal conductivity allows us to consider the heat jump during the transition through the phase transition. Thus, it is possible to determine the zone where the temperature change due to the heat of the phase transition prevails or is comparable to the conductive heat transfer.

Figure 7 shows that the temperature dependencies of the unfrozen water content for chalk and sand are almost the same. In clay, the curve is higher, and therefore, it is more difficult to freeze the porewater in it. It is noteworthy that the dependencies of the effective heat transfer coefficient on the temperatures for clay and chalk are almost the same, and for sand, the curve is higher. With a decrease in temperature, the difference between the dependencies increases, and at a temperature of −30 °C, the effective heat transfer of sand is three times higher than that of clay or chalk. From this, we can conclude that the process of temperature change occurs faster in sand than in clay or chalk. Chalk, despite having approximately the same amount of porewater as sand (see Figure 7b), has a significantly lower effective heat transfer coefficient because chalk solids have a lower thermal conductivity and a higher specific heat capacity than do sand solids.

It can be seen from Figure 7b that the effective heat transfer drops sharply in all three layers near the freezing temperature of the porewater. This is due to a significant increase in the temperature derivative of the enthalpy at the time of transition through the phase transition boundary.

The dependences of thermophysical parameters on salt concentration and temperature obtained in this section were further used as input parameters for numerical simulation of heat and mass transfer during the formation of FW.

3. Numerical Simulation of Heat and Mass Transfer on the FW Formation

3.1. Mathematical Model

The problem of artificial freezing of the soil volume in the form of a circular cylinder was considered. Freezing was technically accomplished using 39 freeze pipes (see Figure 8a). Geometrically, the pipes were considered as circular cylinders with equal radii, at an equal distance relative to each other, and with zero deviation from the vertical direction. The temperature distribution of the brine moving in the freeze pipes was assumed to be homogeneous. In view of these assumptions, a transition was made from a discrete set of heat sources in the form of freeze pipes to a continuous heat source distributed along the freezing contour (see Figure 8b). This transition was made by specifying a distributed source of heat flux density in the form of the ratio of the total specific heat flux from all pipes to the radius of the freezing contour:

$$q = \alpha \frac{N_f r_f}{R_f}\left(T_f - T\right), \quad (19)$$

where q is the heat flux density (W/m^2), α is the heat transfer coefficient between the freeze pipes and the soil (W/(m^2·°C)), N_f is the number of freeze pipes, r_f is the radius of the freeze pipes (m), R_f is the radius of the freezing contour (m), T_f is the temperature on the wall of the freeze pipes (°C), and T is the soil temperature (°C).

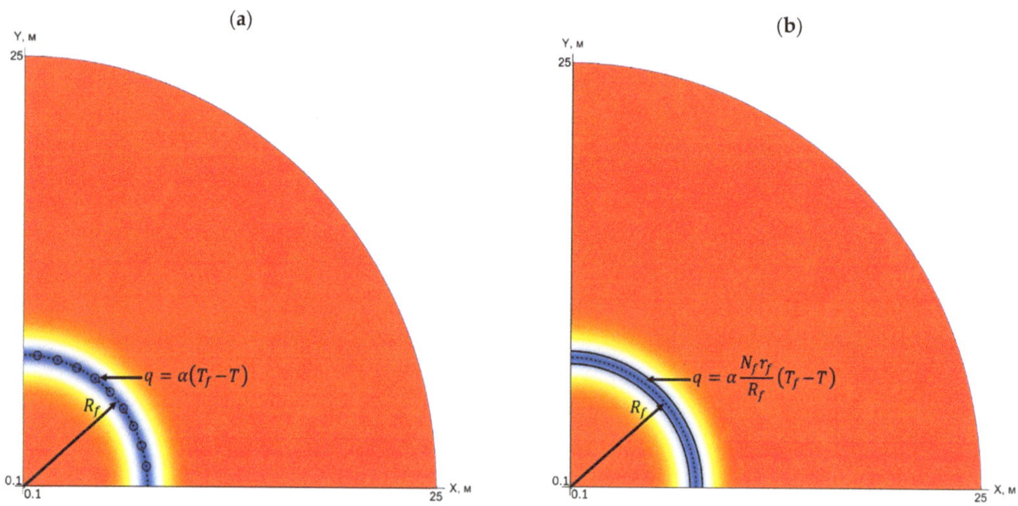

Figure 8. Part of the computational domain with a discrete set of heat sources: with freeze pipes (**a**) and with a distributed heat source (**b**).

It was assumed that the considered soil layers were sufficiently extended in the vertical direction, and therefore, the influence of the vertical heat flows on the temperature field of the median horizontal section of the layers would be negligible over a long period of time. The considered soils were assumed to be homogeneous and isotropic, and heat transfer in them occurs due to thermal diffusion and the convective transport of the porewater. The heat transfer processes between the soil and the freeze pipes were also considered. In addition, the release of latent heat during the phase transition of the porewater was considered. This assumption was made about the local thermodynamic equilibrium of phases at each point of the given cylindrical domain. The effect of supercooling the porewater was not considered.

In view of all the accepted model assumptions, we can say that the problem has angular symmetry, and thermodynamic processes proceed mainly in a radial direction. For these reasons, the problem can be considered using cylindrical coordinates, with its reduction to the one-dimensional case along the radial coordinate. In this case, the transient temperature field in the soil layers is described by the law of energy balance in the following form [32]:

$$\frac{\partial H}{\partial T}\frac{\partial T}{\partial t} + \frac{1}{r}\frac{\partial}{\partial r}(\mathbf{v}rH) = \frac{1}{r}\frac{\partial}{\partial r}\left(r\lambda\frac{\partial T}{\partial r}\right), \tag{20}$$

where H is the soil specific enthalpy (J/m^3), t is the time (s), r is the radial coordinate (m), and $\mathbf{v}$ is the vector of groundwater velocity (m/s).

It was assumed that the flow of porewater is plane-radial. This type of flow is typical, for example, in situations when water bulges from the front of the phase transition due to the expansion of freezing water [33]. The Darcy equation was used to simulate porewater flow [34,35]:

$$\mathbf{v} = -\frac{Kk_r}{\mu_w}\nabla p = -\frac{Kk_r}{\mu_w}\frac{\partial p}{\partial r}\hat{\mathbf{r}}, \tag{21}$$

where K is the soil absolute permeability (m^2), k_r is the soil relative permeability (m^2/m^2), $\mu_w = 0.001236$ Pa·s is the porewater dynamic viscosity, p is the soil density (Pa), and $\hat{\mathbf{r}}$ is the basis vector of the radial coordinate.

The relative permeability depends on the unfrozen water content and is determined through the following ratio [28,29]:

$$k_r = \sqrt{\gamma}\left[1 - \left(1 - \gamma^{\frac{1}{m}}\right)^m\right]^2, \tag{22}$$

The pressure field was calculated based on the assumption of the incompressibility of porewater:

$$\nabla \cdot \mathbf{v} = -\frac{K}{\mu_w}\frac{1}{r}\frac{\partial}{\partial r}\left(rk_r\frac{\partial p}{\partial r}\right) = 0, \tag{23}$$

It is worth noting that the calculation of porewater flow was implemented only for the sand layer. For clay and chalk, the flow of porewater was not considered, because these layers are weakly permeable. Thus, the absolute permeability of sand, K, was assumed to be $0.04 \cdot 10^{-12}$ m^2.

An important feature of the physical process of freezing saline soil is that only pure water crystallizes, while dissolved salt accumulates in unfrozen water. This process can be mathematically expressed by the following expression:

$$s = \frac{s_0}{\gamma}, \tag{24}$$

where s_0 is the initial salinity of porewater in an untouched soil mass (g/100 g).

The dissolved salt accumulates in the unfrozen water until the eutectic point is reached [36,37]. After that, the excess salt precipitates. The solution itself (water + salt) in this case is saturated. At the same time, it is also possible to release the additional heat of the phase transition, which, according to estimates [37,38], is 216–235 J per 1 g of the eutectic solution involved in the chemical reaction. In the case of NaCl, the saturated solution has two ways to reach the eutectic point: (1) when salinity reaches 23.2% [37]; or (2) when water with a salt solution freezes to a temperature of −21.2 °C [37]. However, this is true only in the case where water with a salt solution is in an unstressed state. According to [39], it is difficult to reach the eutectic point in porewater, which is in a confined state. For this reason, the process of salt precipitating upon reaching the eutectic point was not considered in the mathematical model.

To calculate the change in salinity over time, Equation (24) was differentiated by time, and the corresponding term was added to the resulting equation to account for the convective transport of porewater:

$$\frac{\partial s}{\partial t} + \frac{1}{r}\frac{\partial}{\partial r}(\mathbf{v}rs) = -\frac{s_0}{\gamma^2}\frac{\partial \gamma}{\partial t}, \tag{25}$$

The system of Equations (20)–(25) was supplemented by the following initial and boundary conditions:

$$T|_{t=0} = T_0,\ s|_{t=0} = s_0, \tag{26}$$

$$-\lambda \frac{\partial T}{\partial \mathbf{n}}\bigg|_{r=R_f} = q, \tag{27}$$

$$-\lambda \frac{\partial T}{\partial \mathbf{n}}\bigg|_{r=R_1} = 0,\ \frac{\partial s}{\partial \mathbf{n}}\bigg|_{r=R_1} = 0,\ p|_{r=R_1} = p_1, \tag{28}$$

$$T|_{r=R_2} = T_0,\ s|_{r=R_2} = s_0,\ p|_{r=R_2} = p_2, \tag{29}$$

where $\mathbf{n}$ is the normal vector to the boundary (m); R_1 and R_2 are the internal and external radii of the calculated area (see Figure 9), respectively; and p_1 and p_2 are the pressures at the boundaries of the computational domain (Pa).

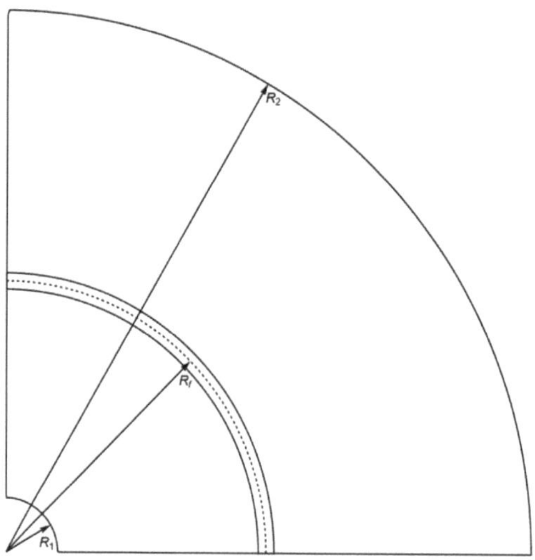

Figure 9. Schematic representation of boundary conditions.

For a numerical discretization of Equations (20)–(29) in space, the finite volume method was used [40]; for discretization in time, an explicit Euler [41] with a dynamic time step was used. The computational mesh was uneven, with a decrease near the freeze pipe (see Figure 10). Table 7 shows the main numerical parameters that were further used in the simulation.

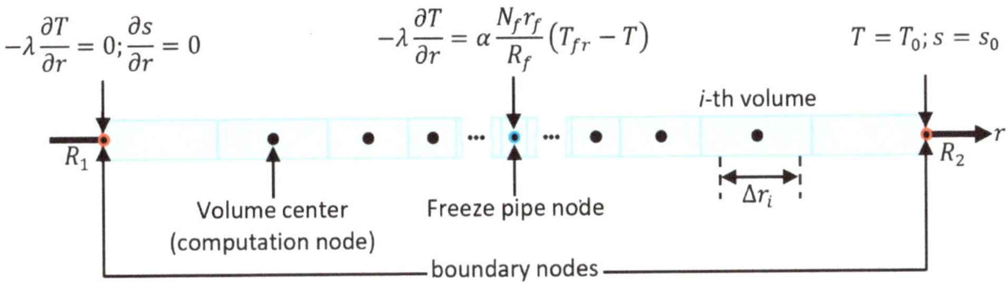

Figure 10. Schematic representation of the computational mesh.

Table 7. Basic numerical parameters.

Parameter	Dimension	Value
Geometry and computational mesh		
Radius, R_1	m	0.1
Radius, R_2	m	25
Radius of the freezing contour, R_f	m	7.7
The characteristic value of the computational mesh around the freeze pipe	m	0.0025
Radius of the freeze pipe, r_f	m	0.073
The ratio of the size of neighboring computing volumes	%	10
Initial and boundary conditions		
Simulation time	day	30
Temperature of the untouched soil, T_0	°C	10
Temperature on the wall of the freeze pipe, T_f	°C	−30
Heat transfer coefficient, α, between the freeze pipe and the soil	W/(m^2·°C)	150
Pressure, p_1	MPa	0
Pressure, p_2	MPa	1.79

3.2. Discussion

Figures 11–13 present the results of the simulation of the dependencies of the distributions of temperature, salinity, and unfrozen water content on the initial salinity, S_0, for clay, sand, and chalk near the heat source. The results are presented for the final moment in time: 30 days. The dashed vertical line in the figure shows the location of the distributed heat source.

From Figure 11, it can be observed that the initial salinity has a weak effect on the temperature distribution in the clay in the range from 0 to 6.9 g/100 g. At the same time, with an increase in initial salinity, the unfrozen water content increases significantly. That is, clay may contain a significant amount of unfrozen porewater despite the weak dependence of temperature on initial salinity, up to 6.9 g/100 g. It may negatively affect the FW's strength properties. However, with an increase in the initial salinity to 13.8 g/100 g, the temperature dependence on the initial salinity begins to manifest itself. In this case, with an increase in the initial salinity above 6.9 g/100 g, the temperature begins to decrease. At the same time, the maximum difference between the curves with initial salinities of 6.9 g/100 g and 13.8 g/100 g is 3 °C, which can be considered significant.

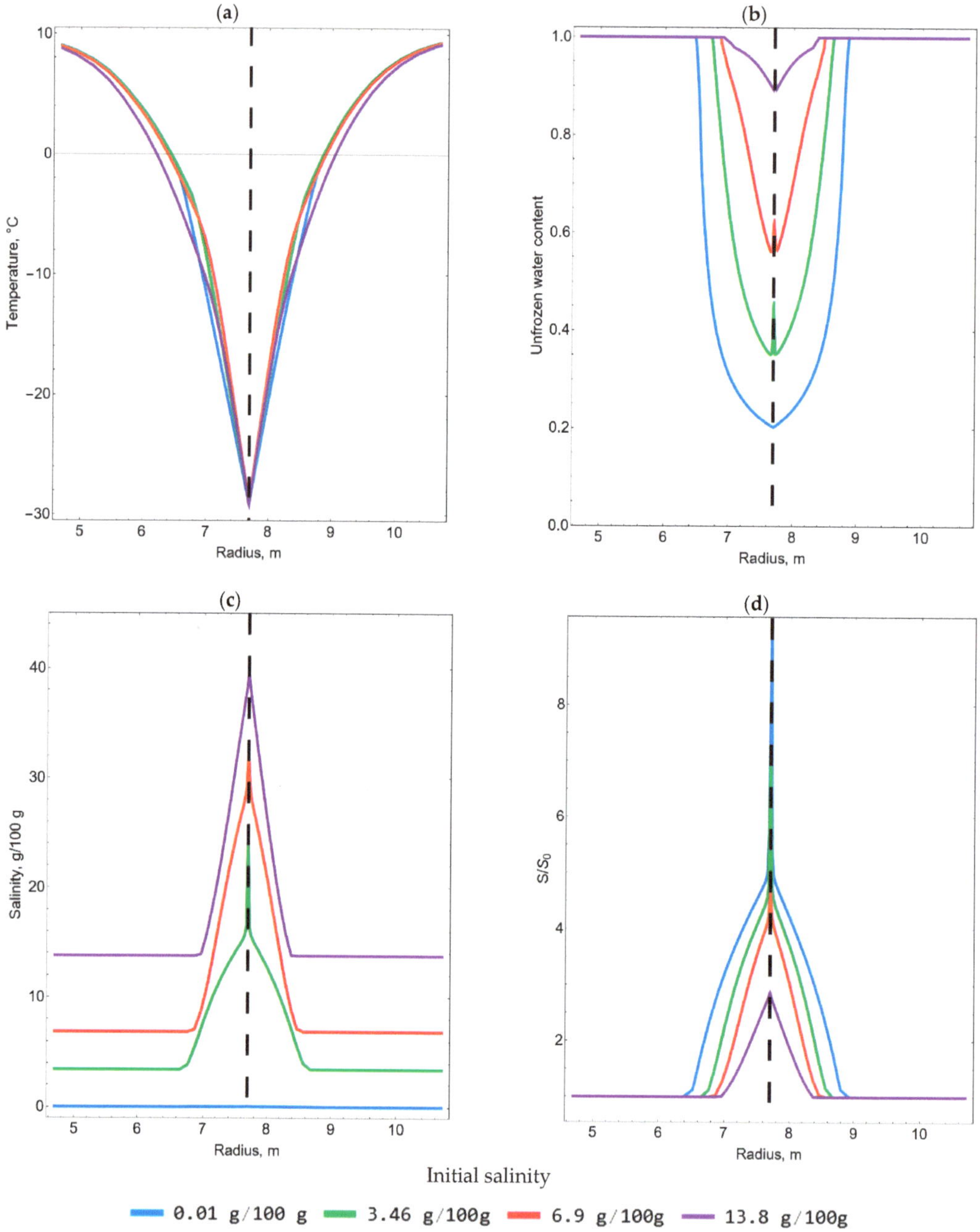

Figure 11. Distribution of temperature (**a**), unfrozen water content (**b**), salinity (**c**), and relative salinity (**d**) for clay.

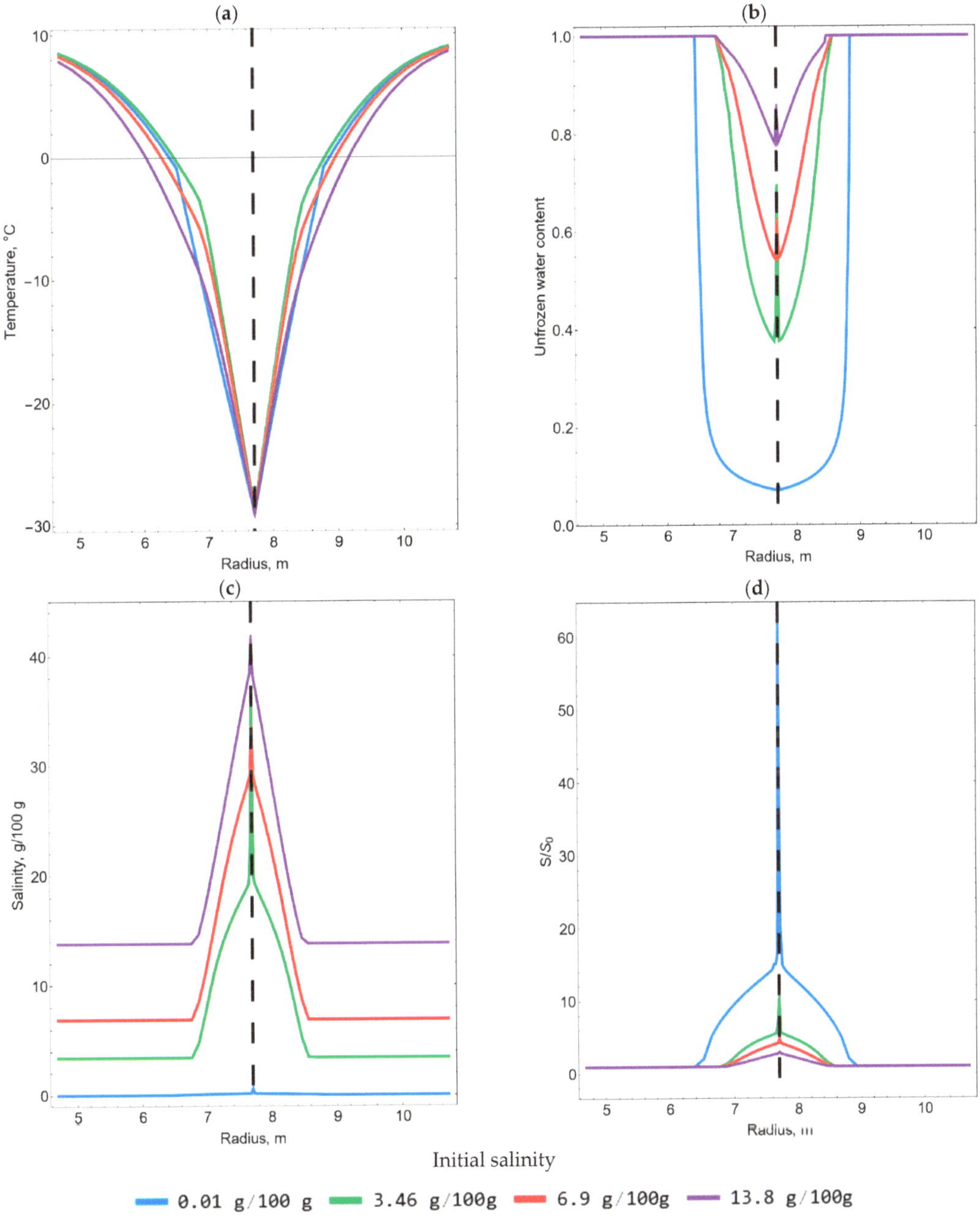

Figure 12. Distribution of temperature (**a**), unfrozen water content (**b**), salinity (**c**), and relative salinity (**d**) for chalk.

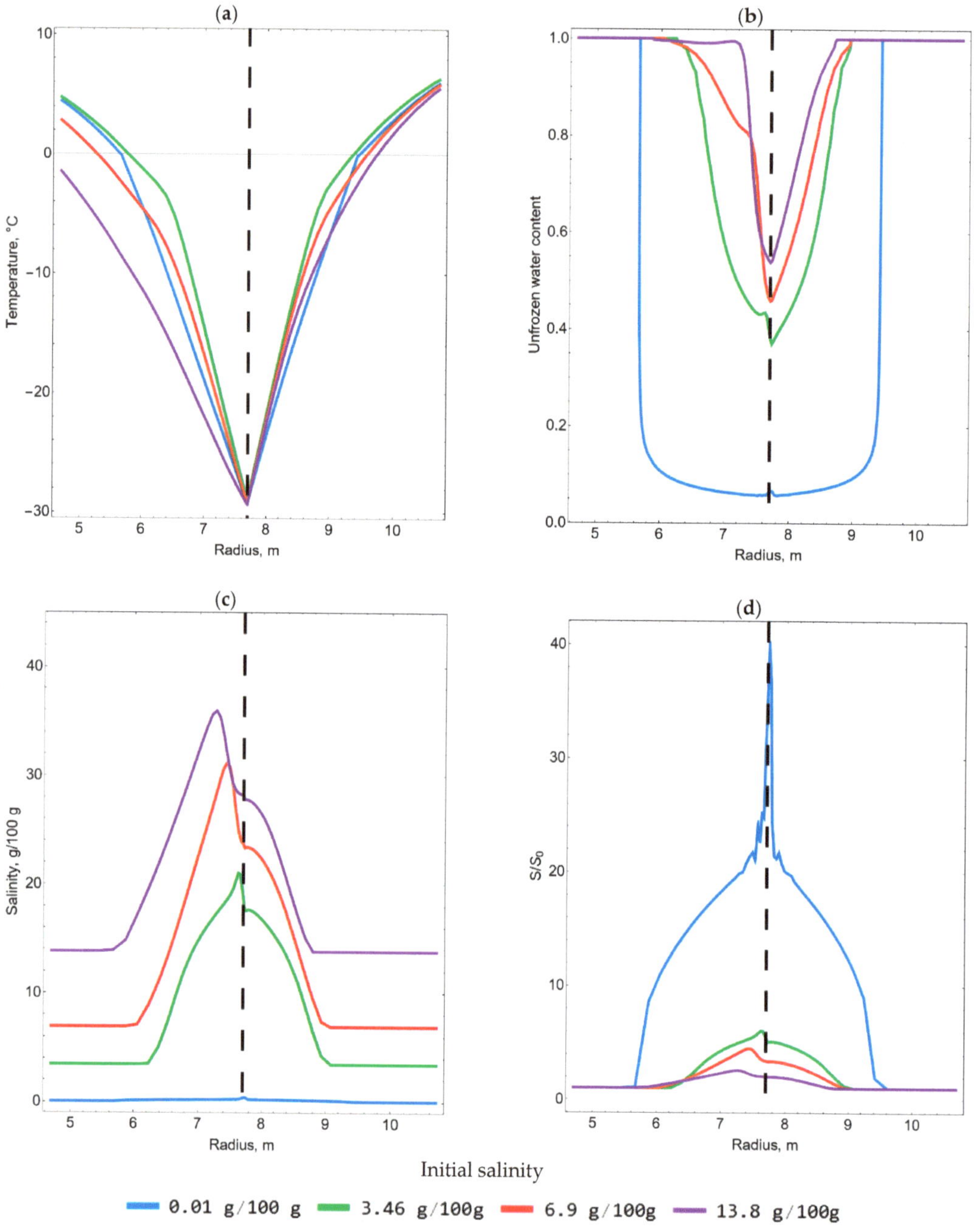

Figure 13. Distribution of temperature (**a**), unfrozen water content (**b**), salinity (c), and relative salinity (**d**) for sand.

It is also worth noting that, since an increase in initial salinity leads to an increase in the unfrozen water content, a natural proportional decrease in relative salinity occurs since the same volume of dissolved salt remains in a larger volume of unfrozen water.

From Figure 11c,d, it can be observed that a significant jump in salinity occurs at the location of the distributed heat source. This is natural because, at the initial moment in time, the temperature of the freeze pipe is rapidly established in the local zone near the heat source. This dramatically reduces the unfrozen water content. As a result, the dissolved salt remains in a smaller unfrozen water content.

For chalk (see Figure 12), one can qualitatively note the same features that were described above for clay. The exceptions are the temperature distributions. For chalk, the effect of the initial salinity on the resulting temperature curves is much stronger and appears earlier than for clay. Figure 12a shows that the temperature curve shifts higher with an increase in the initial salinity from 0.01 g/100 g to 3.46 g/100, but with a further increase in the initial salinity, the temperature curve shifts downward. This can be explained by the fact that an increase in temperature with an increase in salinity is a local effect only for small values of initial salinity. In general, chalk is characterized by a decrease in temperature with an increase in salinity. This is explained by the fact that, if the soil is more saline, then less heat must be removed from the soil to lower its temperature by a given value, ΔT, since less unfrozen water content needs to be frozen. It is worth noting that, in fact, a decrease in temperature with an increase in initial salinity is also characteristic of clay, but this is manifested to a much lesser extent due to more gently sloping curves for unfrozen water content (see Figure 7a).

When comparing the unfrozen water content in clay and chalk, it can be noted that the unfrozen water content near the heat source in chalk is lower than in clay. Therefore, with the same salinity of chalk and clay, it is easier to freeze porewater in chalk.

The case of sand (Figure 13) differs significantly from clay and chalk. First, this is due to the presence of porewater flow. Due to the convective transfer of porewater, the dependence of the temperature distribution on the initial salinity is more pronounced. As in the case of clay and chalk, an increase in temperature can be observed with an increase in initial salinity from 0.01 g/100 g to 3.46 g/100 g, but with a further increase in initial salinity, the temperature decreases significantly. Additionally, in general, it is possible to observe a more complex distribution of temperature, unfrozen water content, and salinity due to the presence of the flow of porewater. For this reason, the design, construction, and monitoring of FWs in highly permeable soils can be complicated because, in this case, there is the problem of selecting a criterion for assessing the thickness of the FW due to a highly asymmetric freezing pattern.

Based on the obtained simulation results, the dependencies of an FW's thickness on time for all three layers were also determined (see Figures 14 and 15). However, as noted above, in the case of sand, the criterion for estimating the FW's thickness is complicated due to the asymmetric distribution of the temperature field. For this reason, we considered three FW thicknesses for sand: (1) the thickness of the inner frozen zone (to the left of the heat source), (2) the thickness of the outer frozen zone (to the right of the heat source), and (3) the total FW's thickness.

It is also worth noting that an FW's thickness is usually determined by reaching the ground temperature of a certain isotherm. In this case, the FW's thickness was calculated from the isotherm of the freezing temperature of the porewater. The choice of this isotherm was based on the fact that it directly varies, depending on the amount of salinity, and it allows one to more accurately assess the variation in the FW's thickness with a change in salinity. If we estimate the FW's thickness using the constant isotherm, then we can arrive at an erroneous conclusion about the beneficial effect of salinity on an FW's growth over time. The fallacy lies in the fact that, as has been shown, with an increase in initial salinity, the soil temperature decreases, and, accordingly, the FW's thickness, as calculated from the constant isotherm, will grow.

Figure 14a shows the expected result for clay; with an increase in initial salinity, the FW's thickness decreases. However, in the case of chalk (Figure 14b), interesting results can be observed; for cases of initial salinity of 3.46 g/100 g and 6.9 g/100 g, the FW's thickness practically does not change, and with a further increase in the initial salinity, the FW's thickness decreases again. This result can be explained by balancing two factors for initial salinities of 3.46 g/100 g and 6.9 g/100 g: a decrease in soil temperature with an increase in salinity, and a decrease in freezing temperature with an increase in salinity. The first factor favorably affects the increase in the FW's thickness; the second, on the contrary, reduces the FW's thickness. However, with a further increase in the initial salinity, the second factor begins to prevail, and the FW's thickness begins to decrease again.

When considering sand (Figure 15), it can be observed that, in the inner frozen zone, the FW's thickness increases with increasing initial salinity. This is a consequence of the influence of the flow of porewater, due to which the temperature of the sand decreases faster to the left of the heat source than its freezing temperature decreases. However, in the outer frozen zone, where the effect of the flow of porewater is not so great, a similar effect can be observed, as in the case of chalk: the FW's thickness at the initial salinity of 3.46 g/100 g and 6.9 g/100 g coincides, and with a further increase in salinity, the FW's thickness decreases. Thus, it can be concluded that the occurrence of the effect of the coincidence of the FW's thicknesses at different initial salinities is determined not only by the thermophysical properties of the soil and porewater, but also by the presence of the flow of pore water.

It should be noted that in the future, it is planned to test this model in the field when monitoring the AFG of potash mine shafts under construction. It is assumed that the obtained dependencies will allow a much better explanation of the observed features of temperature changes in the control wells. At the same time, experimental data on the real process of freezing of soils with known salinity will help to carry out a full-fledged validation of the model proposed by us.

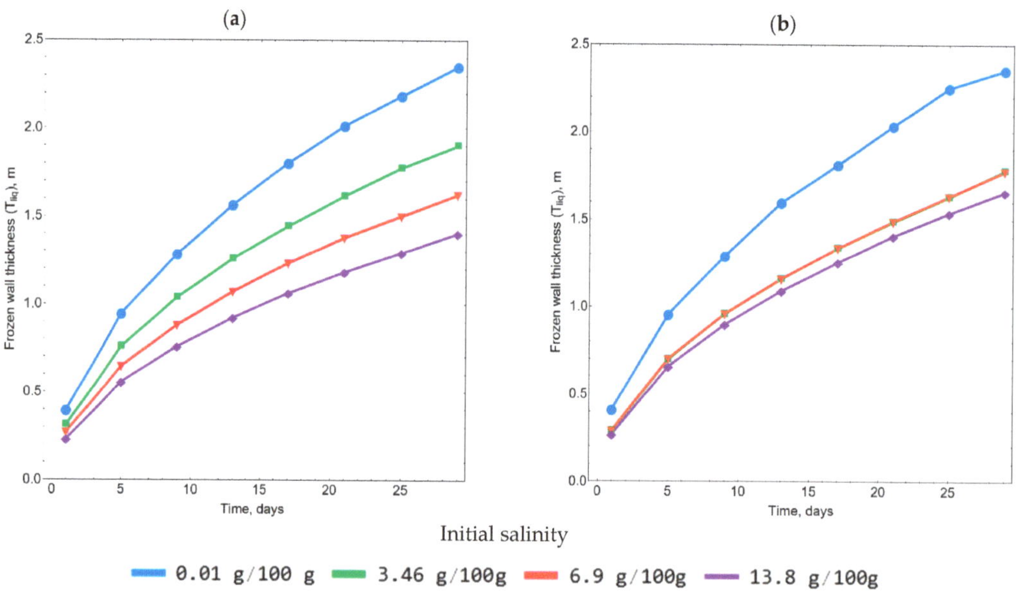

Figure 14. The dependence of the FW's thickness on time for clay (**a**) and chalk (**b**).

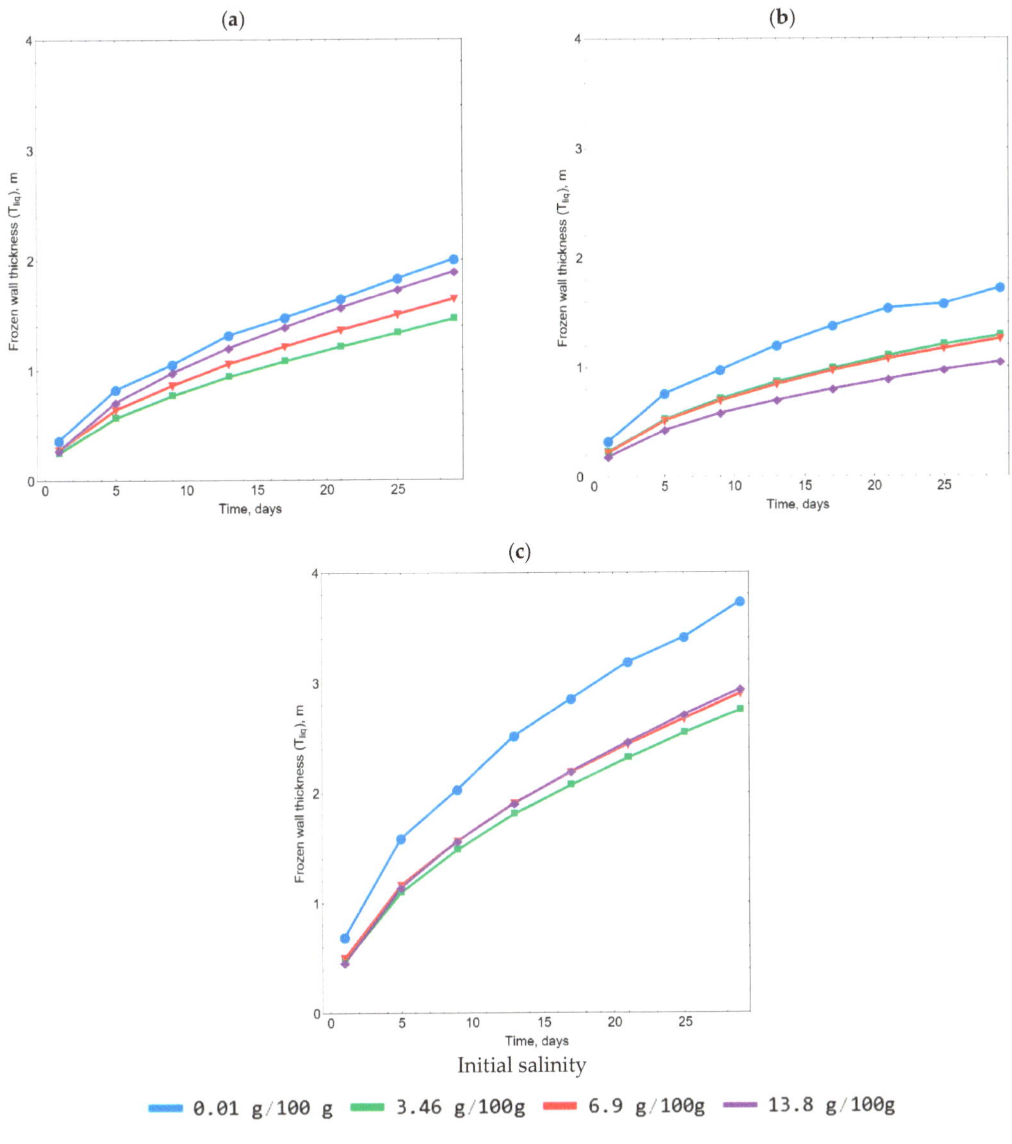

Figure 15. The dependence of the FW's thickness on time for sand in the inner frozen zone (**a**) and in the outer frozen zone (**b**), and on the total thickness (**c**).

4. Conclusions

In the course of this work, laboratory and theoretical studies were implemented on the effects of porewater salinity on the artificial freezing of three soils: clay, chalk, and sand.

Conducting laboratory studies made it possible to determine the dependencies of the freezing temperature of porewater, the unfrozen water content, the specific heat capacity, and the thermal conductivity on the salinity of the considered soils. The experimental data obtained were approximated using analytical functions known in the literature. The obtained dependencies were later used for theoretical research: the numerical simulation of the AGF process for the conditions of the construction of the frozen wall around the designed mineshaft at a potash salt deposit.

The analysis of the simulation results shows that, with an increase in the initial salinity, the temperature of the soil decreases because more saline soils contain a larger quantity of unfrozen water content (which was also shown by the simulation results), and, therefore, to lower their temperature by a given amount, less heat must be removed. However, this effect is less pronounced in clay. In addition, for all the soils under consideration, the opposite local effect was noted: at low values of initial salinity, the soil temperature increases. Since this effect manifests itself only at small initial salinities, we can discuss the insignificance of this effect.

The dependence of the FW's thickness on the initial salinity was found. The thickness was calculated from the isotherm of the freezing temperature of the porewater. The obtained results show that, in the case of clay, a natural process occurs: an increase in initial salinity leads to a decrease in the FW's thickness since the water's freezing temperature decreases. In the case of both chalk and sand, it was found that, at certain values of initial salinity, the FW's thickness becomes independent of the initial salinity. However, with the continued increase in initial salinity, at some point, the FW's thickness begins to decrease. This can be explained by the fact that, at some values of initial salinity, the two processes are balanced: a decrease in ground temperature leads to an increase in the FW's thickness, and a decrease in the freezing temperature of porewater leads to a decrease in the FW's thickness.

The relationships and simulation results acquired in this study hold significant practical importance: they will contribute to the enhancement of existing tools for the design and monitoring of frozen walls under the circumstances involving saline soil layers near the water-protective salt stratum.

Author Contributions: Conceptualization, S.B., M.S., L.L., A.B. and I.D.; methodology, M.S., A.B. and I.D.; software, S.B. and M.S.; validation, S.B., M.S. and L.L.; formal analysis, S.B., M.S. and L.L.; investigation S.B., M.S., L.L., A.B. and I.D.; resources, S.B., M.S., L.L., A.B. and I.D.; data curation, M.S., A.B. and I.D.; writing—original draft preparation, S.B.; writing—review and editing, M.S.; visualization, S.B.; supervision, M.S. and L.L.; project administration, M.S. and L.L. All authors have read and agreed to the published version of the manuscript.

Funding: This research was funded by the Ministry of Science and Higher Education of Russia within the framework of project No. C-26/563.

Institutional Review Board Statement: Not applicable.

Informed Consent Statement: Not applicable.

Data Availability Statement: Not applicable.

Conflicts of Interest: The authors declare no conflict of interest.

References

1. Trupak, N.G. *Ground Freezing in Underground Construction*; Nedra: Moscow, Russia, 1974.
2. Nikolaev, P.V.; Shuplik, M.N. Rationale for choosing heat transfer liquid for artificial ground freezing. *Min. Inf. Anal. Bull. (Sci. Tech. J.)* **2016**, *1*, 320–332.
3. Olkhovikov, Y. *Support of Permanent Openings of Potash and Salt Mines*; Nedra: Moscow, Russia, 1984.
4. Semin, M.; Levin, L.; Bogomyagkov, A.; Pugin, A. Features of Adjusting the Frozen Soil Properties Using Borehole Temperature Measurements. *Model. Simul. Eng.* **2021**, *2021*, 8806159. [CrossRef]
5. Levin, L.; Semin, M.; Parshakov, O. Improving Methods of Frozen Wall State Prediction for Mine Shafts under Construction Using Distributed Temperature Measurements in Test Wells. *J. Min. Inst.* **2019**, *237*, 268–274. [CrossRef]
6. Tsytovich, N.A. *Mechanics of Frozen Soils*; High School: Moscow, Russia, 1973.
7. Semin, M.; Levin, L.; Bublik, S.; Brovka, G.; Brovka, A.; Agutin, K. Parameterization of the Model of Artificial Clay Freezing Considering the Effect of Pore Water Salinity. *Fluids* **2022**, *7*, 186. [CrossRef]
8. Banin, A.; Anderson, D.M. Effects of Salt Concentration Changes During Freezing on the Unfrozen Water Content of Porous Materials. *Water Resour. Res.* **1974**, *10*, 124–128. [CrossRef]
9. Yong, R.N.; Cheung, C.H.; Sheeran, D.E. Prediction of salt influence on unfrozen water content in frozen soils. *Eng. Geol.* **1979**, *13*, 137–155.

10. Lucas, T.; Chourot, J.M.; Bohuon, P.; Flick, D. Freezing of a Porous Medium in Contact with a Concentrated Aqueous Freezant: Numerical Modelling of Coupled Heat and Mass Transport. *Int. J. Heat Mass Transf.* **2001**, *44*, 2093–2106. [CrossRef]
11. Liu, J.; Yang, P.; Yang, Z. (Joey) Electrical Properties of Frozen Saline Clay and Their Relationship with Unfrozen Water Content. *Cold Reg. Sci. Technol.* **2020**, *178*, 103127. [CrossRef]
12. Aleksyutina, D.M.; Motenko, R.G. The Effect of Soil Salinity and the Organic Matter Content on the Thermal Properties and Unfrozen Water Content of Frozen Soils at the West Coast of Baydarata Bay. *Moscow Univ. Geol. Bull.* **2016**, *71*, 275–279. [CrossRef]
13. Qiu, E.; Zhong, C.; Wan, X.; Lu, J.; Chen, H.M.; Pirhadi, N.; Wang, Z.; Chen, Q. Study on Thermal Conductivity Model of Saline Soil Based on Particle Morphology. *Heat Mass Transf.* **2021**, *57*, 2029–2043. [CrossRef]
14. Zhang, J.; Lai, Y.; Li, J.; Zhao, Y. Study on the Influence of Hydro-Thermal-Salt-Mechanical Interaction in Saturated Frozen Sulfate Saline Soil Based on Crystallization Kinetics. *Int. J. Heat Mass Transf.* **2020**, *146*, 118868. [CrossRef]
15. Lyu, C.; Sun, Q.; Zhang, W. Effects of NaCl Concentration on Thermal Conductivity of Clay with Cooling. *Bull. Eng. Geol. Environ.* **2020**, *79*, 1449–1459. [CrossRef]
16. Rong, C.; Sun, S.; Cheng, H.; Duan, Y.; Yang, F. Experimental Study on Temporal and Spatial Evolutions of Temperature Field of Double-Pipe Freezing in Saline Stratum with a High Velocity. *Energies* **2022**, *15*, 1308. [CrossRef]
17. Qin, B.; Rui, D.; Wang, S.; Ji, M.; Zou, Z. Experimental Investigation on Formation and Evolution Characteristics of Frozen Wall under Salty Groundwater Seepage. *J. Eng. Sci. Technol. Rev.* **2020**, *13*, 99–107. [CrossRef]
18. Calderón, D.S.; Palacio, S.; Ibanez García, S.; Bock, S.H. An Adjusted Analytical Solution for Thermal Design in Artificial Ground Freezing. *Int. J. Rock Mech. Min. Sci.* **2023**, *164*, 105310. [CrossRef]
19. Deng, S.; He, Y.; Yang, M.; Zhou, F.; Liu, H.; Zhu, R.; Wan, Z. Numerical Analysis of Shield Tunnelling Breakthrough Working Shaft by Artificial Ground Freezing Method under Extreme Conditions Considering Phase Change Latent Heat. *Appl. Sci.* **2023**, *13*, 3651. [CrossRef]
20. Bi, J.; Wu, Z.; Cao, W.; Zhang, Y.; Wen, H.; Yang, S.; Zhang, Q.; Sun, T.; Wei, T. A Hyperbolic Model for the Thermal Conductivity of Freezing Soils. *Geoderma* **2023**, *436*, 116507. [CrossRef]
21. Wan, X.; Zhong, C.; Mohamed, H.S.; Qiu, E.; Qu, M.; Nkiegaing, F.J. Study on the Thermal Conductivity Model of Sodium Sulfate Soils. *Exp. Heat Transf.* **2021**, *34*, 217–239. [CrossRef]
22. Zhelnin, M.; Kostina, A.; Prokhorov, A.; Plekhov, O.; Semin, M.; Levin, L. Coupled Thermo-Hydro-Mechanical Modeling of Frost Heave and Water Migration during Artificial Freezing of Soils for Mineshaft Sinking. *J. Rock Mech. Geotech. Eng.* **2022**, *14*, 537–559. [CrossRef]
23. Tounsi, H.; Rouabhi, A.; Jahangir, E. Thermo-Hydro-Mechanical Modeling of Artificial Ground Freezing Taking into Account the Salinity of the Saturating Fluid. *Comput. Geotech.* **2020**, *119*, 103382. [CrossRef]
24. Zhang, X.; Wang, Q.; Yu, T.; Wang, G.; Wang, W. Numerical Study on the Multifield Mathematical Coupled Model of Hydraulic-Thermal-Salt-Mechanical in Saturated Freezing Saline Soil. *Int. J. Geomech.* **2018**, *18*, 04018064. [CrossRef]
25. Semin, M.A.; Levin, L.Y.; Zhelnin, M.S.; Plekhov, O.A. Simulation of Artificial Ground Freezing under Conditions of Heterogeneous Mineralization of Pore Water. *High Temp.* **2022**, *60*, 391–398. [CrossRef]
26. Brovka, A.G.; Brovka, G.P.; Dedyulya, I.V. The Influence of Salt Concentration in a Detached Water of Claystone-like Clay on the Ground Freezing Point and the Dependence of the Amount of Unfrozen Water on Temperature. In *Mining Mechanical Engineering and Machine-Building*; JSC «Soligorsk Institute of Resources Saving Problems with Pilot Production»: Soligorsk, Belarus, 2022; pp. 14–23.
27. Rouabhi, A.; Jahangir, E.; Tounsi, H. Modeling Heat and Mass Transfer during Ground Freezing Taking into Account the Salinity of the Saturating Fluid. *Int. J. Heat Mass Transf.* **2018**, *120*, 523–533. [CrossRef]
28. Gens, A.; Nishimura, S.; Jardine, R.; Olivella, S. THM-Coupled Finite Element Analysis of Frozen Soil: Formulation and Application. *Geotechnique* **2009**, *59*, 159–171. [CrossRef]
29. Zhou, M.M.; Meschke, G. A Three-Phase Thermo-Hydro-Mechanical Finite Element Model for Freezing Soils. *Int. J. Numer. Anal. Methods Geomech.* **2013**, *37*, 3173–3193. [CrossRef]
30. Xiao, Z.; Hou, Z.; Zhu, L.; Dong, X. Experimental Investigation of the Influence of Salt on the Phase Transition Temperature in Saline Soil. *Cold Reg. Sci. Technol.* **2021**, *183*, 103229. [CrossRef]
31. Tian, H.; Wei, C.; Wei, H.; Zhou, J. Freezing and Thawing Characteristics of Frozen Soils: Bound Water Content and Hysteresis Phenomenon. *Cold Reg. Sci. Technol.* **2014**, *103*, 74–81. [CrossRef]
32. Voller, V.R.; Swaminathan, C.R.; Thomas, B.G. Fixed Grid Techniques for Phase Change Problems: A Review. *Int. J. Numer. Methods Eng.* **1990**, *30*, 875–898. [CrossRef]
33. Khakimov, K.R. *Artificial Freezing of Soils, Theory and Practice: (Voprosy Teorii i Praktiki Iskusstvennogo Zamorazhivaniya Gruntov)*; Israel Program for Scientific Translations; U.S. Department of Commerce: Washington, DC, USA; Clearinghouse for Federal Scientific and Technical Information: Springfield, VA, USA, 1966.
34. Neuman, S.P. Theoretical Derivation of Darcy's Law. *Acta Mech.* **1977**, *25*, 153–170. [CrossRef]
35. Heinemann, Z.E. *Fluid Flow in Porous Media*; Springer-Verlag: Berlin/Heidelberg, Germany, 2005; pp. 337–362.
36. Wang, M.; Zhu, Y.; Zhao, T.; Cui, L.; Mao, W.; Ye, M.; Wu, J.; Yang, J. Chemical Characteristics of Salt Migration in Frozen Soils during the Freezing-Thawing Period. *J. Hydrol.* **2021**, *606*, 127403. [CrossRef]

37. Leys, J.; Losada-Pérez, P.; Glorieux, C.; Thoen, J. The Melting Behaviour of Water and Water–Sodium Chloride Solutions Studied by High-Resolution Peltier-Element-Based Adiabatic Scanning Calorimetry. *J. Therm. Anal. Calorim.* **2017**, *129*, 1727–1739. [CrossRef]
38. Han, B.; Choi, J.; Dantzig, J.; Bischof, J. A Quantitative Analysis on Latent Heat of an Aqueous Binary Mixture. *Cryobiology* **2006**, *52*, 146–151. [CrossRef]
39. Koniorczyk, M.; Bednarska, D. Kinetics of Water Freezing from Inorganic Salt Solution Confined in Mesopores. *Thermochim. Acta* **2019**, *682*, 178434. [CrossRef]
40. Moukalled, F.; Mangani, L.; Darwish, M. *The Finite Volume Method in Computational Fluid Dynamics: An Advanced Introduction with OpenFOAM® and Matlab*; Fluid Mechanics and Its Applications; Springer International Publishing: Cham, Switzerland, 2016; Volume 113, ISBN 978-3-319-16873-9.
41. Atkinson, K.E.; Han, W.; Stewart, D. *Numerical Solution of Ordinary Differential Equations*; Pure and Applied Mathematics; Wiley: Hoboken, NJ, USA, 2009; ISBN 978-0-470-04294-6.

applied sciences

MDPI

Article

Numerical Analyses of the Effect of the Freezing Wall on Ground Movement in the Artificial Ground Freezing Method

Yazhou Ou [1], Long Wang [1], Hui Bian [1], Hua Chen [1], Shaole Yu [1], Tao Chen [1], Alfrendo Satyanaga [2] and Qian Zhai [3,*]

1 China Construction Eighth Engineering Division Co., Ltd., Shanghai 200112, China; ouyazhou@cscec.com (Y.O.); chentaoone2000@163.com (T.C.)
2 Department of Civil and Environmental Enginering, School of Engineering and Digital Sciences, Nazarbayev University, 53 Kabanbay Batyr Ave, Nur-Sultan 010000, Kazakhstan; alfrendo.satyanaga@nu.edu.kz
3 School of Civil Engineering, Southeast University, Nanjing 211189, China
* Correspondence: zhaiqian@seu.edu.cn

Abstract: The advancement of massive construction in urban subway projects contributes to the increased use of the artificial ground freezing (AGF) method in the construction of cross passages due to its reliability and environmental friendliness. However, the uplift or subsidence of the ground surface induced by the frost heave and thawing settlement of the soil can be a problem for existing buildings, and the current design method places way too much emphasis on the strength requirement of the freezing wall. In this study, FLAC3D was employed to develop a series of state-of-the-art numerical models of the construction of a typical subway cross passage by the AGF method, utilizing freezing walls with different thicknesses. The results of this study can be used to examine the ground deformation arising from the AGF method and the influence of the thickness of the freezing wall on the AGF method.

Keywords: artificial ground freezing; freezing wall; frost heave; thaw settlement; FLAC3D

1. Introduction

Over the last decade, the artificial ground freezing (AGF) method, which was originally applied in mining engineering, has been extensively applied in the massive construction of sub-railways in China due to its sterling performance in terms of waterproofing, structure stabilizing, and environmental friendliness [1]. Among the design factors of the AGF method, the most pivotal is the determination of the thickness of the freezing wall, which is highly related to the stability of the excavation and the deformation of the ground surface [2]. In past mining engineering practice, the prevalent method for the determination of the thickness of the freezing wall was the plane-strain elastic–plastic method proposed by Domke [3]. However, considerable research and field practice have shown that the design method employed in mining engineering seems to be inappropriate for the design of urban tunnel excavations, and more considerations of the displacement aspect are needed [4]. In the water supply project that crosses the Yangtze River at Xinjizhou, Jiangning, Nanjing, diverting water from Xinjizhou Island to the pump station at Jiangning District, the AGF method is planned to be adopted for the construction of tunnels. The effects of freezing walls on ground movement need to be evaluated.

This study aims to first investigate the influence of the thickness of the freezing wall on the displacement development characteristics of the freezing and thawing process, examine the temperature distribution within freezing walls with different thicknesses at the selected indicative time points of the characteristic period, and finally evaluate the influence of the thickness of the freezing wall on the ground displacement characteristics. Due to its accuracy and powerful ability to carry out thermal–mechanical coupling analyses, the finite volume program FLAC3D utilizing the three-dimensional explicit Lagrangian method was employed in this study to simulate a set of numerical analyses to model the construction of a subway cross passage using the AGF method with freezing walls of different thicknesses. These numerical analyses are expected to shed light on the thermal–mechanical behavior of freezing walls with different thicknesses, i.e., the frost heave and thaw settlement induced by the freezing wall.

2. Literature Review

Over the past two decades, a great number of researchers have sought new ways to efficiently design freezing walls for the AGF method of subway construction. Yue et al. [5] introduced the widely employed structural mechanics method, which simplifies the freezing wall into an elastic, statically indeterminate structure. Modifying the assumptions, many researchers made improvements to these methods [6–9]. Despite its safety and efficacy, the structural mechanics method suffers from several major drawbacks: (1) this method is limited to the elastic regime of the material, and the cohesion between unfrozen soil and frozen soil is omitted, which leads to a more conservative design of freezing wall thickness; (2) the thickness of the freezing wall is primarily determined by the strength requirement, with little concern for the deformation of the ground surface caused by freezing heave or thaw settlement; and (3) little attention has been paid to thermal dynamics problems during the development of freezing walls [2,10–13]. Noting the limitations of the structural analytical method, some scholars tried to develop analytical solutions for predicting the deformation of the ground surface. However, the equations derived are only applicable to limited configurations and have relatively complicated forms [14,15]. Thus, many researchers made an attempt to examine freezing walls using numerical simulation methods. Zhang and He [16], employing the three-dimensional finite element numerical simulation method, carried out research on the mechanical behaviors of the freezing wall during the excavation. Their observation showed that the bottom of the freezing wall had the largest tension zone area. Employing finite element modeling, Fu et al. [17] evaluated temperature field variations during artificial freezing and observed the potential factors affecting the changes in temperature. They found that thermal conductivity had the most significant impact on the development of the temperature field, while the phase-change latent heat had little impact. Huang et al. [18] undertook a study utilizing the COMSOL multi-physical platform to optimize the distribution of the freezing pipes. While the numerical simulation method has been applied to either solely mechanical or purely thermal dynamic problems when modeling freezing walls, research works on the thermal–mechanical problems induced by freezing walls, i.e., frost heave and thaw settlement, remain scarce. Based on a metro entrance construction project, Zhang et al. [19] investigated time variations in frost heave and found that intermittent freezing could effectively control surface deformation. Through laborary model tests in combination with numerical simulations, Cai et al. [20] examined frost heave during twin-tunnel construction; the results showed that sequential freezing could bring about less frost heave displacement than that of simultaneous freezing.

3. Basic Theory for Thermal–Mechanical Calculations

In FLAC3D, thermal dynamic problems are solved by coupling the energy balance equation and transport laws at the specific boundaries and initial conditions given. The equations and boundaries are summarized as follows:

3.1. Energy Balance Equation

The temperature calculation process follows the law of conservation of energy, and the energy balance differential equation has the following form, as shown in Equation (1):

$$-q_{i,i} + q_v = \frac{\partial \zeta}{\partial t} \tag{1}$$

where $q_{i,i}$ is the heat-flux vector, q_v is the volumetric heat-source intensity, ζ is the heat stored per unit volume, and t is the time.

In general, changes in both energy storage and volumetric strain, ϵ, can result in a change in the temperature, and the thermal constitutive equation describing this relationship can be expressed as Equation (2):

$$\frac{\partial T}{\partial t} = M_{th}\left(\frac{\partial \zeta}{\partial t} - \beta_{th}\frac{\partial \epsilon}{\partial t}\right) \tag{2}$$

where M_{th} and β_{th} are material constants, T is temperature, ϵ is the volumetric strain, and the other variables are the same as those mentioned above.

In FLAC3D, the strain changes are thought to have a minimal impact on temperature, which is valid for solid and liquid matter. This assumption leads to a simplification of Equation (2) into Equation (3):

$$\frac{\partial \zeta}{\partial t} = \rho C_v \frac{\partial T}{\partial t} \tag{3}$$

where ρ is the mass density of the medium, C_v is the specific heat at constant volume, and the other variables are the same as those mentioned above.

Substituting Equation (3) into Equation (1) yields Equation (4):

$$-q_{i,i} + q_v = \rho C_v \frac{\partial T}{\partial t} \tag{4}$$

where the variables are the same as those in the equations above.

3.2. Transport Law

Fourier's law links the heat-flux vector to the temperature gradient. For a stationary, homogeneous, isotropic solid, this law can be expressed as Equation (5):

$$q = -k\nabla T \tag{5}$$

where q is the heat-flux vector, k is the thermal conductivity, and ∇T is the temperature gradient.

3.3. Thermal Boundary Conditions

In FLAC3D, four types of conditions are considered: (1) given temperature; (2) given component of the flux normal to the boundary; (3) convective boundaries; and (4) insulated (adiabatic) boundaries. In FLAC3D, boundaries are adiabatic by default.

3.4. Mechanical–Thermal Coupling

Solutions to thermal stress problems are accomplished by subtracting the portion due to temperature change from the total strain increment since free thermal expansion does not contribute to any angular distortion in an isotropic material, and thus no impact is exerted on the shearing strain increments. The thermal strain increments of the free expansion at a given temperature increment can be expressed in Equation (6):

$$\Delta\epsilon_{ij} = \alpha_t \Delta T \delta_{ij} \tag{6}$$

where $\Delta\epsilon_{ij}$ is the thermal strain increment, α_t is the coefficient of linear thermal expansion, and δ_{ij} is the Kronecker delta.

3.5. Numerical Formulation

In FLAC3D, the solution of a problem is gained using the finite volume approach, which means that the 3D geometric model is first meshed into volumetric finite elements of a tetrahedron, among which the space and time derivatives of a variable are considered constant. Specifically, the temperature gradient, by the Gauss divergence theorem, could be calculated from the nodal temperature values by Equation (7), as illustrated in Figure 1:

$$T_{,j} = -\frac{1}{3V}\sum_{l=1}^{4} T^{l} n_j^{(l)} S^{(l)} \tag{7}$$

where $T_{,j}$ is the temperature gradient, V is the tetrahedron volume, T^l is the temperature of face l (which is the average of the four nodal temperature), $n_j^{(l)}$ is the exterior unit vector normal to face l, and $S^{(l)}$ is the face surface area.

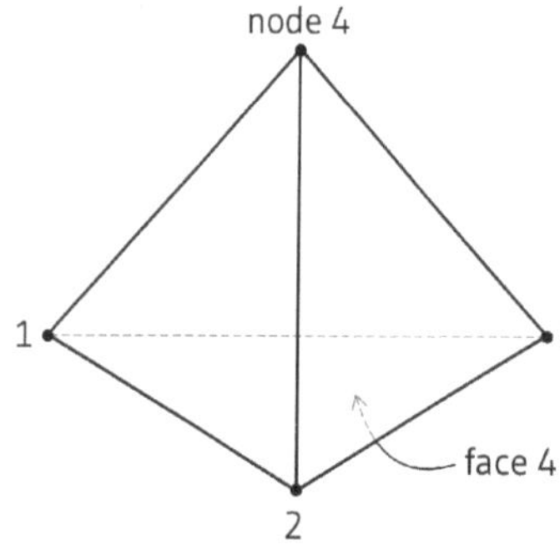

Figure 1. The schematic diagram of the tetrahedron element used in FLAC3D.

In addition to temperature, the energy balance equation can be rewritten as Equation (8):

$$Q_e^n = Q_t^n - \frac{q_v V}{4} + m^n C_v^n \frac{dT^n}{dt} \tag{8}$$

where n is the index of each node, Q_e^n is the equivalent nodal heat, Q_t^n is the nodal contribution and can be calculated from Equation (9), m^n is a parameter and can be obtained through Equation (10), and the other variables are the same as those mentioned above.

$$Q_t^n = \frac{q_i n_i^{(n)} S^{(n)}}{3} \tag{9}$$

$$m^n = \frac{\rho V}{4} \tag{10}$$

4. Model Setting

4.1. Model Geometry and Boundaries

The numerical model employed in this study, as shown in Figure 2, was built to simulate a typical application of the AGF method in the construction of a cross passage between two tunnels of an urban subway project. In this study, the focus is on the displacement of the ground surface; the groundwater dynamics, as suggested by Zheng [13], have little effect on it. Therefore, the effect of groundwater migration and interaction with the construction is not considered in the model. To reduce the model calculation time, only one layer of saturated clay was considered. The model geometry boundaries, as shown in Figure 2a, was set as 80 m × 80 m × 40 m, and the tunnels were located 20 m under the ground surface. As presented in Figure 2b, the two tunnels were 14 m away from each other with a 0.3 m thick shield tunnel segment. The arch-shaped cross passage, as illustrated in Figure 2c, was 4 m wide and high, and the liner within it was set as 0.1 m thick. When considering the differences in the thickness of the liners, the tunnel liners were simulated using volumetric elements, while the liners of the cross passage were modeled using shell elements. In addition, to examine the impact of the thickness of the freezing wall on ground surface deformation, five models with different freezing wall thicknesses, i.e., 1.0 m, 1.5 m, 2.0 m, 2.5 m, and 3.0 m, were developed.

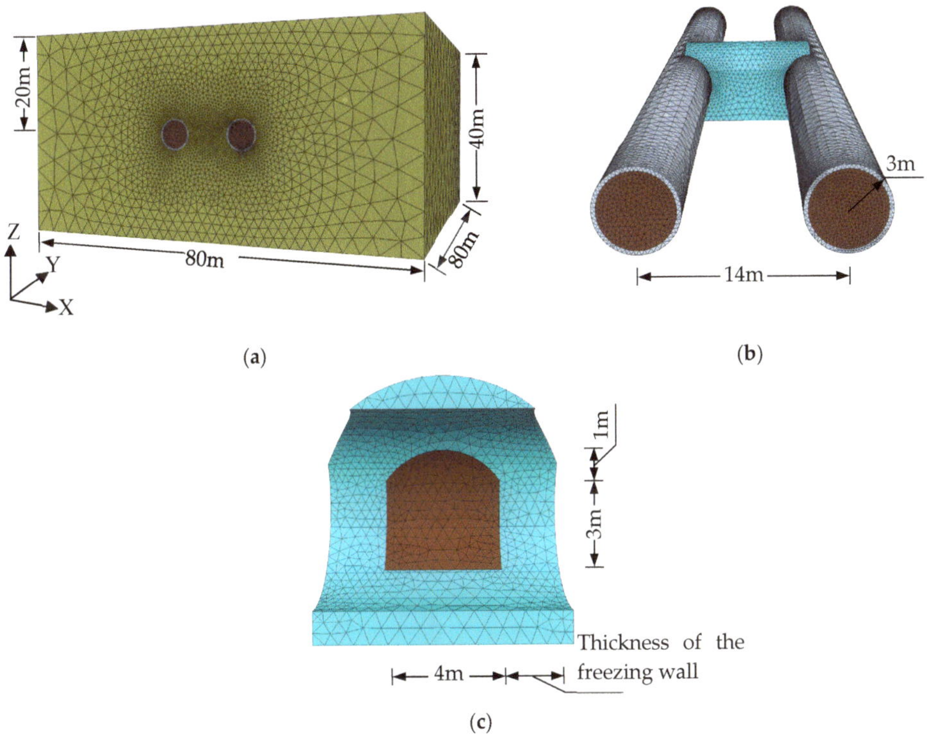

Figure 2. Dimensions of the model. (**a**) Dimensions of the overall model; (**b**) dimensions of the tunnel; (**c**) dimensions of the cross-passage.

For the mechanical boundaries, the top of the model was set as a free boundary, the bottom of the model was set as completely constrained, and the normal displacement of the four vertical faces of the model was set as constrained. For thermal conditions, the temperature field was initialized uniformly at 20 °C. Considering that the upper face of the model would be open to the external environment and exchange heat with it, the constant temperature boundary of 20 °C was fixed for all faces except for the top face, while the others were considered as adiabatic boundaries.

4.2. Simulation of Construction Procedures

The simulation of the construction procedures involved several activities, such as (1) stress initialization; (2) excavation of the tunnel and installation of the concrete liners; (3) artificial freezing of the surrounding soil of the cross passage; (4) excavation of the cross-passage and the installation of the liner; and (5) thawing of the freezing wall.

In phase (1), the initial stress was modeled by applying the gravity of 9.81 N/kg, with a coefficient of earth pressure at rest of 0.71. In phase (2), the excavation of the tunnel was simulated in 10 steps, with 8 m excavated in each step. While excavating the tunnel, the liners are installed simultaneously. In phase (3), the installation of the freezing pipes surrounding the cross passage was modeled, equivalent to the additional volumetric heat-absorbing source in the freezing wall. This volumetric source was added after the excavation of the tunnel, and it was removed before the excavation of the cross passage. In real-world engineering practices, the total power available for artificial freezing always has a limit, and thus, to control the variables of the freezing walls with different thicknesses, the total freezing power (in the unit of W) was set as the same. However, in FLAC3D, the valid input of freezing power is in unit volumetric form (in the unit of W/m^3); therefore, as recommended by Zheng [13], the volumetric heat-absorbing source corresponding to a freezing wall with a thickness of 3 m was set as -25 W/m^3, and the others were determined by keeping the total freezing power constant. In addition, the freezing period was determined by ensuring that the temperature throughout the freezing wall was below -10 °C. The source strength and freezing periods corresponding to the freezing walls with different thicknesses are summarized in Table 1. In phase (4), the excavation of the cross-passage was modeled in four steps.

Table 1. Source strengths and freezing periods of different freezing wall thicknesses.

Thickness (m)	1.0	1.5	2.0	2.5	3.0
Source strength (W/m^3)	-105	-64	-44	-32	-25
Freezing period (day)	18	26	36	48	60

4.3. Material Parameters

Material properties can be affected by temperature. When the temperature is below the freezing point, studies have shown that the physical parameters of reinforced concrete are insensitive to temperatures as low as -10 °C [21]. For soil, however, although properties such as friction angle are also less affected by temperature when below the freezing point [22], considerable research has shown that many properties of the soil, such as bulk modulus, shear modulus, cohesion, thermal conductivity, and specific heat capacity, are highly affected by the freezing and thawing process [23–26]. Moreover, these properties of the soil could be thought of as linear functions of temperature [27,28]. When the temperature falls into the range of freezing point to 20 °C, the physical properties of both soil and reinforced concrete are barely affected by the temperature change [25,29]. In addition, the properties of soils at the same temperature but under different processes (i.e., freezing or thawing process) have different values [13].

The data of the soil properties were retrieved from Zheng et al. [30] as shown in Tables 2 and 3. According to Zheng et al. [30], the freezing temperature of the soil was set as -0.56 °C. Following the discussion above, the properties of liners were treated as constant throughout the whole process, and the value of some properties, i.e., the bulk modulus, shear modulus, cohesion, thermal conductivity, and specific heat capacity of the soil under different processes, were determined by linear interpolation of the corresponding test results separately when the temperature of the soil was below the freezing point, but set as constant when the temperature was above the freezing point. The friction angle, on the other hand, was thought to be irrelevant to the changes in temperature within certain confining pressures, which is consistent with the experimental results obtained by Liu et al. [22]. In the numerical models, by utilizing the user-defined fish function in the FLAC3D, the properties of the soil were automatically adjusted with the change in temperature.

Table 2. Mechanical properties of the materials.

Material	Temperature (°C)	Density (kg/m³)	Bulk Modulus (MPa)	Shear Modulus (MPa)	Constitutive Model	Friction Angle (°)	Cohesion (kPa)
Soil	20 (Original soil) −10 20 (Thawed soil)	1880 1880 1880	16.67 106 40	7.69 57.9 18.46	Mohr–Coulomb Mohr–Coulomb Mohr–Coulomb	17 * 17 * 17 *	22 330 5.5
Liner of the tunnel	-	2500	2×10^4	1.5×10^4	Elastic	-	-
Liner of the cross passage	-	2400	1.56×10^4	1.17×10^4	Elastic	-	-

* When the temperature is above the freezing temperature, the thermal expansion coefficient of soils is positive and constant.

Table 3. Thermal properties of the materials.

Material	Temperature (°C)	Thermal Conductivity (W/m·°C)	Specific Heat Capacity (J/kg·°C)	Thermal Expansion Coefficient (°C⁻¹)	Heave Rate (°C⁻¹)	Thaw Settlement Rate (°C⁻¹)
Soil	20 −10	1.23 1.38	1040 1110	1×10^{-5} * -	- -9.64×10^{-4}	- -1.03×10^{-3}
Liner of the tunnel	-	0.5	1000	1×10^{-5}	-	-
Liner of the cross passage	-	- #	- #	1×10^{-5}	-	-
Air in the tunnel	-	0.025	1000	-	-	-

* When the temperature is above the freezing temperature, the thermal expansion coefficient of soils is positive and constant. # Because the liners of the cross passage were simulated using shell element, these properties were currently not available for setting in FLAC3D.

5. Results and Discussions

To present the results more clearly, the coordinates were constructed as illustrated in Figure 3 on the ground surface. The Y-axis is aligned with the longitudinal direction of the tunnel, while the X-axis is set as the transverse direction of the tunnel. In addition, the origin is right above the center of the cross passage.

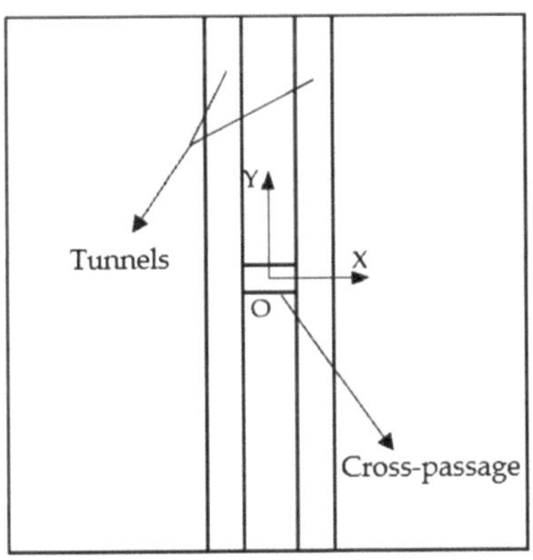

Figure 3. The schematic diagram of the coordinate used in the presentation of the results.

5.1. Development of Frost Heave and Thaw Settlement and How the Thickness of the Freezing Wall Affected Them

Figure 4 illustrates the frost heaves (the vertical displacements arising from the freezing process) and thaw settlements (the vertical displacements caused by the thawing process) at the origin O in Figure 3 induced by freezing walls of different thicknesses with respect to time. As illustrated in Figure 4a, there is a lag in the frost heave of the ground surface during the freezing process after the heat-absorbing source is added to the freezing wall. This lag corresponds to a slight settlement of the surface ground (the shrinkage stage), which is most likely caused by a small shrinkage in the soil before the temperature is below the freezing temperature of the soil. After passing the shrinkage stage, the frost heave begins to develop approximately linearly with respect to time (the fast-freezing stage). In addition, as shown in Figure 4b, the thawing process, the reversed process of the frost heave, shows 'inverted' curves compared with their counterparts of the frost heaving process. Similar to the the frost heave curve having a fast-freezing stage, the thaw settlement curve has a 'fast-thawing stage' just after the removal of the heat-absorbing source. Similar to the frost heave at the fast-freezing stage, the thaw settlement at the fast-thawing stage develops approximately linearly with a rather large slope. After the fast-thawing stage, the settlement becomes stable. These results are similar to that reported by Zheng et al. [2].

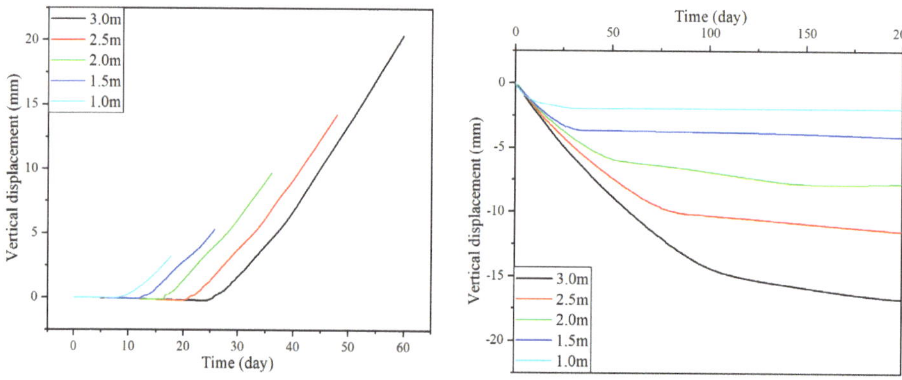

(a) (b)

Figure 4. Development of the frost heave and thaw settlement induced by the freezing walls with different thicknesses. (**a**) Development of the frost heave; (**b**) development of the thaw settlement.

An increase in the thickness of the freezing wall in the freezing process, as illustrated in Figure 5, can prolong the period of freezing. In addition, the shrinkage period of the frost heave extends linearly with an increase in the thickness of the freezing wall, while the fast-freezing stage lengthens faster and nonlinearly. In the thawing process, as shown in Figure 6, with an increase in the thickness of the freezing wall, the fast-thawing stage also extends, and the effect on it is much more apparent than that shown on the fast-freezing stage.

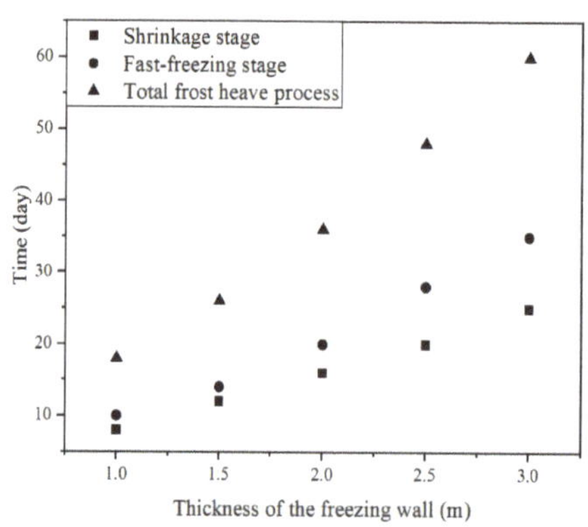

Figure 5. Changes in the time corresponding to the shrinkage stage, the fast-freezing stage, and the total freezing process with an increase in the thickness of the freezing wall.

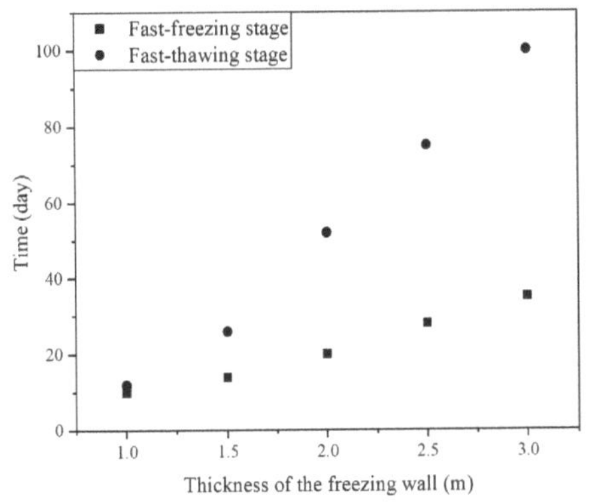

Figure 6. Changes in the time corresponding to the fast-freezing stage and the fast-thawing stage with an increase in the thickness of the freezing wall.

5.2. Temperature Distribution within the Freezing Wall at Two Characteristic Time Points

As mentioned in the Section 1, problems concerning temperature distribution within freezing walls have long been of great interest. In this study, the temperature distribution within the freezing wall at two characteristic time points, i.e., the starting point of the fast-freezing stage and the finishing point of the fast-thawing stage, were investigated. These two points were selected to shed light on the identification and explanation of the beginning of the fast-freezing stage and the end of the fast-thawing stage with temperature distribution. The distribution results are presented in Figures 7 and 8. Figure 7 illustrates the temperature distribution within freezing walls of different thicknesses at the cross-section of the Y-axis right before the fast-freezing stage. Before entering the fast-freezing stage, the temperature of the area right around the cross passage would already have been below freezing temperature. This may be traced to the shrinkage of the unfrozen soil due to the drop in the temperature. The shrinkage of the soil cancels part of the frost heave. Therefore, part of the soil froze ahead of the commencement of the frost heave of the ground surface. In addition, as shown in Figure 7, the lowest temperature within the freezing wall at this time point first drops and then increases with an increase in the thickness of the freezing wall. Figure 8 demonstrates the temperature distribution within freezing walls of different thicknesses at the cross-section of the Y-axis right after the fast-thawing stage. Like its counterpart in the freezing process, when achieving a stable settlement, the temperature of some areas in the middle of the freezing wall would still have been below freezing temperature. There are several potential explanations for this. There was only a very thin layer of frozen soil within the freezing wall, and most of the soil was in a thawed state. Therefore, the displacement arising from thawing would not contribute too much to the entire thawing settlement. In addition to this, when the temperature rises during the thawing process, the unfrozen soil expands, and this also counteracts the thaw settlement of the ground surface. Therefore, part of the soil remains frozen when the settlement is about to become stable. Figure 8 also shows that the lowest temperature within the freezing wall at this time point drops with an increase in the thickness of the freezing wall in a gradual, gentle way. Apart from this, it can be seen that with an increase in the thickness of the freezing wall, the center of the temperature contour moves downward. This may arise from the setting of the temperature thermal boundary at the top of the model. The temperature at the top of the model is fixed and can be thought of as a heat source which transfers the heat downward to the freezing wall, where the temperature is much lower than the surrounding soil. This heat transfer movement makes the temperature of the upper side of the cross passage higher than that of the lower side. This effect might be more apparent as the prolongation of the fast-thawing stage with an increase in the thickness of the freezing wall.

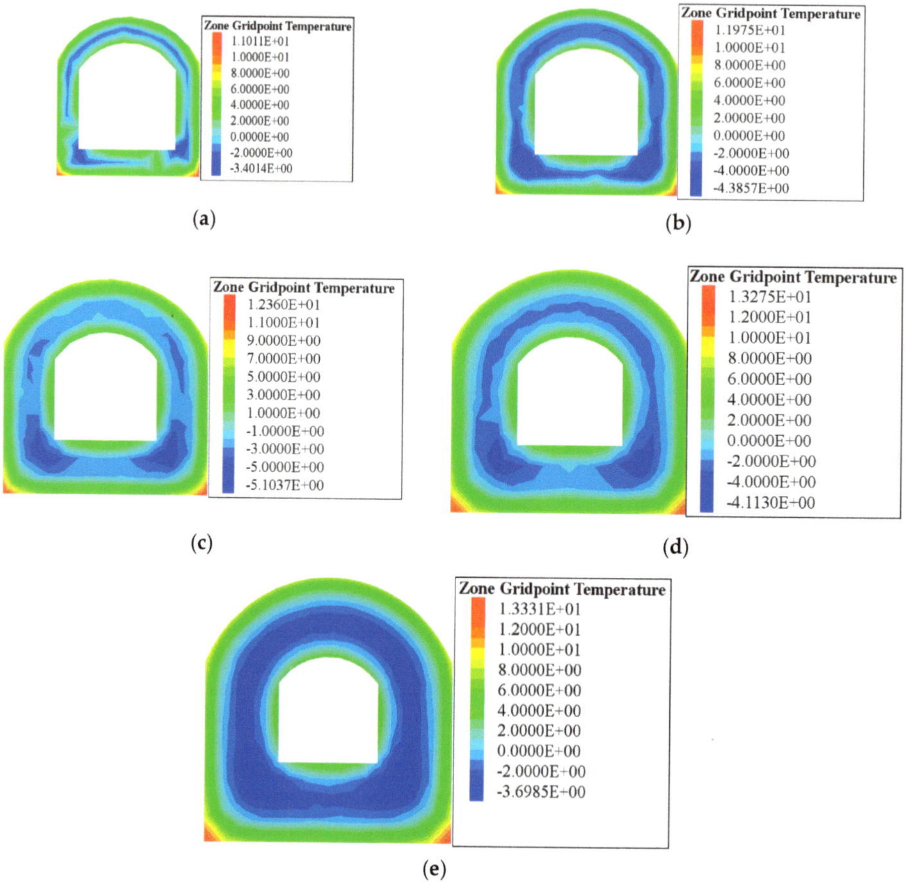

Figure 7. The temperature distribution within freezing walls of different thicknesses at the cross-section of the Y-axis at the starting point of the fast-freezing stage: (**a**) 1.0 m on the 8th day; (**b**) 1.5 m on the 12th day; (**c**) 2.0 m on the 16th day; (**d**) 2.5 m on the 20th day; (**e**) 3.0 m on the 25th day.

5.3. Spatial Distribution of Frost Heave and Thaw Settlement and How the Thickness of the Freezing Wall Affected It

Figures 9 and 10 reveal that the frost heave distribution along the X-axis is wider than that along the Y-axis. This reveals that the frost heave has a gentler variation in the transverse direction of the tunnel, yet a more marked variation in the longitudinal direction of the tunnel. On the contrary, as illustrated by Figures 11 and 12, except for that corresponding to the freezing wall with a thickness of 1 m, thaw settlement distribution along the X-axis has a similar size to that along the Y-axis, which represents that the fluctuation of the thaw settlement along the longitudinal direction of the tunnel is approximately as large as that along the transverse direction of the tunnel. Furthermore, Figure 11 conveys that the settlement is distributed uniformly among different directions. In addition, as presented by Figures 10 and 12, with the increase in the thickness of the freezing wall, the frost heave and the thaw settlement grow accordingly. In addition, to qualify the effects of the thickness of the freezing wall on the maximum ground surface displacements, an attempt was made to fit the simulation results into mathematical equations. It was found that the exponential equation (Equation (11)) can best represent the trends with respect to the thickness of the freezing wall, which consequently implies that the increase in the thickness of the freezing wall might give rise to an exponential growth in the displacements of the ground surface. The fitting parameters are summarized in Table 4, and the fitting results are shown in Figure 13.

$$y = y_0 + Ae^{kx} \tag{11}$$

where y_0, A, and k are the fitting parameters.

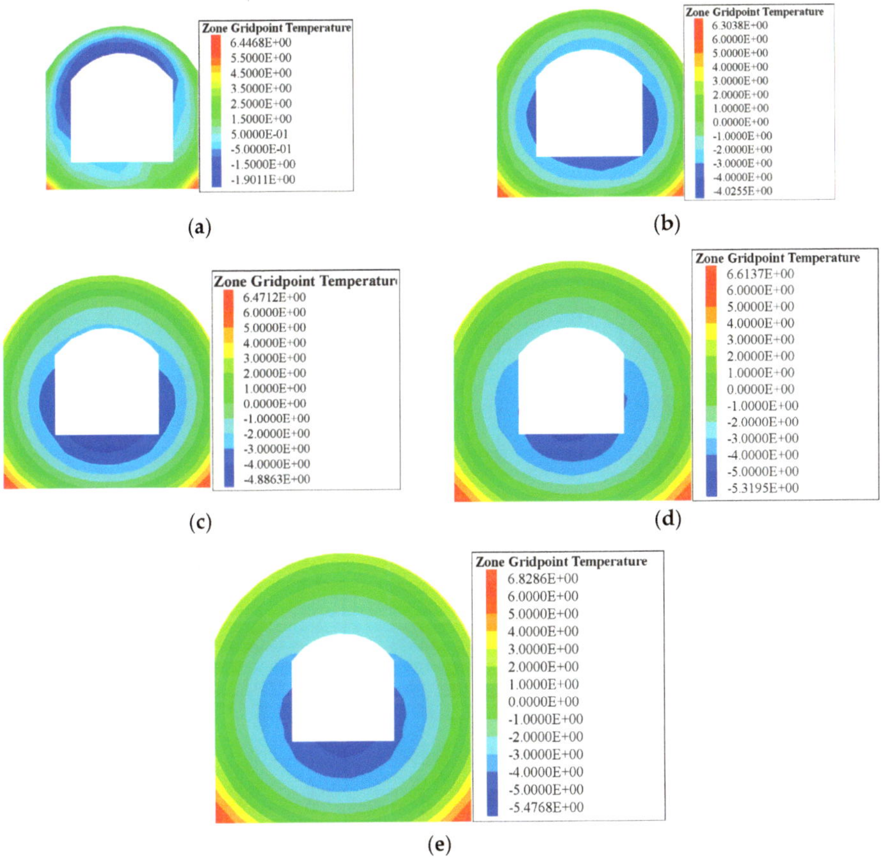

Figure 8. The temperature distribution within freezing walls of different thicknesses at the cross-section of the Y-axis at the finishing point of the fast-thawing stage: (**a**) 1.0 m on the 12th day; (**b**) 1.5 m on the 26th day; (**c**) 2.0 m on the 52nd day; (**d**) 2.5 m on the 75th day; (**e**) 3.0 m on the 100th day.

Table 4. Fitting results of the maximum vertical displacements.

	Fitting Parameters			R^2
	y_0	A	k	
Maximum frost heave	−6.83	5.80	0.52	1.00
Maximum thaw settlement	−9.70	7.75	0.41	1.00

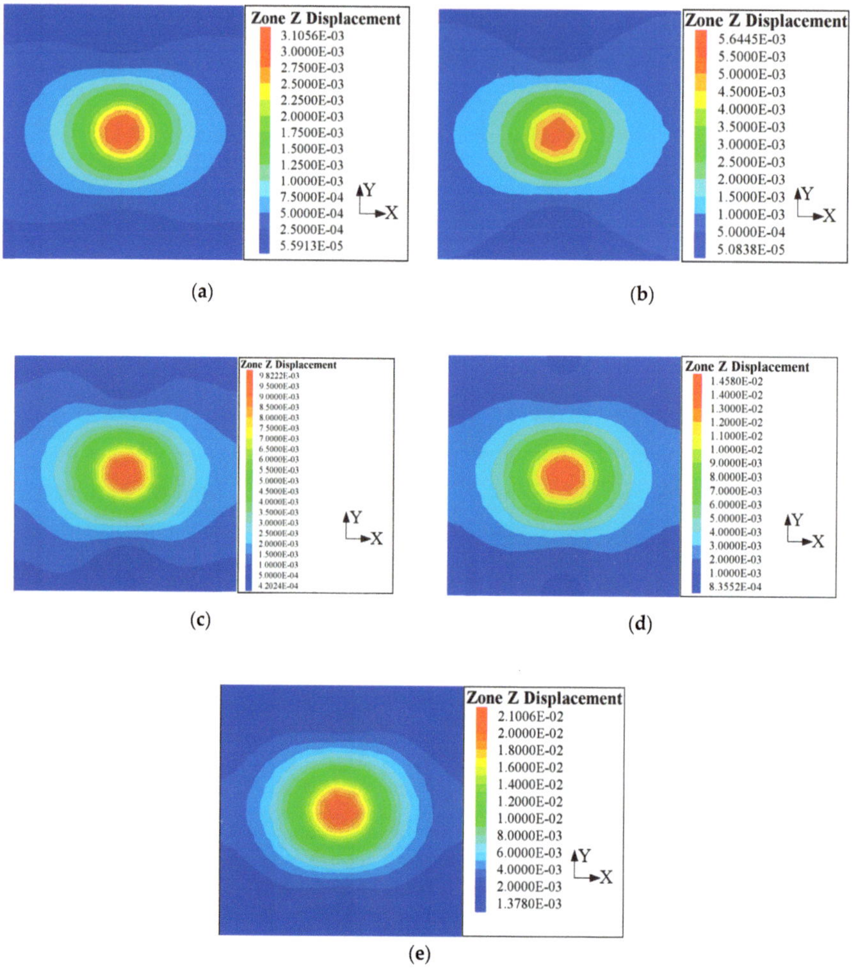

Figure 9. The displacement contours of the frost heave induced by freezing walls of different thicknesses: (**a**) 1.0 m; (**b**) 1.5 m; (**c**) 2.0 m; (**d**) 2.5 m; (**e**) 3.0 m.

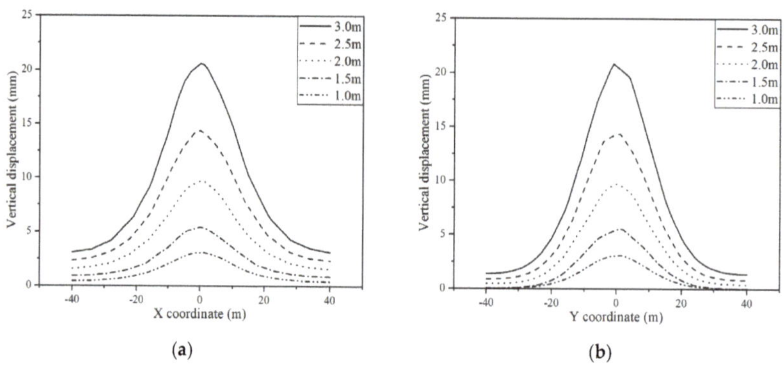

Figure 10. The displacement distribution curves of the frost heave of the ground surface (**a**) along the X-axis and (**b**) along the Y-axis.

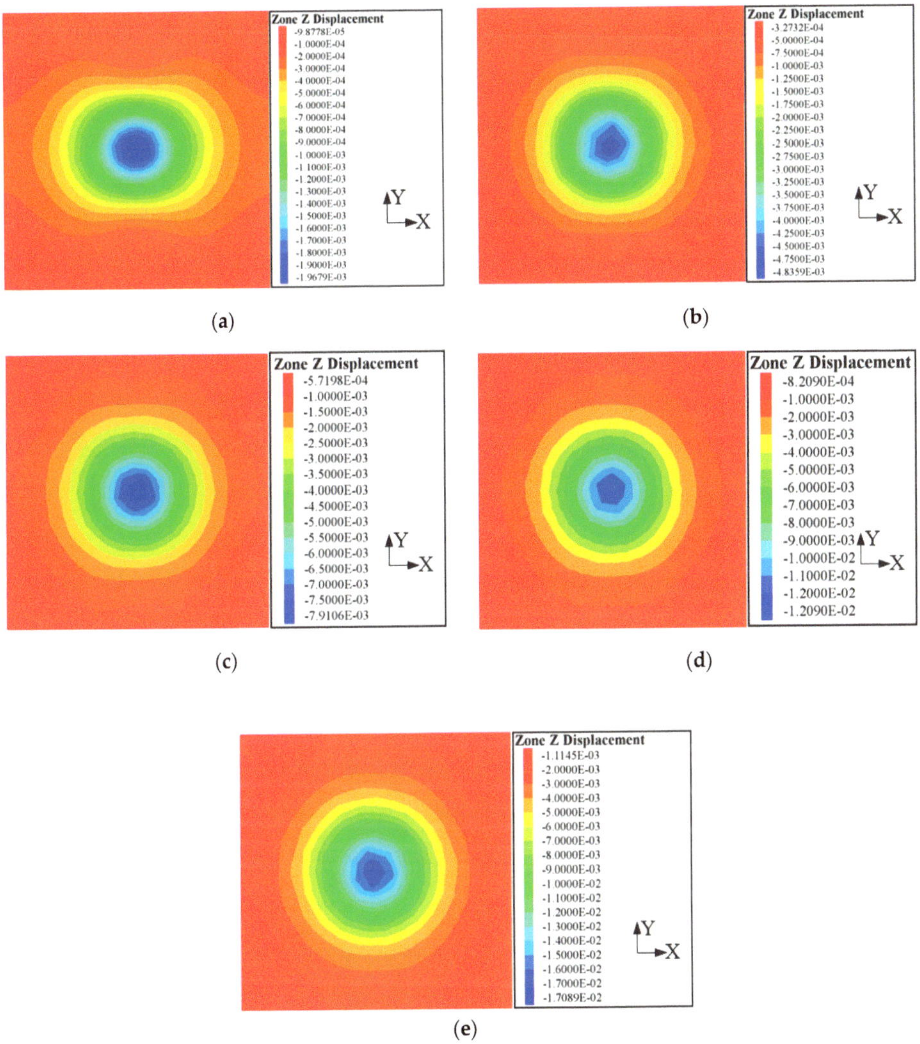

Figure 11. The displacement contours of the thaw settlement induced by freezing walls of different thicknesses: (**a**) 1.0 m; (**b**) 1.5 m; (**c**) 2.0 m; (**d**) 2.5 m; (**e**) 3.0 m.

To evaluate the displacement distribution of frost heave and thaw settlement quantitively, the absolute values of the displacement data of the frost heave and thaw settlement gained from the numerical simulation were fitted to the amplitude version of the Gaussian peak function in the form of Equation (12). The results are summarized in Table 5. As illustrated in Table 5, the parameter w, governing the width of the distribution curve, is larger for the curves of the X-axis than those of the Y-axis, which indicates that the curves along the X-axis are wider than those along the X-axis. This may stem from the geometrical shape of the freezing wall, which is longer in the direction of the X-axis than the Y-axis. The longer side might give rise to a wider settlement, while the shorter side causes a narrower one. However, the difference between them is rather sharp in the curves of frost heave, yet relatively minor in those of thaw settlement, which is consistent with what is shown in Figures 9 and 10. It is also noteworthy that the values of the w of the thaw settlement curves are larger than those of the frost heave curves, which indicates that the thaw settlement curves are generally gentler than the frost heave curves. In addition, the value of w slightly grows with an increase in the thickness of the freezing wall, which implies that the displacement distribution of the frost heave and thaw settlement becomes wider with an increase in the thickness of the freezing wall.

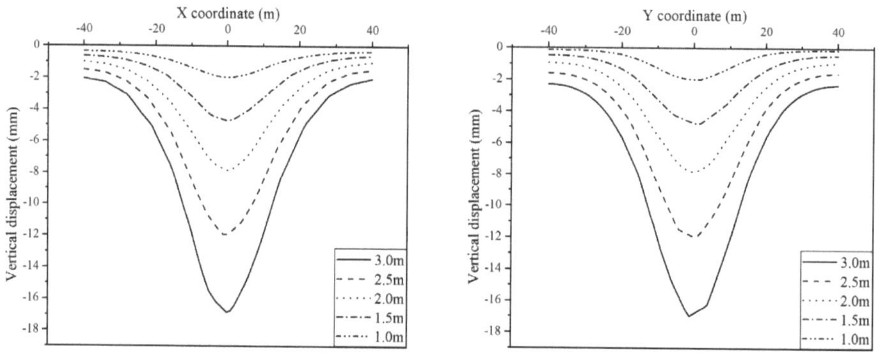

(a) (b)

Figure 12. The displacement distribution curves of the thaw settlement of the ground surface (**a**) along the X-axis and (**b**) along the Y-axis.

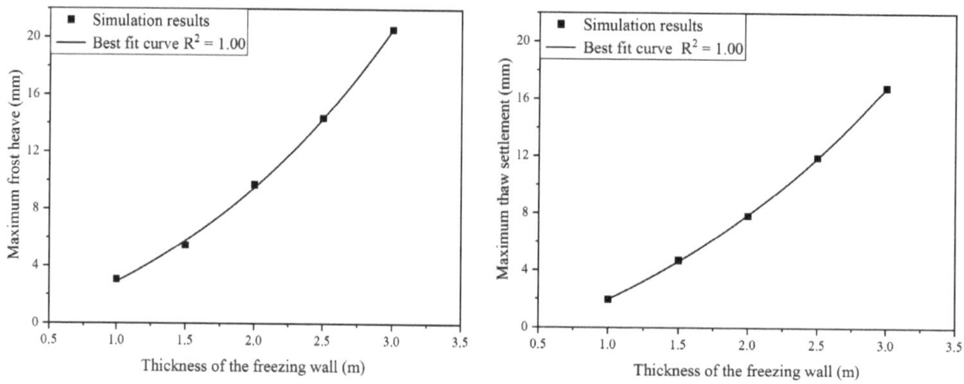

(a) (b)

Figure 13. The variation in the maximum vertical displacements with respect to freezing wall thickness: (**a**) frost heave; (**b**) thaw settlement.

Table 5. The fitting results of the displacement data of the frost heave and thaw settlement.

Thickness of the Freezing Wall (m)	Displacement Type	Direction of the Cross-Section	y_0	x_c	w	A	$y_0 + A$	R^2
1.0 m	Frost heave	X-axis	0.48	−0.03	10.42	2.52	3.00	1.00
		Y-axis	0.08	0.05	9.81	2.97	3.05	1.00
1.5 m		X-axis	0.98	−0.15	10.32	4.29	5.27	1.00
		Y-axis	0.06	0.03	9.99	5.34	5.40	1.00
2.0 m		X-axis	1.75	0.03	10.92	7.71	9.47	1.00
		Y-axis	0.48	−0.01	10.16	9.09	9.57	1.00
2.5 m		X-axis	2.51	0.11	11.37	11.54	14.05	1.00
		Y-axis	0.91	−0.15	10.31	13.34	14.25	1.00
3.0 m		X-axis	3.37	0.15	11.30	16.82	20.19	1.00
		Y-axis	1.49	0.09	10.48	18.91	20.40	1.00

Table 5. Cont.

Thickness of the Freezing Wall (m)	Displacement Type	Direction of the Cross-Section	y_0	x_c	w	A	$y_0 + A$	R^2
1.0 m	Thaw settlement	X-axis	0.35	0.11	11.14	1.57	1.92	1.00
		Y-axis	0.12	−0.35	10.09	1.83	1.95	1.00
1.5 m		X-axis	0.69	0.04	11.19	3.88	4.57	1.00
		Y-axis	0.48	0.06	11.19	4.17	4.65	1.00
2.0 m		X-axis	1.12	0.02	11.51	6.51	7.64	1.00
		Y-axis	0.97	0.00	11.40	6.72	7.70	1.00
2.5 m		X-axis	1.61	0.10	11.69	10.08	11.69	1.00
		Y-axis	1.63	−0.07	11.43	10.19	11.82	1.00
3.0 m		X-axis	2.24	0.08	11.64	14.24	16.48	1.00
		Y-axis	2.38	0.11	11.50	14.29	16.68	1.00

For the maximum absolute values of the displacement curve, which are indicated by $y_0 + A$, although the thaw settlement rate is slightly larger than the frost heave rate, those of the frost heave are larger than those corresponding to the thaw settlement. This may be attributed to the excavation of the cross passage and the installation of the liners. When freezing the surrounding soil of the cross passage, part of the soil within the cross passage area will also freeze. However, when excavating the cross passage, this part of the soil will be removed, which will reduce the volume of the soil which has the potential to settle, along with the restriction of the settlement imposed by the liners. These two factors might work together to lead to a smaller displacement of the thawing process.

$$y = y_0 + Ae^{-\frac{(x-x_c)^2}{2w^2}} \tag{12}$$

where A is the parameter related to the amplitude of the curve, where a larger A will result in a higher peak; x_c determines the location of the center of the curve; w has an impact on the width of the curve, where a greater w will yield a wider curve; and y_0 is the parameter indicating the base of the curve, where a larger y_0 will yield a higher base.

6. Conclusions

In this paper, the frost heave and thaw settlement induced by freezing walls in the AGF method, which are inadequately investigated in the current literature, are systematically examined by a set of state-of-the-art numerical models built by FLAC3D with the data retrieved from the literature. Similar to the results reported by Zheng et al. [2], for the characteristics of the freezing and thawing processes, our results show that a shrinkage stage occurs before entering the linear fast-freezing stage and the settlement becomes stable after a linear fast settlement during the thaw settlement. For temperature distribution within a freezing wall, it is found that the temperature of a certain area within the freezing wall is below freezing temperature prior to the fast-freezing stage and after the fast-thawing stage. For ground displacement, the results indicate that during the frost heave process, the displacement curve along the transverse direction of the tunnel is wider than that along the longitudinal direction of the tunnel, while that of the thaw settlement is comparatively uniform in different directions.

In terms of the effects of the thickness of the freezing wall on the freezing and thawing characteristics, it is observed that (1) the increase in the thickness of the freezing wall gives rise to the extension of the freezing process and the fast-thawing stage; (2) similar to that pointed out by Zheng [13], the frost heave and thaw settlement of the ground surface might increase exponentially with growth in the thickness of the freezing wall, which can be informative when it comes to the determination of the thickness of the freezing wall; and (3) the thickness of the freezing wall is also a contributory factor to a more gentle displacement distribution of the surface ground.

This study provided insights into the ground displacements during the freezing and thawing processes of artificial freezing methods in the construction of subway cross passages. However, in this study, the groundwater movement during freezing and thawing process was not considered. More research is needed to investigate the interaction of groundwater dynamics with the freezing and thawing processes.

Author Contributions: Methodology, Y.O.; Software, H.C.; Validation, S.Y.; Formal analysis, H.B.; Data curation, T.C.; Writing—original draft, Y.O. and L.W.; Writing—review & editing, A.S. and Q.Z. All authors have read and agreed to the published version of the manuscript.

Funding: This research received no external funding.

Institutional Review Board Statement: Not Applicable.

Informed Consent Statement: Not Applicable.

Data Availability Statement: The raw data supporting the conclusions of this article will be made available by the authors on request.

Conflicts of Interest: Authors Yazhou Ou, Long Wang, Hui Bian, Hua Chen, Shaole Yu and Tao Chen were employed by the company China Construction Eighth Engineering Division Co., Ltd. The remaining authors declare that the re-search was conducted in the absence of any commercial or financial relationships that could be construed as a potential conflict of interest.

References

1. Lu, X.; Chen, X.; Chen, X. Risk Prevention and Control of Artificial Ground Freezing (AGF). *Chin. J. Geotech. Eng.* **2021**, *43*, 2308–2314.
2. Zheng, L.; Gao, Y.; Zhou, Y.; Liu, T.; Tian, S. A Practical Method for Predicting Ground Surface Deformation Induced by the Artificial Ground Freezing Method. *Comput. Geotech.* **2021**, *130*, 103925. [CrossRef]
3. Domke, O. Über Die Beanspruchung Der Frostmauer Beim Schachtabteufen Nach Dem Gefrierverfahren. *Glückauf* **1915**, *51*, 1129–1135.
4. Zhou, X.; Jiang, G.; Li, F.; Gao, W.; Han, Y.; Wu, T.; Ma, W. Comprehensive Review of Artificial Ground Freezing Applications to Urban Tunnel and Underground Space Engineering in China in the Last 20 Years. *J. Cold Reg. Eng.* **2022**, *36*, 04022002. [CrossRef]
5. Yue, F.; Qiu, P.; Yang, G.; Shi, R. Design and Practice of Freezing Method Applied to Connected Aisle in Tunnel under Complex Conditions. *Chin. J. Geotech. Eng.* **2006**, *28*, 660–663.
6. Fang, L.; Li, F.; Cui, H.; Ding, H. Research on Freezing Wall Thickness Design of Metro Cross Passage Based on Structural Mechanics Method. *Urban Mass. Transit.* **2020**, *23*, 117–121.
7. Liu, Z. *Study on the Thickness Design of the Freezing Wall of the Subway Liaison Channel in the Water Rich Gravel Stratum*; Xi'an University of Science and Technology: Xi'an, China, 2018.
8. Zhou, F.; Zhou, P.; Li, J.; Ge, T.; Lin, J.; Wang, Z. Key Parameters Design Method of AGF Method for Metro Connecting Passage in Water-Rich Coastal Area. *KSCE J. Civ. Eng.* **2022**, *26*, 5301–5317. [CrossRef]
9. Yan, S.; Li, J.; Sun, L.; Wu, K. Study on Design and Calculating Methods of Artificial Freezing Curtain on Metro Connected Passage-Way Project. *Site Investig. Sci. Technol.* **2014**, *2014*, 5–10.
10. Tang, Y.; Xiao, S.; Zhou, J. Deformation Prediction and Deformation Characteristics of Multilayers of Mucky Clay under Artificial Freezing Condition. *KSCE J. Civ. Eng.* **2019**, *23*, 1064–1076. [CrossRef]
11. Li, J.; Li, J.; Cai, Y.; Wu, D.; Guo, C.; Zhao, W.; Tang, K.; Liu, Y. Application of Artificial Freezing Method in Deformation Control of Subway Tunnel. *Adv. Mater. Sci. Eng.* **2022**, *2022*, 3251318. [CrossRef]
12. Zhou, J.; Tang, Y. Practical Model of Deformation Prediction in Soft Clay after Artificial Ground Freezing under Subway Low-Level Cyclic Loading. *Tunn. Undergr. Space Technol.* **2018**, *76*, 30–42. [CrossRef]
13. Zheng, L. *Research on Optimization Method and Engineering Application of Frozen Wall Thickness for Cross-Passage in Urban Rail Transit*; University of Science and Technology Beijing: Beijing, China, 2021.
14. Cai, H.; Peng, L.; Zheng, T. Prediction Method of Surface Frost Heave Based on Stochastic Medium Theory in Tunnel Freezing Period. *J. Cent. South Univ.* **2014**, *45*, 4251–4257.
15. Cai, H.; Liu, Z.; Li, S.; Zheng, T. Improved Analytical Prediction of Ground Frost Heave during Tunnel Construction Using Artificial Ground Freezing Technique. *Tunn. Undergr. Space Technol.* **2019**, *92*, 103050. [CrossRef]
16. Zhang, Z.; He, C. Study on Construction of Cross Connection of Shield Tunnel and Connecting Aisle by Freezing Method. *Chin. J. Rock Mech. Eng.* **2005**, *24*, 3211–3217.
17. Fu, Y.; Hu, J.; Wu, Y. Finite Element Study on Temperature Field of Subway Connection Aisle Construction via Artificial Ground Freezing Method. *Cold Reg. Sci. Technol.* **2021**, *189*, 103327. [CrossRef]
18. Huang, S.; Guo, Y.; Liu, Y.; Ke, L.; Liu, G.; Chen, C. Study on the Influence of Water Flow on Temperature around Freeze Pipes and Its Distribution Optimization during Artificial Ground Freezing. *Appl. Therm. Eng.* **2018**, *135*, 435–445. [CrossRef]
19. Zhang, S.; Zhou, X.; Zhang, J.; Sun, T.; Ma, W.; Liu, Y.; Yang, N. A Case Study of Energy-Saving and Frost Heave Control Scheme in Artificial Ground Freezing Project. *Geofluids* **2022**, *2022*, 1004735. [CrossRef]

20. Cai, H.; Li, S.; Liang, Y.; Yao, Z.; Cheng, H. Model Test and Numerical Simulation of Frost Heave during Twin-Tunnel Construction Using Artificial Ground-Freezing Technique. *Comput. Geotech.* **2019**, *115*, 103155. [CrossRef]
21. Lee, G.C.; Shih, T.S.; Chang, K.-C. Mechanical Properties of Concrete at Low Temperature. *J. Cold Reg. Eng.* **1988**, *2*, 13–24. [CrossRef]
22. Liu, X.; Liu, E.; Zhang, D.; Zhang, G.; Song, B. Study on Strength Criterion for Frozen Soil. *Cold Reg. Sci. Technol.* **2019**, *161*, 1–20. [CrossRef]
23. Liu, J.; Chang, D.; Yu, Q. Influence of Freeze-Thaw Cycles on Mechanical Properties of a Silty Sand. *Eng. Geol.* **2016**, *210*, 23–32. [CrossRef]
24. Wang, D.; Ma, W.; Niu, Y.; Chang, X.; Wen, Z. Effects of Cyclic Freezing and Thawing on Mechanical Properties of Qinghai–Tibet Clay. *Cold Reg. Sci. Technol.* **2007**, *48*, 34–43. [CrossRef]
25. Özgan, E.; Serin, S.; Ertürk, S.; Vural, I. Effects of Freezing and Thawing Cycles on the Engineering Properties of Soils. *Soil Mech. Found. Eng.* **2015**, *52*, 95–99. [CrossRef]
26. Kok, H.; McCool, D.K. *CRREL Special Report 90-1*; CRREL Special and Products: Hanover, NH, USA, 1990; pp. 70–76.
27. Christ, M.; Kim, Y.-C.; Park, J.-B. The Influence of Temperature and Cycles on Acoustic and Mechanical Properties of Frozen Soils. *KSCE J. Civ. Eng.* **2009**, *13*, 153–159. [CrossRef]
28. Wang, D.; Zhu, Y.; Ma, W.; Niu, Y. Application of Ultrasonic Technology for Physical–Mechanical Properties of Frozen Soils. *Cold Reg. Sci. Technol.* **2006**, *44*, 12–19. [CrossRef]
29. Shoukry, S.N.; William, G.W.; Downie, B.; Riad, M.Y. Effect of Moisture and Temperature on the Mechanical Properties of Concrete. *Constr. Build. Mater.* **2011**, *25*, 688–696. [CrossRef]
30. Zheng, L.; Gao, Y.; Zhou, Y.; Tian, S. Research on Surface Frost Heave and Thaw Settlement Law and Optimization of Frozen Wall Thickness in Shallow Tunnel Using Freezing Method. *Rock Soil Mech.* **2020**, *41*, 10.

applied sciences

MDPI

Article

Sensitivity Analysis of Different Hydrothermal Characteristics in the Variable Thermodynamic Processes of Soft Clay Rock

Tao Wang [1,2,3,4,*], Huixi Lin [1,2], Kexiong Ren [3], Jian Gao [1,2] and Di Wang [4]

[1] SINOPEC Key Laboratory of Geology and Resources in Deep Stratum, Beijing 102206, China
[2] SINOPEC Petroleum Exploration and Production Research Institute, Beijing 102206, China
[3] The Natural Resources and Planning Bureau of Yiling District, Yichang 443100, China
[4] State Key Laboratory for Geomechanics and Deep Underground Engineering, School of Mechanics and Civil Engineering, China University of Mining and Technology, Xuzhou 221116, China
* Correspondence: taowang@cumt.edu.cn

Abstract: Artificial ground freezing technology is the most important construction method of complex water-bearing soft clay rock. The thermodynamic properties of soft clay rock are important evidence for the design and construction of space resources development, and the variable hydrothermal parameter can directly affect the uncertain thermodynamic properties of soft clay rock. In this work, an array of field experiments on the soft clay rock are carried out, and the anisotropic spatial variations of hydrothermal parameters of soft clay rock are obtained. The statistical variability characteristics of variable hydrothermal parameters are estimated. A stochastic coupling model of soft clay rock with heat conduction and porous flow is proposed, and the uncertain thermodynamic properties of soft clay rock are computed by the self-compiled program. Model validation with the experimental and numerical temperatures is also presented. According to the relationship between anisotropic spatial variations and statistical variability characteristics for the different random field correlation models, the effects of the autocorrelation function, coefficient of variation, and autocorrelation distance of variable hydrothermal parameters on the uncertain thermodynamic properties of soft clay rock are analyzed. The results show that the proposed stochastic analysis model for the thermal characteristics of soft clay rock, considering the spatial variability of frozen soil layers, is scientifically reasonable. The maximum standard deviation of average thickness is 2.33 m, and the maximum average temperature is 2.25 °C. For the autocorrelation function, the most significant impact comes from DBIN. For the coefficient of variation, the most significant impact comes from thermal conductivity. Different variations of hydrothermal parameters have different effects on the standard deviation of soft clay rock temperature. The biggest influence is the thermal conductivity, while the lowest influence is the specific heat capacity.

Keywords: soft clay rock; uncertain thermodynamic properties; stochastic analysis method; spatial variability; influencing factor

1. Introduction

In the underground space resources development within highly permeable and weakly consolidated formations, sustainable hazards often arise. These include groundwater inundation, leading to tunnel flooding and potential structural instability, soil subsidence due to water seepage, risking tunnel collapse, and environmental impacts like land settlement and groundwater contamination [1–3]. Addressing these challenges requires comprehensive engineering solutions, including artificial ground freezing, soil stabilization techniques, and effective groundwater management strategies, to ensure the sustainable and safe development of underground space in such geological conditions [4–6]. Artificial ground freezing (AGF) is a sophisticated engineering method extensively employed in underground space resources development. The freezing tube temperature can be reduced to −30 °C. The soil temperature can be lowered to −15 °C. The permeability can be reduced by about three times. The strength can be increased by about three times. By manipulating groundwater levels and soil temperatures, AGF creates a protective frozen barrier underground, effectively shielding underground space excavation sites from water and soil infiltration [7,8]. This technique is particularly invaluable in areas with complex geological compositions and high groundwater levels, where traditional construction methods are prone to safety hazards like ground settlement and underground space collapse due to water seepage. AGF involves a series of meticulously executed steps, including pre-freezing to lower groundwater temperatures, primary freezing utilizing freezing pipes, and subsequent underground space excavation within the frozen support layer. Post-freezing curing is then carried out to ensure the long-term stability and integrity of the frozen barrier [9]. AGF not only guarantees high safety standards by mitigating the risks associated with groundwater ingress but also enhances construction efficiency by enabling uninterrupted excavation processes. Furthermore, its minimal environmental impact compared to conventional support structures underscores its significance in sustainable urban development initiatives. The temperature of underground space construction is crucial for ensuring the effectiveness and safety of AGF techniques [10]. It directly influences the formation and stability of the frozen barrier underground, which is essential for controlling groundwater inflow and stabilizing surrounding soil during excavation. By studying the optimal freezing temperature, engineers can enhance the efficiency of this method, minimize construction risks such as underground space collapse and ground settlement, and ultimately ensure the long-term structural integrity and safety of underground space [11,12].

The temperature field of the soft clay rock in underground space resources development is influenced by several factors. Firstly, the properties of the surrounding soft clay rock, such as thermal conductivity and moisture content, impact heat transfer rates and the extent of freezing. Secondly, the type and temperature of the refrigerant used for ground freezing play a critical role in controlling the temperature distribution within the soft clay rock. Additionally, environmental factors like ambient temperature and groundwater flow rates can affect the effectiveness of the freezing process. Furthermore, the geometry and design of the freezing system, including the spacing and arrangement of freezing pipes, also influence the temperature field [13–15]. Soft clay rock thermal conductivity determines the rate of heat transfer within the ground, impacting the freezing process's efficiency and the uniformity of temperature distribution in the underground space. Additionally, the moisture content of soft clay rock affects thermal conductivity and heat capacity, influencing the heat absorption and release capabilities. The permeability coefficient of soft clay rock governs the flow of groundwater, which can affect the effectiveness of ground freezing by influencing the distribution and stability of the frozen barrier [16,17]. The horizontal fluctuation range of clay rock is generally about 5 m, and the vertical fluctuation range is generally about 2 m. Variations in mineral composition, organic content, and moisture distribution lead to differences in thermal conductivity and heat capacity across soil layers. Similarly, variations in pore size, shape, and connectivity influence permeability coefficients, affecting the movement of fluids within the soft clay rock. Geological factors such as fractures, faults, and bedding planes further contribute to spatial heterogeneity in soft clay rock properties [18–20]. The mechanism behind quantifying spatial variability in geotechnical properties through stochastic field methods lies in capturing the inherent randomness and spatial correlation of geotechnical attributes [21,22]. These methods utilize statistical techniques to model the spatial distribution of soft clay rock parameters, considering factors such as geological heterogeneity, depositional processes, and environmental influences. By characterizing the spatial dependence structure of geotechnical properties, stochastic field approaches provide insights into the underlying processes governing variability across landscapes [23,24]. This includes accounting for factors such as soil texture, moisture content, and compaction, which influence the spatial distribution of properties like permeability, porosity, and shear strength [25,26]. Hence, the spatial variability of hydrothermal parameters needs to be clarified, and the effects of variable hydrothermal parameters on the thermodynamic properties of soft clay rock need to be evaluated.

In this paper, the anisotropic spatial variations of the hydrothermal parameters of soft clay rock are counted by test data results. The statistical variability characteristics of variable hydrothermal parameters are estimated. A stochastic coupling model of thermodynamic characteristics for soft clay rock is proposed, and the uncertain temperature properties of underground space are computed. The reliability of the model is verified by test results. According to the relationship between anisotropic spatial variations and statistical variability characteristics for the different random field correlation models, the effects of autocorrelation function (ACF), coefficient of variation (COV), and autocorrelation distance (ACD) of variable hydrothermal parameter on the uncertain thermodynamic properties of soft clay rock are discussed. The results can provide a basis for the sustainable development of underground space.

2. Spatial Variability of Hydrothermal Parameter

The thermodynamic parameters of soft clay rock in underground space resource development are paramount for successful construction in soft clay rock layers. Ground freezing techniques rely on precise control of parameters like freezing temperature, freezing rate, and latent heat of fusion. These parameters dictate the extent and stability of frozen zones around underground space excavations, ensuring temporary ground support, preventing groundwater ingress, and minimizing settlement risks [27]. A thorough understanding and accurate management of thermodynamic parameters are crucial for maintaining underground space stability, protecting surrounding structures, and ensuring worker safety during construction. Improper control or deviation from optimal parameters can lead to ground instability, structural damage, and delays in project completion [28,29]. Therefore, meticulous attention to thermodynamic parameters is essential for the efficient and safe execution of underground space projects in soft clay rock formations.

2.1. Anisotropic Spatial Variations

The spatial variability of soft clay rock exhibits significant anisotropy, characterized by distinct directional variations in geotechnical properties. This anisotropy arises due to the orientation and arrangement of clay particles, which are influenced by factors such as sedimentation processes, depositional history, and mechanical loading conditions. In soft clay rock, the arrangement of clay particles tends to align along the principal stress directions, resulting in anisotropic behaviors in terms of strength, stiffness, and permeability. Moreover, the presence of soil structure and fabric further enhances the directional dependence of geotechnical properties. For instance, the orientation of clay platelets and micro-fabrics within the soil matrix can lead to pronounced variations in shear strength along different directions. Additionally, external factors such as loading history and environmental conditions can induce changes in the anisotropic behavior of soft clay rock over time. Understanding and characterizing the spatial variability and anisotropy of soft clay rock is essential for AGF applications [30–32]. In this study, the project background of this paper is that the section from Mengcheng Road Station to North Second Ring Road station of Hefei Rail Transit Line 5 is about 979 m long, the section roof is about 9.7~18.5 m, and the section floor is about 15.7~24.5 m buried. The shield method is intended to be used for construction; the tunnel diameter is about 6 m, and the formation constructed by the freezing method is soft clay rock. A series of soft clay rocks were collected, and thermodynamic tests were carried out. The range of DSC is 0~ ±500 mW, the temperature range is from −60 °C to 200 °C, the heating rate and cooling rates are 1~80 °C/min and 1~20 °C/min, respectively, the temperature resolution is 0.1 °C, the temperature fluctuation and repeatability are ±0.1 °C, the noise is 0.01 μW, the resolution and accuracy are 0.01 μW and 0.1 μW. The sensitivity is 0.1 μW. Our DSC equipment comes from Germany. A known temperature difference is applied across a sample of clay, and the rate of heat transfer is measured. By calculating the heat flux and knowing the geometry of the sample, the thermal conductivity can be determined. Differentiating scanning calorimetry was used to measure the specific heat capacity. The sample is subjected to controlled temperature changes, and the heat flow into or out of the sample is measured. By analyzing the heat flow data, the specific heat capacity of the soft clay rock can be determined. The permeameter was used to assess the permeability. The constant head test is used in this paper. The inner diameter of the metal sealing cylinder, metal orifice plate, filter screen, pressure measuring pipe, and metal cylinder for the water supply bottle is 10 cm, and the height is 40 cm. The clay sample is placed in a permeameter cell, and water is allowed to flow through it under controlled conditions. By measuring the flow rate and applying Darcy's law, the permeability of the clay can be calculated. The depth of the soil sample was 10 m, and 10 sets of samples were obtained. The test results of the hydrothermal parameters of soft clay rock for ten different sampling points are shown in Figure 1. It can be seen that the discrete features are very evident because of the heterogeneity of the soft clay rock. The relationship between spatial variation in anisotropy and heterogeneous characteristics of soft clay rock is significant. Soft clay rock exhibits anisotropic behavior due to variations in its mineral composition, fabric orientation, and depositional history. These variations lead to spatial heterogeneity in properties such as permeability, compressibility, and thermophysical properties. The alignment of clay particles along certain directions can result in preferential pathways for fluid flow, causing anisotropic permeability. Additionally, localized concentrations of specific minerals or organic matter can create heterogeneities in thermodynamic behavior. Therefore, it is essential to analyze the statistical characteristics and variability of these data.

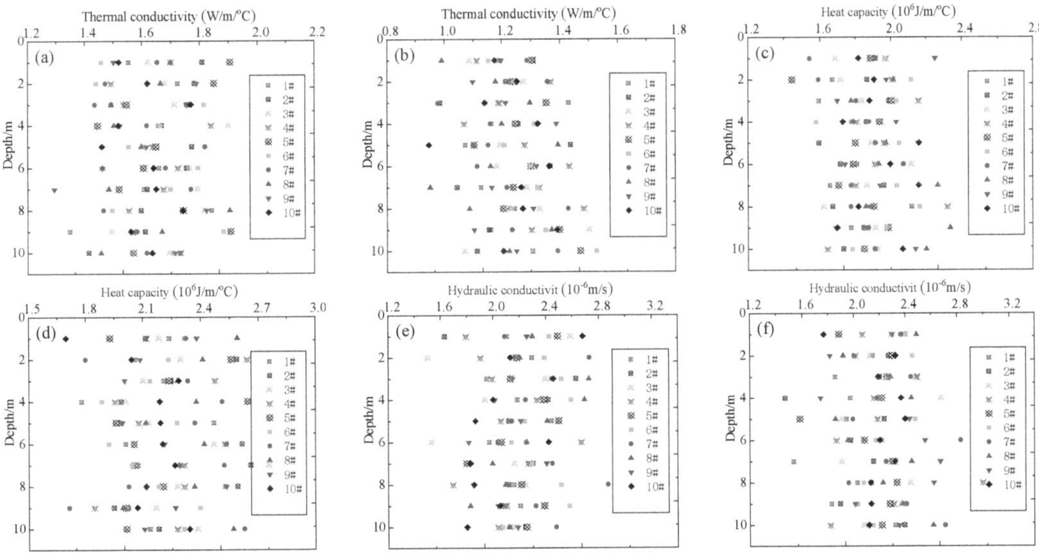

Figure 1. Hydrothermal parameter for 10 different sampling points: (**a**) Thermal conductivity of frozen state, (**b**) thermal conductivity of unfrozen state, (**c**) heat capacity of frozen state, (**d**) heat capacity of unfrozen state, (**e**) hydraulic conductivity of horizontal direction, (**f**) hydraulic conductivity of vertical direction.

2.2. Statistical Distribution Characteristics

The results of thermodynamic parameter testing for soft clay rock reveal a distinct statistical distribution pattern, highlighting the inherent variability in clay properties. Factors such as mineral composition, pore structure, and moisture content contribute to this variability, leading to diverse values of hydrothermal parameters. Rather than displaying a singular fixed value, these parameters exhibit a range of values distributed probabilistically. This statistical distribution reflects the heterogeneous nature of soft clay rock, where different regions within a sample may possess varying properties [33,34]. Understanding this statistical distribution is essential for accurate modeling and prediction of clay behavior in engineering applications, as it allows for the incorporation of uncertainty and variability into predictive models. Advanced statistical techniques are often employed to analyze and interpret these results, providing valuable insights into the spatial variability and complexity of soft clay rock. Recognizing the statistical distribution of thermodynamic parameters in clay enhances our understanding of its behavior and facilitates more informed decision-making in engineering and environmental contexts. Based on the test results of the hydrothermal parameters of soft clay rock for ten different sampling points, the statistical value was calculated. The results are shown in Figure 2. For the thermal conductivity of the frozen state, the maximum is 1.907 W/m/°C while the minimum is 1.294 W/m/°C. The mean and standard deviation (SD) are 1.640 W/m/°C and 0.134 W/m/°C, respectively. For the thermal conductivity of the unfrozen state, the maximum is 1.526 W/m/°C while the minimum is 0.947 W/m/°C. The mean and SD are 1.241 W/m/°C and 0.124 W/m/°C, respectively. For the heat capacity of the frozen state, the maximum is 2.336 $\times$ 10^6 J/m^3/°C while the minimum is 1.445 $\times$ 10^6 J/m^3/°C. Mean and SD are 1.893 $\times$ 10^6 J/m^3/°C and 0.1764 $\times$ 10^6 J/m^3/°C, respectively. For the heat capacity of the unfrozen state, the maximum is 2.755 $\times$ 10^6 J/m^3/°C while the minimum is 1.697 $\times$ 10^6 J/m^3/°C. Mean and SD are 2.227 $\times$ 10^6 J/m^3/°C and 0.230 $\times$ 10^6 J/m^3/°C, respectively. For the hydraulic conductivity of horizontal direction, the maximum is 2.869 $\times$ 10^{-6} m/s, while the minimum is 1.504 $\times$ 10^{-6} m/s. Mean and SD are 2.203 $\times$ 10^{-6} m/s and 0.266 $\times$ 10^{-6} m/s, respectively. For the hydraulic conductivity of vertical direction, the maximum is 2.999 $\times$ 10^{-6} m/s while the minimum is 1.471 $\times$ 10^{-6} m/s. Mean and SD are 2.204 $\times$ 10^{-6} m/s and 0.265 $\times$ 10^{-6} m/s, respectively. It can be seen that the statistical distribution characteristics of hydraulic conductivity in the horizontal direction and vertical direction are approximate. Thus, it can be concluded that the permeability of soft clay rock in different directions is the same. In general, the statistical characteristics of the hydrothermal parameters show a normal distribution.

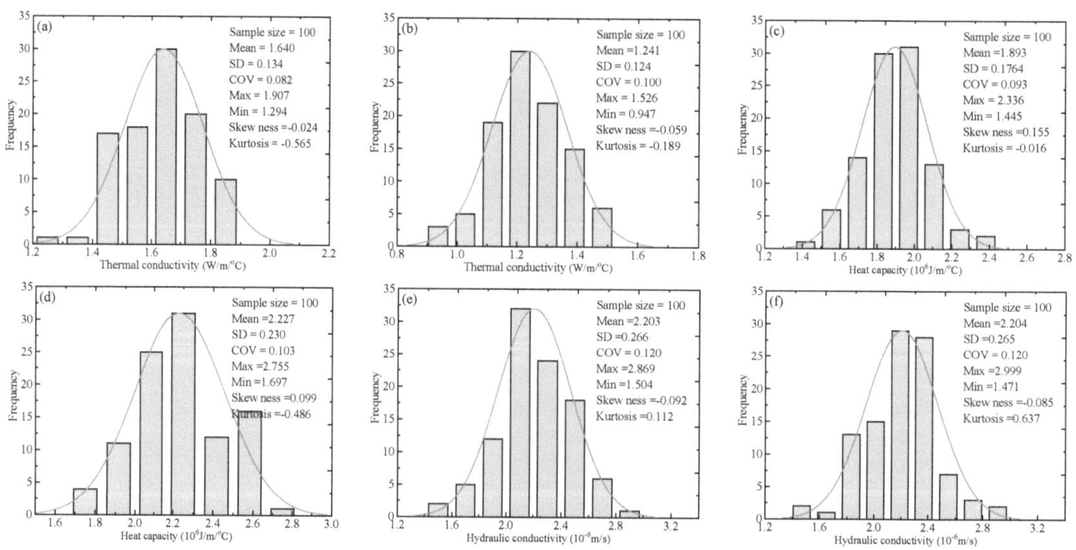

Figure 2. Statistical characteristics of hydrothermal properties: (**a**) Thermal conductivity of frozen state, (**b**) thermal conductivity of unfrozen state, (**c**) heat capacity of frozen state, (**d**) heat capacity of unfrozen state, (**e**) hydraulic conductivity of horizontal direction, (**f**) hydraulic conductivity of vertical direction.

2.3. Variability Distribution Characteristics

The spatial variability of thermodynamic parameters in soft clay rock is caused by a complex interplay of several underlying mechanisms. One primary factor contributing to this variability is the heterogeneous nature of clay at both micro and macro scales. At the microscale, clay minerals exhibit intricate structures and compositions, resulting in variations in thermal properties such as thermal conductivity and specific heat capacity. Variations in mineral composition, including the presence of different types of clay minerals and accessory minerals, can significantly influence thermal conductivity. Additionally, the arrangement of clay particles and their orientation within the clay matrix can create anisotropic behavior, further contributing to spatial variability. At the macroscale, geological processes such as sedimentation, weathering, and diagenesis play a crucial role in shaping the spatial distribution of clay properties. Variations in depositional environments, including factors like sedimentary facies and depositional conditions, can lead to spatial heterogeneity in clay properties across different geological formations. Furthermore, post-depositional processes such as compaction, cementation, and deformation can further modify the spatial distribution of soft clay rock properties, introducing additional variability. Environmental factors also influence the spatial variability of thermodynamic parameters. Variations in moisture content, temperature gradients, and chemical composition of pore fluids can affect thermal conductivity and specific heat capacity. Additionally, hydraulic gradients and flow pathways within clay formations can lead to preferential fluid flow and transport, influencing permeability distribution [35,36]. The spatial variability of clay thermodynamic parameters is further influenced by scale-dependent phenomena. At smaller scales, such as the pore scale, interactions between clay particles, pore fluid properties, and surface chemistry play a significant role in determining thermal and hydraulic properties. At larger scales, geological heterogeneity, structural features, and boundary conditions become more pronounced, influencing the overall spatial distribution of clay properties. Therefore, the spatial variability of thermodynamic parameters in soft clay rock is governed by a combination of factors, including mineralogical composition, geological processes, environmental conditions, and scale-dependent phenomena. Understanding these mechanisms is essential for accurately characterizing and predicting the behavior of soft clay rock in various engineering and environmental applications. Based on test data and random field theory, the COV, scale of fluctuation (SOF), and ACD can be calculated. The variability results of hydrothermal parameters are shown in Table 1.

Table 1. Variability of the hydrothermal parameters for the soft clay rock.

NO.	Thermal Conductivity				Heat Capacity				Hydraulic Conductivity			
	Frozen State		Unfrozen State		Frozen State		Unfrozen State		Horizontal Direction		Vertical Direction	
	COV	SOF	COV	SOF	COV	SOF	COV	SOF	COV	SOF	COV	SOF
1#	0.19	1.84	0.12	2.63	0.24	3.34	0.14	1.65	0.12	3.17	0.14	2.32
2#	0.23	2.38	0.18	2.32	0.13	2.75	0.14	1.99	0.16	3.48	0.17	2.84
3#	0.20	2.09	0.15	3.02	0.12	2.52	0.12	1.91	0.17	2.21	0.14	1.89
4#	0.09	2.56	0.18	1.81	0.18	3.60	0.12	1.73	0.15	3.05	0.15	3.53
5#	0.16	1.61	0.21	1.94	0.22	2.39	0.13	2.16	0.17	2.81	0.13	2.47
6#	0.20	2.32	0.18	3.60	0.19	3.27	0.10	1.49	0.08	2.10	0.13	1.66
7#	0.10	1.48	0.14	2.87	0.18	2.55	0.15	2.25	0.15	2.73	0.18	2.90
8#	0.16	1.48	0.19	3.50	0.15	3.26	0.14	1.27	0.16	2.75	0.18	3.70
9#	0.19	2.24	0.14	2.70	0.13	3.31	0.14	2.36	0.17	2.25	0.11	2.91
10#	0.07	2.03	0.13	2.91	0.17	2.38	0.05	2.32	0.16	1.70	0.12	2.96

Random field theory is a powerful framework for characterizing the spatial variability of hydrothermal parameters [37]. In this theory, soft clay rock properties are treated as random fields, exhibiting inherent variability across space. The essence of this approach lies in recognizing that soft clay rock properties vary stochastically due to diverse geological processes, environmental influences, and heterogeneity in soil composition and structure. By representing soft clay rock parameters as random fields, stochastic field theory captures the complex and nonlinear spatial patterns of variability observed in soft clay rock properties. Statistical descriptors such as mean, variance, and spatial correlation length are used to quantify the average behavior, degree of variability, and spatial dependence of soft clay rock parameters, respectively. Variogram analysis is often employed to characterize the spatial correlation structure, providing insights into the range over which parameter values exhibit spatial correlation. Through stochastic field theory, spatial realizations of soil parameter fields can be generated, simulating multiple plausible spatial distributions of parameters. Based on the variability results of the hydrothermal parameters and random field, the COV, SOF, and ACD can be obtained. The variability distribution characteristics of the hydrothermal parameters for the soft clay rock are shown in Figure 3. The observation reveals that the COV varies between 0.05 and 0.25, the SOF varies between 1.2 m and 4.0 m, and the ACD varies between 0.4 m and 1.4 m, respectively.

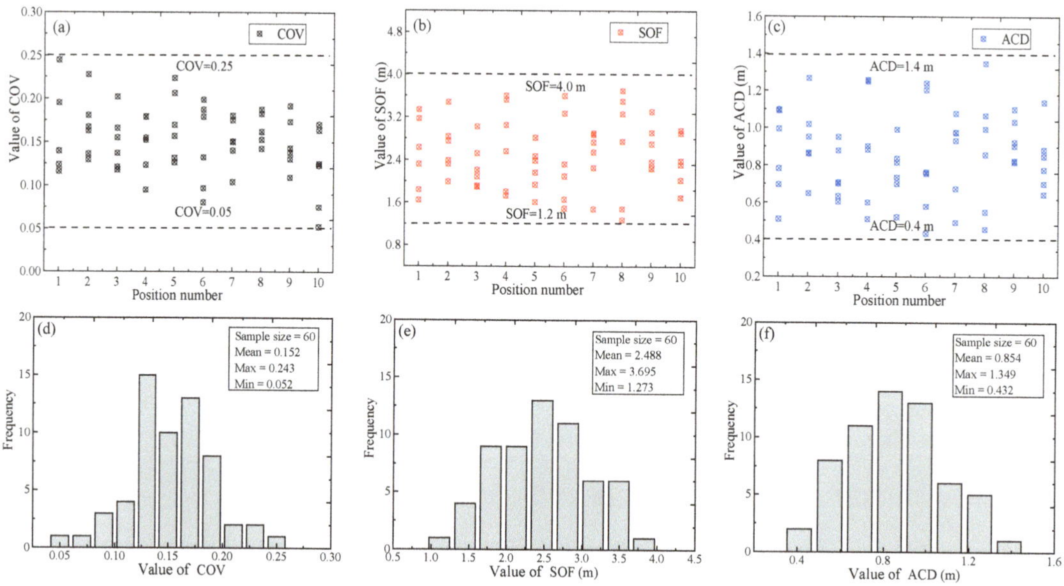

Figure 3. Variability characteristics of hydrothermal properties: (**a**) Discrete distribution of COV, (**b**) discrete distribution of SOF, (**c**) discrete distribution of ACD, (**d**) statistical characteristics of COV, (**e**) statistical characteristics of SOF, (**f**) statistical characteristics of ACD.

3. Uncertain Analysis Method of Thermodynamic Properties

3.1. Coupled Governing Equations

The mechanism of temperature-permeability coupling in frozen soft clay rock elucidates the intricate interplay between thermal dynamics and fluid flow. At its core, the mechanism involves the manipulation of temperature to induce freezing in rock pores, altering their permeability and hydraulic conductivity. As temperature decreases, pore water freezes, leading to a reduction in pore size and an increase in rock matrix strength. Consequently, the permeability of the soft clay rock decreases, restricting fluid flow. However, the coupling is not solely determined by temperature. Factors such as ice formation rate, soil composition, and boundary conditions also influence the process. Understanding this mechanism is critical for optimizing freezing techniques, designing effective containment systems, and predicting the behavior of soft clay rock under various conditions. By comprehending the temperature-permeability coupling mechanism in underground space, engineers can develop sustainable solutions for infrastructure development in cold regions, mitigate environmental impacts, and ensure the long-term stability of engineered structures in frozen soft clay rock environments [38]. The coupled equations can be expressed as follows:

In the frozen area Ω_{f}:

$$C_{\mathrm{f}}\frac{\partial T_{\mathrm{f}}}{\partial t} = \frac{\partial}{\partial x}\left(\lambda_{\mathrm{f}}\frac{\partial T_{\mathrm{f}}}{\partial x}\right) + \frac{\partial}{\partial y}\left(\lambda_{\mathrm{f}}\frac{\partial T_{\mathrm{f}}}{\partial y}\right) - C_w\rho_w\left(V_x^{\mathrm{f}}\frac{\partial T_{\mathrm{f}}}{\partial x} + V_y^{\mathrm{f}}\frac{\partial T_{\mathrm{f}}}{\partial y}\right) \tag{1}$$

$$S_{\mathrm{f}}\frac{\partial H_{\mathrm{f}}}{\partial t} = \frac{\partial}{\partial x}\left(K_x^{\mathrm{f}}\frac{\partial H_{\mathrm{f}}}{\partial x}\right) + \frac{\partial}{\partial y}\left(K_y^{\mathrm{f}}\frac{\partial H_{\mathrm{f}}}{\partial y}\right) + Q \tag{2}$$

$$V_x^{\mathrm{f}} = -K_x^{\mathrm{f}}\frac{\partial H_{\mathrm{f}}}{\partial x},\ V_y^{\mathrm{f}} = -K_y^{\mathrm{f}}\frac{\partial H_{\mathrm{f}}}{\partial y} \tag{3}$$

In the unfrozen area Ω_{u}:

$$C_{\mathrm{u}}\frac{\partial T_{\mathrm{u}}}{\partial t} = \frac{\partial}{\partial x}\left(\lambda_{\mathrm{u}}\frac{\partial T_{\mathrm{u}}}{\partial x}\right) + \frac{\partial}{\partial y}\left(\lambda_{\mathrm{u}}\frac{\partial T_{\mathrm{u}}}{\partial y}\right) - C_w\rho_w\left(V_x^{\mathrm{u}}\frac{\partial T_{\mathrm{u}}}{\partial x} + V_y^{\mathrm{u}}\frac{\partial T_{\mathrm{u}}}{\partial y}\right) \tag{4}$$

$$S_{\mathrm{u}}\frac{\partial H_{\mathrm{u}}}{\partial t} = \frac{\partial}{\partial x}\left(K_x^{\mathrm{u}}\frac{\partial H_{\mathrm{u}}}{\partial x}\right) + \frac{\partial}{\partial y}\left(K_y^{\mathrm{u}}\frac{\partial H_{\mathrm{u}}}{\partial y}\right) + Q \tag{5}$$

$$V_x^{\mathrm{u}} = -K_x^{\mathrm{u}}\frac{\partial H_{\mathrm{u}}}{\partial x},\ V_y^{\mathrm{u}} = -K_y^{\mathrm{u}}\frac{\partial H_{\mathrm{u}}}{\partial y} \tag{6}$$

In Equations (1)–(6), f and u represent the frozen and the unfrozen states, respectively. T_{f}, C_{f}, and λ_{f} are the temperature, volumetric heat capacity, and thermal conductivity of foundation soils in the frozen area Ω_{f}, respectively. H_{f}, S_{f}, K_x^f and K_y^f are the water head, specific yield, and permeability coefficients in the frozen area Ω_{f}, respectively. Q is the source or collection of the seepage field. Variables V_x and V_y are the seepage rates in the x and y directions, respectively. C_w and ρ_w are the volumetric heat capacity and bulk density of water. Parameters with subscript u are the corresponding physical components in the unfrozen area Ω_{u}. t is time, and x, y are distances.

At the freezing-front position $s\ (\boldsymbol{x};t)$, in which $\boldsymbol{x} = \{x, y\}$, The following equation needs to be satisfied.

$$T_{\mathrm{f}}(s(\boldsymbol{x},t),t) = T_{\mathrm{u}}(s(\boldsymbol{x},t),t) = T_{\mathrm{m}} \tag{7}$$

$$\lambda_{\mathrm{f}}\frac{\partial T_{\mathrm{f}}}{\partial n} - \lambda_{\mathrm{u}}\frac{\partial T_{\mathrm{u}}}{\partial n} = L\frac{\mathrm{d}s(\boldsymbol{x},t)}{\mathrm{d}t} \tag{8}$$

where L is the latent heat per unit volume, n is the direction vector of the moving boundary $s\ (\boldsymbol{x};t)$, and T_{m} is the phase change temperature.

The hydrothermal parameters can be expressed as

$$C^* = \begin{cases} C_{\mathrm{f}} & T < T_{\mathrm{m}} - \Delta T \\ \frac{C_{\mathrm{u}}+C_{\mathrm{f}}}{2} + \frac{L}{2\Delta T} & T_{\mathrm{m}} - \Delta T \le T \le T_{\mathrm{m}} + \Delta T_{\mathrm{u}} \\ C_{\mathrm{u}} & T > T_{\mathrm{m}} + \Delta T_{\mathrm{u}} \end{cases} \tag{9}$$

$$\lambda^* = \begin{cases} \lambda_{\mathrm{f}} & T < T_{\mathrm{m}} - \Delta T \\ \lambda_{\mathrm{f}} + \frac{\lambda_{\mathrm{u}}-\lambda_{\mathrm{f}}}{2\Delta T}[T - (T_{\mathrm{m}} - \Delta T)] & T_{\mathrm{m}} - \Delta T \le T \le T_{\mathrm{m}} + \Delta T \\ \lambda_{\mathrm{u}} & T > T_{\mathrm{m}} + \Delta T \end{cases} \tag{10}$$

The hydraulic conductivity can be written as follows:

$$K_x^* = \begin{cases} \left(\frac{W_{\mathrm{u}}}{W}\right)^{\beta} \cdot K_x^{\mathrm{u}} & T \le T_{\mathrm{m}} \\ K_x^{\mathrm{u}} & T > T_{\mathrm{m}} \end{cases},\ K_y^* = \begin{cases} \left(\frac{W_{\mathrm{u}}}{W}\right)^{\beta} \cdot K_y^{\mathrm{u}} & T \le T_{\mathrm{m}} \\ K_y^{\mathrm{u}} & T > T_{\mathrm{m}} \end{cases} \tag{11}$$

where W and W_{u} denote the water content and unfrozen water content of soil, respectively. The constant β is determined experimentally.

Using Equations (9)–(11), Equations. (1)–(8) can be written simply as

$$C^* \frac{\partial T}{\partial t} = \frac{\partial}{\partial x}\left(\lambda^* \frac{\partial T}{\partial x}\right) + \frac{\partial}{\partial y}\left(\lambda^* \frac{\partial T}{\partial y}\right) - C_w \rho_w \left(V_x^* \frac{\partial T}{\partial x} + V_y^* \frac{\partial T}{\partial y}\right) \tag{12}$$

$$S^* \frac{\partial H}{\partial t} = \frac{\partial}{\partial x}\left(K_x^* \frac{\partial H}{\partial x}\right) + \frac{\partial}{\partial y}\left(K_y^* \frac{\partial H}{\partial y}\right) + Q \tag{13}$$

where

$$S^* = \begin{cases} S_{\mathrm{f}} & T \leq T_{\mathrm{m}} \\ S_{\mathrm{u}} & T > T_{\mathrm{m}} \end{cases} \tag{14}$$

$$V_x^* = -K_x^* \frac{\partial H}{\partial x},\ V_y^* = -K_y^* \frac{\partial H}{\partial y} \tag{15}$$

3.2. Uncertainty Characterization Method

The principle behind quantifying soft clay rock parameter uncertainty in spatial random fields lies in understanding and modeling the inherent spatial variability of soft clay rock properties. Geotechnical parameters, such as permeability, porosity, and shear strength, exhibit spatial heterogeneity due to geological processes, depositional patterns, and environmental factors. Spatial random field theory provides a framework to represent this variability by characterizing the statistical properties of geotechnical parameters across a given area. This involves analyzing spatial autocorrelation structures using tools like variograms and covariance functions. Geostatistical techniques, including kriging and stochastic simulation, are then employed to interpolate or simulate geotechnical parameters, accounting for spatial dependency and uncertainty. The local averaging method in random fields, also known as the moving window technique, is a powerful approach used in geostatistics and spatial analysis to simulate spatially correlated data. This method involves partitioning a random field into a series of overlapping windows and computing the local average within each window. The process begins by defining a window size and shape, typically a square or circular area, and then moving this window across the field, calculating the average of the data points within each window. The center of the window is then assigned the average value obtained. This process is repeated for each window, resulting in a new simulated field with spatially correlated values. The local averaging method offers several advantages, including simplicity, computational efficiency, and the ability to preserve local spatial patterns [39,40]. It is particularly useful for simulating spatially correlated phenomena with complex spatial structures, such as geotechnical properties, groundwater levels, or geological formations. Parameterization of the local averaging method involves determining the appropriate window size and shape based on the spatial characteristics of the data and the desired level of spatial correlation. The choice of parameters can significantly impact the simulated results, with larger windows capturing broader spatial trends and smaller windows preserving finer-scale variability. To optimize parameters, practitioners often employ cross-validation techniques or sensitivity analyses to assess the performance of different parameter combinations. By adjusting the parameters and comparing the simulated results to observed data or known spatial patterns, the most suitable parameter values can be identified for a given application. The three hydrothermal parameters of the freezing curtain are simulated as three two-dimensional random fields. It can be expressed as follows:

$$X_e = \frac{1}{A_e} \int_{\Omega_e} X(x, y) dx dy \tag{16}$$

The covariance can be expressed as

$$Cov(X_e, X_{e'}) = \frac{\sigma^2}{A_e A_{e'}} \sum_{i=1}^{4} \sum_{j=1}^{4} \sum_{m=1}^{4} \sum_{l=1}^{4} H_i H_j H_m H_l \rho(\Delta x, \Delta y) |J| |J'| \tag{17}$$

where A_e is the area of e; $A_{e'}$ is the area of e'; H_i, H_j, H_m, and H_l are the weighting coefficients; $\rho\,(\Delta x, \Delta y)$ is the correlation structure of original random field; $|J|$ and $|J'|$ are the Jacobian determinants of coordinate transformation.

It can be seen that the covariance can be calculated as long as $\rho\,(\Delta x, \Delta y)$ is given. The autocorrelation function in stochastic fields quantifies the degree of similarity between a random variable and a lagged version of itself across space or time. It is a fundamental tool in analyzing spatial or temporal dependence patterns within stochastic processes. The autocorrelation function describes how the correlation between points changes as the distance between them varies. For example, in spatial applications, the autocorrelation function often exhibits decay with increasing distance, indicating a diminishing correlation between points farther apart. Various mathematical forms can represent the autocorrelation function, such as exponential, Gaussian, or spherical models, each suited to different spatial or temporal contexts. Understanding and accurately modeling the autocorrelation function are crucial for tasks like spatial interpolation, prediction, and simulation within stochastic field analysis. In this study, the symbols θ_x and θ_y are used to represent horizontal ACD and vertical ACD. The symbols δ_x and δ_y are used to represent horizontal and vertical SOF. The mathematical expressions and parametric relationships of five common autocorrelation functions are shown in Table 2. Five kinds of two-dimensional random fields have different related structures, and the parameter relationships between SOF and ACD are different.

Table 2. Autocorrelation functions for characterizing the spatial variability of hydrothermal properties.

Name	Autocorrelation Functions	Parameter Relationship
DSNX	$\rho(\tau) = \exp\left[-\left(\frac{\Delta x}{\theta_x} + \frac{\Delta y}{\theta_y}\right)\right]$	$\theta_x = \frac{\delta_x}{2}, \theta_y = \frac{\delta_y}{2}$
DSQX	$\rho(\tau) = \exp\left\{-\left[\left(\frac{\Delta x}{\theta_x}\right)^2 + \left(\frac{\Delta y}{\theta_y}\right)^2\right]\right\}$	$\theta_x = \frac{\delta_x}{\sqrt{\pi}}, \theta_y = \frac{\delta_y}{\sqrt{\pi}}$
DSMK	$\rho(\tau) = \exp\left[-\left(\frac{\Delta x}{\theta_x} + \frac{\Delta y}{\theta_y}\right)\right]\left[\left(1 + \frac{\Delta x}{\theta_x}\right)\left(1 + \frac{\Delta y}{\theta_y}\right)\right]$	$\theta_x = \frac{\delta_x}{4}, \theta_y = \frac{\delta_y}{4}$
DCSX	$\rho(\tau) = \exp\left[-\left(\frac{\Delta x}{\theta_x} + \frac{\Delta y}{\theta_y}\right)\right]\cos\left(\frac{\Delta x}{\theta_x}\right)\cos\left(\frac{\Delta y}{\theta_y}\right)$	$\theta_x = \delta_x, \theta_y = \delta_y$
DBIN	$\rho(\tau) = \begin{cases} \left(1 - \frac{\Delta x}{\theta_x}\right)\left(1 - \frac{\Delta y}{\theta_y}\right) & \Delta x \le \theta_x, \Delta y \le \theta_y \\ 0 & \Delta x > \theta_x, \Delta y > \theta_y \end{cases}$	$\theta_x = \delta_x, \theta_y = \delta_y$

3.3. Hydrothermal Parameters and Boundary Conditions

Boundary and initial conditions significantly influence the thermal characteristics analysis of freeze curtains in underground space engineering. Boundary conditions dictate the exchange of heat between the soft clay rock and the surrounding environment, affecting temperature distribution and freeze front progression. Similarly, initial conditions determine the starting temperature profile within the ground, impacting the rate and extent of freezing. Accurate consideration of these conditions is vital for assessing the effectiveness of soft clay rock in controlling ground temperatures and preventing settlement during underground space construction [41]. Understanding their influence aids in optimizing underground space resource development design and operational parameters to ensure efficient heat extraction and ground stabilization. Moreover, it facilitates the prediction of potential thermal impacts on adjacent structures and utilities, enhancing safety and sustainability in underground infrastructure projects. According to the coupling equation of the soft clay rock, the content of unfrozen water in soft clay rock is an important index to evaluate the water transport characteristics, and it is also a common parameter in thermal calculation. The experimental fitting curve equation can be expressed as

$$W_u = \begin{cases} P|T|_Q & T \le T_\mathrm{m} \\ W & T > T_\mathrm{m} \end{cases} \tag{18}$$

Based on field test data, parameters P and Q are 0.348 and -0.172, respectively.

Water head boundary condition:

$$H = H_0 \tag{19}$$

Flow boundary condition:

$$K_x \frac{\partial H}{\partial x} \cos(n, x) + K_y \frac{\partial H}{\partial y} \cos(n, y) = -q_0 \tag{20}$$

Temperature conditions:

$$T(r_0, t) = T_c \tag{21}$$

$$T(\infty, t) = T_0 \tag{22}$$

where r_0 is outer diameter; T_c is surface temperature; T_0 is initial temperature.

Initial temperature conditions:

$$T(r, 0) = T_0 \tag{23}$$

In this paper, underground space engineering is implemented in soft clay rock layers with three rings of freezing pipes. The inner ring has a diameter of 3.5 m, the middle ring has a diameter of 4.1 m, and the outer ring has a diameter of 4.7 m. There are 18 freezing pipes in both the inner and middle rings, while the outer ring has 32 freezing pipes. The outer diameter of the freezing pipes, r_0, is 0.06 m, and the average surface temperature, *Tc*, is −30 °C. The initial temperature, T_0, of the ground soil is 25 °C. The water head, H_0, is about 1.5 m. From the spatial variability of the hydrothermal parameters of soft clay rock (Figures 2 and 3), the statistical distribution patterns and the range of variation of COV and ACD are known. Based on the uncertain analysis method, the effect of variable parameters on the uncertain thermodynamic properties of soft clay rock can be evaluated.

4. Results and Analyses

4.1. Verification of Stochastic Coupling Model

Validating analytical models for analyzing the thermal characteristics of soft clay rock in underground space engineering is essential for several reasons. Firstly, it ensures the reliability and accuracy of predictions concerning temperature distributions and the progression of the freezing front within the surrounding soil. Accurate predictions are critical for optimizing freeze designs and operational parameters to effectively control ground temperatures and prevent soil settlement during underground space engineering, ensuring the safety and stability of the infrastructure. Secondly, validation enables the assessment of the model's applicability across various geological and climatic conditions. Subsurface conditions can vary significantly from one location to another, and validating the model across different scenarios helps ensure its robustness and generalizability. Thirdly, the validation process helps identify any limitations or discrepancies between model predictions and real-world observations. This feedback loop guides improvements and refinements to the analytical model, enhancing its accuracy and reliability over time. Ultimately, validated analytical models serve as valuable tools for engineers and planners, providing insights into the thermal behavior of soft clay rock and supporting informed decision-making in underground space engineering. Points A, B, and C represent three sample points vertically spaced 1 m apart within the first ring of freezing pipes. Figure 4a shows the results of the experimental and numerical temperatures of soft clay rock for Points A, B, and C. From Figure 4a, it can be observed that the calculated mean temperature results of the uncertain analysis model of thermodynamic properties for the soft clay rock are close to the experimental results at Points A, B, and C. This is consistent with the law of large numbers in statistics, indicating that the Monte Carlo random simulation yields mean temperatures that are approximately equal to the statistically measured temperatures. As freezing time increases, the temperature of Points A, B, and C gradually decreases. When the mean temperature approaches 0 °C, there is a noticeable plateau in the curve. This is due to the phase transition between ice and water, releasing latent heat, which slows down the cooling trend. Similarly, Points D, E, and F represent three sample points horizontally spaced 1 m apart within the first ring of freezing pipes. Figure 4b shows the results with the experimental temperatures and numerical temperatures of soft clay rock for Points D, E, and F. From Figure 4b, the calculated mean temperature results of the uncertain analysis model of thermodynamic properties for the soft clay rock are close to the experimental results at Points D, E, and F. This is consistent with the Law of Large Numbers in statistics, indicating that the average temperature generated by random simulation is approximately equal to the statistically measured temperature. As freezing time increases, the temperature of Points D, E, and F gradually decreases. When the mean temperature approaches 0 °C, there is a noticeable plateau in the curve. This is due to the phase transition between ice and water, releasing latent heat, which slows down the cooling trend. Therefore, the proposed stochastic analysis model for the thermal characteristics of soft clay rock, considering the spatial variability of variable hydrothermal parameters, is scientifically reasonable.

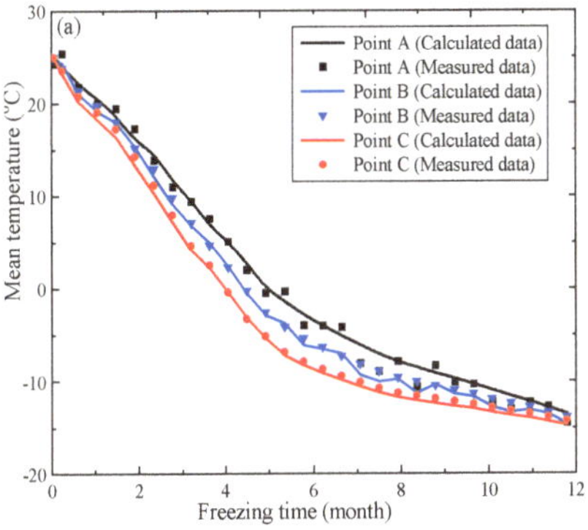

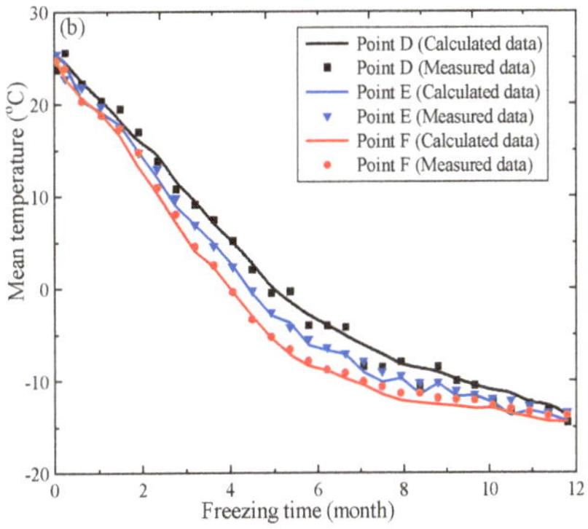

Figure 4. Experimental temperatures and numerical temperatures of soft clay rock: (**a**) measured data and calculated data of Points A, B, and C; (**b**) measured data and calculated data of Points D, E, and F.

4.2. Mean and SD of Thermodynamic Properties

Analyzing the evolution of average temperature and average thickness of soft clay rock in underground space engineering is crucial for several reasons. Firstly, it provides insights into the thermal behavior of the surrounding soft clay rock and the effectiveness of the soft clay rock in controlling ground temperatures. Understanding how the average temperature changes over time helps in optimizing freeze designs and operational parameters to ensure adequate ground freezing and prevent soil settlement during tunnel construction. Secondly, monitoring the average thickness evolution allows engineers to assess the progression of the freezing front within the soil. By tracking changes in thickness, they can identify potential areas of inadequate freezing and take corrective measures to enhance freeze-soft clay rock performance. Moreover, analyzing the evolution of both temperature and thickness provides valuable data for validating analytical models and simulation tools used in underground space engineering. Comparing model predictions with observed data enhances the accuracy and reliability of these tools, ultimately improving decision-making and risk management during underground space engineering. Furthermore, understanding the long-term evolution of average temperature and thickness helps in assessing the stability and durability of underground space engineering over time. By predicting how these parameters will evolve under different conditions, engineers can implement preventive maintenance measures and ensure the long-term integrity of underground space engineering. In summary, analyzing the evolution of average temperature and thickness of soft clay rock is essential for optimizing design, validating models, ensuring construction safety, and maintaining the long-term stability of underground infrastructure. Sections A, B, C, D, and E represent five sections through the center of the inner ring of freezing pipes, with azimuth angles of 0°, 45°, 90°, 135°, and 180°, respectively. Figure 5 shows the variability of temperature characteristic value of soft clay rock. From Figure 5a, the overall trend of the average thickness of the soft clay rock increases with the increase of freezing time. From the second month to the 10th month, it increases at a faster rate, and there are differences in the thickness of different sections, with the largest difference being 1.35 m, which is due to the fact that the frozen curtain has not formed a stable form. When the freezing time reaches 12 months, the average thickness stabilizes, indicating that the stable formation of the frozen curtain is essentially achieved at this point. From Figure 5c, the overall trend of the average temperature of the frozen curtain decreases with the increase of freezing time. From the second month to the 10th month, it decreases at a relatively fast rate, and there are differences in the temperature of different sections, with the largest difference being 0.58 °C, which is due to the fact that the frozen curtain has not formed a stable form. When the freezing time reaches 12 months, the average temperature stabilizes, indicating that the stable formation of the frozen curtain is essentially achieved at this point. From Figure 5b,c, the overall trend of SD of the freeze curtain increases with time. When the freezing time reaches 12 months, the SD of the average thickness and temperature stabilizes.

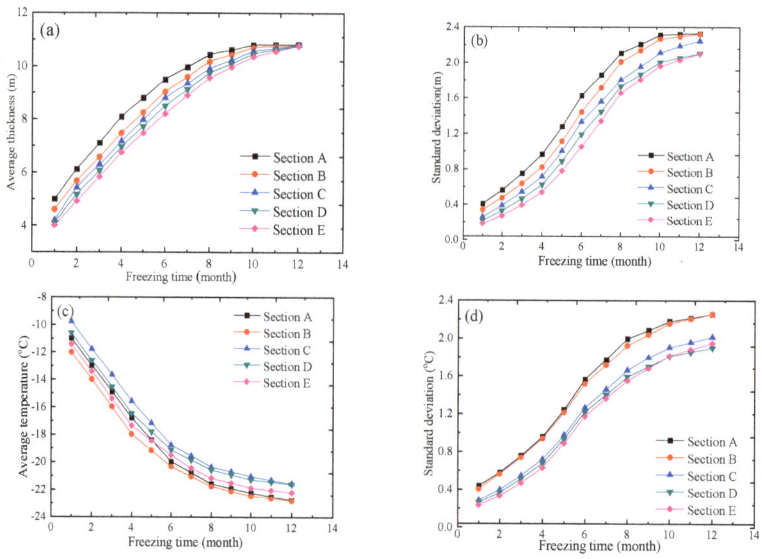

Figure 5. Variability of temperature characteristic value of soft clay rock: (**a**) Average thickness of freezing curtain; (**b**) standard deviation of average thickness; (**c**) average temperature of freezing curtain; (**d**) standard deviation of average temperature.

4.3. Effect of ACF on Hydrothermal Parameters

Evaluating the ACF of stochastic fields of hydrothermal parameters is crucial for understanding the impact on temperature distribution of soft clay rock in underground space engineering. The ACF characterizes the spatial dependency of hydrothermal parameters for the soft clay rock. This information is vital for predicting the spatial variation of temperature within the soil mass and, consequently, the performance of freeze curtains. By analyzing the autocorrelation function, engineers can assess how thermal properties vary over different distances within the soft clay rock. This understanding enables the estimation of the range at which thermal influences propagate, guiding the design and placement of freeze curtain elements for optimal temperature control. Additionally, it helps identify areas of high spatial variability where additional monitoring or intervention may be necessary to ensure uniform freezing and prevent ground settlement. Furthermore, incorporating the ACF into thermal analysis models enhances their accuracy and reliability. By capturing the spatial correlation of thermal parameters, these models can provide more realistic predictions of temperature distributions within the soil mass and along the soft clay rock. This is essential for optimizing freeze curtain designs, ensuring adequate ground freezing, and minimizing the risk of soil instability during underground space engineering. Overall, evaluating the ACF of hydrothermal parameters is essential for understanding the spatial variability of temperature within underground space engineering environments. It facilitates informed decision-making in freeze curtain design and construction, ultimately ensuring the safety, stability, and longevity of underground infrastructure. To elucidate the ACF on the variability of temperature characteristics of frozen curtains, this paper selects the DSNX, DSQX, DSMK, DCSX, and DBIN models to represent the ACF of frozen soft clay rock layers and conducts sensitivity grouping (Table 3) calculations for correlation structure models. The results are shown in Figure 6.

Table 3. Parametric influence of different ACFs of hydrothermal properties.

Case ID	Thermal Conductivity		Heat Capacity		Hydraulic Conductivity	
	λ_u (W/m/°C)	λ_f (W/m/°C)	C_u (10^6 J/m³/°C)	C_f (10^6 J/m³/°C)	$K_x{}^u$ (10^{-6} m/s)	$K_y{}^u$ (10^{-6} m/s)
Reference case	DSQX	DSQX	DSQX	DSQX	DSQX	DSQX
ACF of thermal conductivity	DSNX	DSNX	DSQX	DSQX	DSQX	DSQX
	DSQX	DSQX	DSQX	DSQX	DSQX	DSQX
	DSMK	DSMK	DSQX	DSQX	DSQX	DSQX
	DCSX	DCSX	DSQX	DSQX	DSQX	DSQX
	DBIN	DBIN	DSQX	DSQX	DSQX	DSQX
ACF of heat capacity	DSQX	DSQX	DSNX	DSNX	DSQX	DSQX
	DSQX	DSQX	DSQX	DSQX	DSQX	DSQX
	DSQX	DSQX	DSMK	DSMK	DSQX	DSQX
	DSQX	DSQX	DCSX	DCSX	DSQX	DSQX
	DSQX	DSQX	DBIN	DBIN	DSQX	DSQX
ACF of hydraulic conductivity	DSQX	DSQX	DSQX	DSQX	DSNX	DSNX
	DSQX	DSQX	DSQX	DSQX	DSQX	DSQX
	DSQX	DSQX	DSQX	DSQX	DSMK	DSMK
	DSQX	DSQX	DSQX	DSQX	DCSX	DCSX
	DSQX	DSQX	DSQX	DSQX	DBIN	DBIN

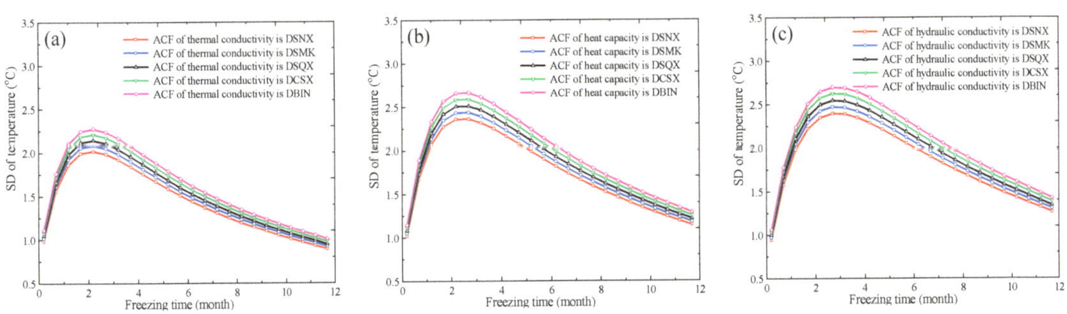

Figure 6. Effects of different ACFs of variable hydrothermal parameters on uncertain thermodynamic properties: (**a**) Thermal conductivity; (**b**) heat capacity; (**c**) hydraulic conductivity.

Figure 6 shows that the five different random field-related structures have different effects on the SD of freezing soft clay rock temperature. When DBIN is used, the SD of freezing curtain temperature is the largest. When DSQX is used, the SD of the freezing curtain temperature is in the middle. When DSNX is used, the SD of the freezing curtain temperature is the smallest. For the thermal conductivity, the maximum and minimum SD of freezing curtain temperature during the freezing process are 2.27 °C and 0.89 °C, respectively. For the spatial variability of heat capacity, the maximum and minimum SD of freezing curtain temperature during the freezing process are 2.66 °C and 1.02 °C, respectively. For the hydraulic conductivity, the maximum and minimum SD of freezing curtain temperature during freezing is 2.70 °C and 0.94 °C, respectively. Therefore, different correlation functions have different variability effects on the freezing curtain temperature. The most significant impact comes from DBIN, followed by DSQX, with DSNX having the smallest effect.

4.4. Effect of COV of Hydrothermal Parameter

The COV of thermal parameters in soft clay rock plays a significant role in influencing the temperature distribution of freeze curtains. As the COV increases, it indicates greater variability or heterogeneity in thermal properties within the frozen soft clay rock. This variability directly affects the spatial distribution of temperature within the soil mass and along the freeze curtain. In regions with high COV values, there is greater uncertainty in thermal properties, leading to more pronounced temperature variations. This can result in non-uniform freezing along the freeze curtain, with some areas experiencing higher or lower temperatures than others. In contrast, lower COV values indicate more consistent thermal properties, resulting in a more uniform temperature distribution along the freeze curtain. Understanding the impact of COV on freeze curtain temperature is crucial for optimizing freeze curtain designs and ensuring effective ground stabilization during underground space engineering. Engineers can use this information to identify regions of high COV, where additional measures may be needed to achieve uniform freezing and prevent ground settlement. Additionally, incorporating COV data into thermal analysis models improves their accuracy and reliability, enabling better predictions of temperature distributions within the soft clay rock. Overall, the relationship between the COV of thermal parameters and freeze curtain temperature highlights the importance of considering spatial variability in thermal properties when designing and implementing freeze curtain systems in underground space engineering. By accounting for COV values, engineers can make informed decisions to enhance the safety, stability, and efficiency of underground infrastructure projects. To elucidate the effect of COV on the temperature characteristics of soft clay rock, this paper selects the 0.05, 0.10, 0.15, 0.20, and 0.25 to represent the COV of hydrothermal parameters and conducts sensitivity grouping (Table 4) calculations for parameter variability. The results are shown in Figure 7.

Figure 7 shows that the five different COVs have different effects on the SD of freezing temperature. For the thermal conductivity, When the COV is 0.05, the SD of the freezing curtain temperature is the smallest. When the COV is 0.15, the SD of the freezing curtain temperature is in the middle. When the COV is 0.25, the SD of the freezing curtain temperature is the largest. The maximum and minimum SD of freezing curtain temperature during the freezing process is 3.09 °C and 0.73 °C, respectively. For the heat capacity, When the COV is 0.05, the SD of the freezing curtain temperature is the smallest. When the COV is 0.15, the SD of the freezing curtain temperature is in the middle. When the COV is 0.25, the SD of the freezing curtain temperature is the largest. The maximum and minimum SD of the freezing curtain temperature during the freezing process are 1.91 °C and 0.59 °C, respectively. For the hydraulic conductivity, When the COV is 0.05, the SD of the freezing curtain temperature is the smallest. When the COV is 0.15, the SD of the freezing curtain temperature is in the middle. When the COV is 0.25, the SD of the freezing curtain temperature is the largest. The maximum and minimum SD of freezing curtain temperature during the freezing process are 2.56 °C and 0.79 °C, respectively. Different variations of hydrothermal parameters have different effects on the standard deviation of freezing curtain temperature. Different COV parameters have varying impacts on the temperature of the freeze-soft clay rock. The most significant impact comes from thermal conductivity, followed by permeability coefficient, with specific heat capacity having the smallest effect. Additionally, the larger the COV of the hydrothermal parameters, the greater the SD of the freezing curtain temperature.

Table 4. Parametric influence of different COVs of hydrothermal properties.

Case ID	Thermal Conductivity		Heat Capacity		Hydraulic Conductivity	
	λ_u (W/m/°C)	λ_f (W/m/°C)	C_u(10^6 J/m^3/°C)	C_f (10^6 J/m^3/°C)	$K_x{}^u$ (10^{-6} m/s)	$K_y{}^u$ (10^{-6} m/s)
Reference case	0.15	0.15	0.15	0.15	0.15	0.15
COV of thermal conductivity	0.05	0.05	0.15	0.15	0.15	0.15
	0.10	0.10	0.15	0.15	0.15	0.15
	0.15	0.15	0.15	0.15	0.15	0.15
	0.20	0.20	0.15	0.15	0.15	0.15
	0.25	0.25	0.15	0.15	0.15	0.15
COV of heat capacity	0.15	0.15	0.05	0.05	0.15	0.15
	0.15	0.15	0.10	0.10	0.15	0.15
	0.15	0.15	0.15	0.15	0.15	0.15
	0.15	0.15	0.20	0.20	0.15	0.15
	0.15	0.15	0.25	0.25	0.15	0.15
COV of hydraulic conductivity	0.15	0.15	0.15	0.15	0.05	0.05
	0.15	0.15	0.15	0.15	0.10	0.10
	0.15	0.15	0.15	0.15	0.15	0.15
	0.15	0.15	0.15	0.15	0.20	0.20
	0.15	0.15	0.15	0.15	0.25	0.25

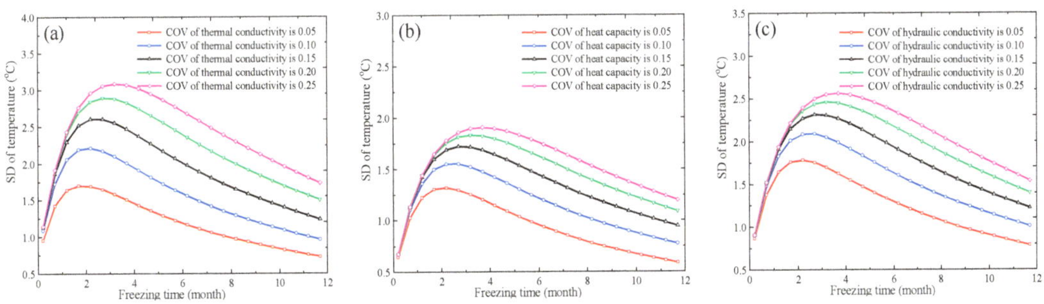

Figure 7. Effects of different COV of variable hydrothermal parameters on uncertain thermodynamic properties: (**a**) Thermal conductivity; (**b**) heat capacity; (**c**) hydraulic conductivity.

4.5. Effect of ACD on Hydrothermal Parameter

Analyzing the autocorrelation distance of thermal parameters in soft clay rock is essential for assessing its impact on freeze curtain temperatures. The autocorrelation distance delineates how thermal properties fluctuate spatially, directly influencing temperature distribution. Understanding this distance aids in optimizing freeze curtain designs to ensure uniform ground freezing and mitigate soil settlement risks during underground space engineering. Incorporating autocorrelation distance into analytical models enhances accuracy, enabling precise predictions of temperature variations along the freeze curtain. This information is vital for informed decision-making in underground space engineering, ensuring the safety, stability, and effectiveness of freeze curtain systems. By considering the autocorrelation distance, engineers can strategically deploy freeze curtain elements, minimizing temperature differentials and promoting reliable ground stabilization. Overall, analyzing the autocorrelation distance of thermal parameters is imperative for comprehensive thermal management in underground space engineering, enhancing project outcomes and safeguarding underground infrastructure. To elucidate the ACD on the temperature characteristics of frozen soft clay rock, this paper selects the 0.4 m, 0.6 m, 0.8 m, 1.0 m, and 1.2 m to represent the ACD of hydrothermal parameters and conducts sensitivity grouping (Table 5) calculations for parameter variability. The results are shown in Figure 8.

Table 5. Parametric influence of different ACDs of hydrothermal properties.

Case ID	Thermal Conductivity		Heat Capacity		Hydraulic Conductivity	
	λ_u (W/m/°C)	λ_f (W/m/°C)	C_u (10^6 J/m^3/°C)	C_f (10^6 J/m^3/°C)	$K_x{}^u$ (10^{-6} m/s)	$K_y{}^u$ (10^{-6} m/s)
Reference case	0.8	0.8	0.8	0.8	0.8	0.8
ACD of thermal conductivity	0.4	0.4	0.8	0.8	0.8	0.8
	0.6	0.6	0.8	0.8	0.8	0.8
	0.8	0.8	0.8	0.8	0.8	0.8
	1.0	1.0	0.8	0.8	0.8	0.8
	1.2	1.2	0.8	0.8	0.8	0.8
ACD of heat capacity	0.8	0.8	0.4	0.4	0.8	0.8
	0.8	0.8	0.6	0.6	0.8	0.8
	0.8	0.8	0.8	0.8	0.8	0.8
	0.8	0.8	1.0	1.0	0.8	0.8
	0.8	0.8	1.2	1.2	0.8	0.8
ACD of hydraulic conductivity	0.8	0.8	0.8	0.8	0.4	0.4
	0.8	0.8	0.8	0.8	0.6	0.6
	0.8	0.8	0.8	0.8	0.8	0.8
	0.8	0.8	0.8	0.8	1.0	1.0
	0.8	0.8	0.8	0.8	1.2	1.2

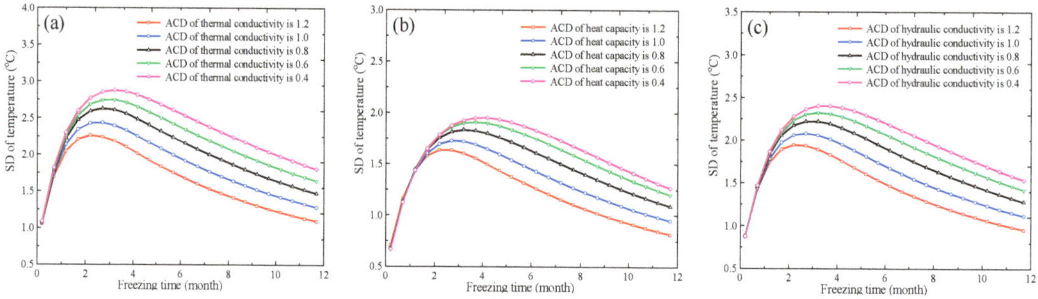

Figure 8. Effects of different ACDs of variable hydrothermal parameters on uncertain thermodynamic properties: (**a**) Thermal conductivity; (**b**) heat capacity; (**c**) hydraulic conductivity.

Figure 8 shows that the five different ACDs have different effects on the SD of freezing curtain temperature. For the thermal conductivity, when the ACD is 0.4 m, the SD of the freezing curtain temperature is the smallest. When the ACD is 0.8 m, the SD of the freezing curtain temperature is in the middle. When the ACD is 1.2 m, the SD of the freezing curtain temperature is the largest. The maximum and minimum SD of freezing curtain temperature during the freezing process are 2.88 °C and 1.05 °C, respectively. For the heat capacity, When the ACD is 0.4 m, the SD of the freezing curtain temperature is the smallest. When the ACD is 0.8 m, the SD of the freezing curtain temperature is in the middle. When the

ACD is 1.2 m, the SD of the freezing curtain temperature is the largest. The maximum and minimum SD of freezing curtain temperature during the freezing process are 1.95 °C and 0.67 °C, respectively. For the hydraulic conductivity, When the ACD is 0.4 m, the SD of the freezing curtain temperature is the smallest. When the ACD is 0.8 m, the SD of the freezing curtain temperature is in the middle. When the ACD is 1.2 m, the SD of the freezing curtain temperature is the largest. The maximum and minimum SD of freezing curtain temperature during the freezing process are 2.40 °C and 0.87 °C, respectively. Different variations of hydrothermal parameters have different effects on the standard deviation of freezing curtain temperatures. Different ACDs of parameters have varying impacts on the temperature of the freeze curtain. The biggest influence is the thermal conductivity, the second is the permeability coefficient, and the least influence is the specific heat capacity. Additionally, the larger the ACD of the hydrothermal parameter, the smaller the SD of the freezing curtain temperature.

5. Conclusions

In this study, the spatial variations of hydrothermal parameters of soft clay rock were calculated based on experimental data. The statistical variability characteristics of variable hydrothermal parameters are estimated. A stochastic coupling model of soft clay rock with heat conduction and porous flow is proposed. According to the relationship between anisotropic spatial variations and statistical variability characteristics for the different random field correlation models, the effects of the ACF, COV, and ACD of variable hydrothermal parameters on underground space engineering are analyzed. The main conclusions obtained are as follows:

(1) The average temperature calculation results of the freezing soft clay rock are close to the experimental results. When the mean temperature approaches 0 °C, there is a noticeable plateau in the curve. The average thickness and temperature of the freezing curtain show an overall increasing trend with increasing freezing time;

(2) From the second month to the tenth month, both the average thickness and temperature increase at a faster rate. When the freezing time reaches twelve months, it tends to stabilize. The overall trend of the SD of the freezing curtain increases with increasing freezing time;

(3) The greater the COV of hydrothermal parameters, the larger the SD of the freezing curtain temperature. For the ACF, the most significant impact comes from DBIN, followed by DSQX, with DSNX having the smallest effect. For COV, the most significant impact comes from thermal conductivity, followed by permeability coefficient, with specific heat capacity having the smallest effect;

(4) Different ACD parameters have varying impacts on the temperature of the freeze-soft clay rock. The biggest influence is the thermal conductivity, the second is the permeability coefficient, and the least influence is the specific heat capacity. The larger the ACD of the hydrothermal parameter, the smaller the SD of freezing soft clay rock temperature.

The variable hydrothermal parameter of soft clay rock holds profound implications for urban underground space sustainable development. As cities continue to expand, the construction and maintenance of underground space infrastructure play a pivotal role in enhancing transportation efficiency, reducing carbon emissions, and fostering economic growth. However, the effectiveness of freeze curtains in controlling water ingress and stabilizing underground space structures depends on the accurate characterization of thermodynamic parameters such as thermal conductivity, permeability coefficient, and specific heat capacity. In the future, we can focus on the dynamic evolution relationship between parameter spatial variability and temperature. By considering the variability in these parameters, urban planners and engineers can make informed decisions that contribute to the sustainability of underground space engineering. Understanding the variable hydrothermal parameters enables the design of freeze curtains that are resilient to changing environmental conditions and geological characteristics. Furthermore, considering variability fosters innovation in sustainable infrastructure development. This aligns with the broader goals of sustainable urban development, which seek to create cities that are livable, equitable, and environmentally conscious. Therefore, the consideration of variability in thermodynamic parameters of soft clay rock is instrumental in promoting urban sustainable development.

Author Contributions: Formal analysis, T.W.; Investigation, D.W. and H.L.; Resources, J.G.; Data curation, T.W. and K.R.; Project administration, H.L. and J.G.; Funding acquisition, T.W. All authors have read and agreed to the published version of the manuscript.

Funding: This research was supported by the National Natural Science Foundation of China (Grant No. U20B6001, 92255302), the Open Fund of State Key Laboratory of Coal Mining and Clean Utilization (China Coal Research Institute) (Grant No. 2021-CMCU-KF019).

Institutional Review Board Statement: Not applicable.

Informed Consent Statement: Not applicable.

Data Availability Statement: The original contributions presented in the study are included in the article, further inquiries can be directed to the corresponding author.

Acknowledgments: The authors wish to thank three anonymous reviewers and editors for their comments and advice.

Conflicts of Interest: The authors declare no conflicts of interest.

References

1. Long, Y.Y.; Tan, Y. Soil arching due to leaking of tunnel buried in water-rich sand. *Tunn. Undergr. Space Technol.* **2020**, *95*, 103158. [CrossRef]
2. Zhang, S.M.; Li, X.; Li, D.H.; Ding, Z.; Wei, G. Study on failure mode of underground diaphragm wall in soft soil area. *Appl. Mech. Mater.* **2013**, *405*, 1375–1382. [CrossRef]
3. Sousa, R.L.; Einstein, H.H. Lessons from accidents during tunnel construction. *Tunn. Undergr. Space Technol.* **2021**, *113*, 103916. [CrossRef]
4. Ma, E.; Lai, J.; Xu, S.; Shi, X.; Zhang, J.; Zhong, Y. Failure analysis and treatments of a loess tunnel being constructed in ground fissure area. *Eng. Fail. Anal.* **2022**, *134*, 106034. [CrossRef]
5. Liu, J.; Tan, Z.; Zhao, Q.; Liu, B.; Wang, X. Mechanical properties and durability analysis of CS-CG stabilized soil: Towards sustainable subgrade soil enhancement. *Constr. Build. Mater.* **2024**, *442*, 137634. [CrossRef]
6. Ouyang, Z.; Li, P.; Cui, J.; Luo, R.; Yuan, D. Shaking table test study on flexible and rigid immersed tube tunnel in liquefiable soil layer. *Math. Probl. Eng.* **2020**, *2020*, 4980549. [CrossRef]
7. Pimentel, E.; Papakonstantinou, S.; Anagnostou, G.E. Numerical interpretation of temperature distributions from three ground freezing applications in urban tunnelling. *Tunn. Undergr. Space Technol.* **2012**, *28*, 57–69. [CrossRef]
8. Fu, Y.; Hu, J.; Wu, Y. Finite element study on temperature field of subway connection aisle construction via artificial ground freezing method. *Cold Reg. Sci. Technol.* **2021**, *189*, 103327. [CrossRef]
9. Zhao, J.; Yang, P.; Li, L. Investigating influence of metro jet system hydration heat on artificial ground freezing using numerical analysis. *KSCE J. Civ. Eng.* **2021**, *25*, 724–734. [CrossRef]
10. Yue, F.T.; Lv, S.G.; Shi, R.J.; Zhang, Y.; Lu, L. Numerical analysis and spot-survey study of horizontal artificial ground freezing in tunnel connecting passage construction. *Appl. Mech. Mater.* **2015**, *744*, 969–977. [CrossRef]
11. Mauro, A.; Normino, G.; Cavuoto, F.; Marotta, P.; Massarotti, N. Modeling artificial ground freezing for construction of two tunnels of a metro station in Napoli (Italy). *Energies* **2020**, *13*, 1272. [CrossRef]
12. Li, Z.; Chen, J.; Sugimoto, M.; Ge, H. Numerical simulation model of artificial ground freezing for tunneling under seepage flow conditions. *Tunn. Undergr. Space Technol.* **2019**, *92*, 103035. [CrossRef]
13. Liu, X.; Shen, Y.; Zhang, Z.; Liu, Z.; Wang, B.; Tang, T.; Liu, C. Field measurement and numerical investigation of artificial ground freezing for the construction of a subway cross passage under groundwater flow. *Transp. Geotech.* **2022**, *37*, 100869. [CrossRef]
14. Liu, X.; Nowamooz, H.; Shen, Y.; Liu, Y.; Han, Y.; An, Y. Heat transfer analysis in artificial ground freezing for subway cross passage under seepage flow. *Tunn. Undergr. Space Technol.* **2023**, *133*, 104943. [CrossRef]
15. Ma, J.; Huang, K.; Zou, B.; Li, X.; Deng, Y. The influence of tunnel insulation measures on the temperature spatiotemporal variation of frozen soil during artificial ground freezing. *Cold Reg. Sci. Technol.* **2023**, *214*, 103942. [CrossRef]
16. Zhou, J.; Tang, Y. Practical model of deformation prediction in soft clay after artificial ground freezing under subway low-level cyclic loading. *Tunn. Undergr. Space Technol.* **2018**, *76*, 30–42. [CrossRef]
17. Fan, W.; Yang, P. Ground temperature characteristics during artificial freezing around a subway cross passage. *Transp. Geotech.* **2019**, *20*, 100250. [CrossRef]

18. Gao, J.; Li, X.; Cheng, G.; Luo, H.; Zhu, H. Structural evolution and characterization of organic-rich shale from macroscopic to microscopic resolution: The significance of tectonic activity. *Adv. Geo-Energy Res.* **2023**, *10*, 84–90. [CrossRef]
19. Alibeikloo, M.; Khabbaz, H.; Fatahi, B. Random field reliability analysis for time-dependent behaviour of soft soils considering spatial variability of elastic visco-plastic parameters. *Reliab. Eng. Syst. Saf.* **2022**, *219*, 108254. [CrossRef]
20. Elachachi, S.M.; Breysse, D.; Denis, A. The effects of soil spatial variability on the reliability of rigid buried pipes. *Comput. Geotech.* **2012**, *43*, 61–71. [CrossRef]
21. Sun, G.; Peng, F.; Mu, M. Uncertainty assessment and sensitivity analysis of soil moisture based on model parameter errors–results from four regions in China. *J. Hydrol.* **2017**, *555*, 347–360. [CrossRef]
22. Annabi, M.; Raclot, D.; Bahri, H.; Bailly, J.S.; Gomez, C.; Le Bissonnais, Y. Spatial variability of soil aggregate stability at the scale of an agricultural region in Tunisia. *Catena* **2017**, *153*, 157–167. [CrossRef]
23. Wang, F.; Huang, H.; Yin, Z.; Huang, Q. Probabilistic characteristics analysis for the time-dependent deformation of clay soils due to spatial variability. *Eur. J. Environ. Civ. Eng.* **2022**, *26*, 6096–6114. [CrossRef]
24. Kalantari, A.R.; Johari, A.; Zandpour, M.; Kalantari, M. Effect of spatial variability of soil properties and geostatistical conditional simulation on reliability characteristics and critical slip surfaces of soil slopes. *Transp. Geotech.* **2023**, *39*, 100933. [CrossRef]
25. Zhou, Z.; Li, X.; Zhao, H. Probabilistic study of offshore monopile foundations considering soil spatial variability. *ASCE-ASME J. Risk Uncertain. Eng.Syst. Part A Civ. Eng.* **2022**, *8*, 04022035. [CrossRef]
26. Nguyen, T.S.; Likitlersuang, S. Influence of the spatial variability of soil shear strength on deep excavation: A case study of a Bangkok underground MRT station. *Int. J. Geomech.* **2021**, *21*, 04020248. [CrossRef]
27. Wang, T.; Cao, J.; Liu, J.; Xu, J.; Zhou, G. Characterizing anisotropic spatial variations of uncertain mechanical parameters for clay layer using incomplete probability data. *Probabilistic Eng. Mech.* **2024**, *76*, 103623. [CrossRef]
28. Luo, Z.; Hu, B.; Wang, Y.; Di, H. Effect of spatial variability of soft clays on geotechnical design of braced excavations: A case study of Formosa excavation. *Comput. Geotech.* **2018**, *103*, 242–253. [CrossRef]
29. Viviescas, J.C.; Griffiths, D.V.; Osorio, J.P. Geological influence on the spatial variability of soils. *Int. J. Geotech. Eng.* **2022**, *16*, 382–390. [CrossRef]
30. Ching, J.; Uzielli, M.; Phoon, K.K.; Xu, X. Characterization of autocovariance parameters of detrended cone tip resistance from a global CPT database. *J. Geotech. Geoenviron. Eng.* **2023**, *149*, 04023090. [CrossRef]
31. Cai, Y.; Bransby, M.F.; Gaudin, C.; Tian, Y. Accounting for soil spatial variability in plate anchor design. *J. Geotech. Geoenviron. Eng.* **2022**, *148*, 04021178. [CrossRef]
32. Luo, Z.; Di, H.; Kamalzare, M.; Li, Y. Effects of soil spatial variability on structural reliability assessment in excavations. *Undergr. Space* **2020**, *5*, 71–83. [CrossRef]
33. Bong, T.; Stuedlein, A.W. Efficient methodology for probabilistic analysis of consolidation considering spatial variability. *Eng. Geol.* **2018**, *237*, 53–63. [CrossRef]
34. Zhang, W.; Han, L.; Gu, X.; Wang, L.; Chen, F.; Liu, H. Tunneling and deep excavations in spatially variable soil and rock masses: A short review. *Undergr. Space* **2022**, *7*, 380–407. [CrossRef]
35. Cami, B.; Javankhoshdel, S.; Phoon, K.K.; Ching, J. Scale of fluctuation for spatially varying soils: Estimation methods and values. *ASCE-ASME J. Risk Uncertain. Eng.Syst. Part A Civ.Eng.* **2020**, *6*, 03120002. [CrossRef]
36. Yoshida, I.; Tomizawa, Y.; Otake, Y. Estimation of trend and random components of conditional random field using Gaussian process regression. *Comput. Geotech.* **2021**, *136*, 104179. [CrossRef]
37. Gan, X.; Gong, X.; Liu, N.; Yu, J.; Li, W. Random analysis method for nonlinear interaction between shield tunnel and spatially variable soil. *Comput. Geotech.* **2024**, *166*, 105964. [CrossRef]
38. Tounsi, H.; Rouabhi, A.; Jahangir, E. Thermo-hydro-mechanical modeling of artificial ground freezing taking into account the salinity of the saturating fluid. *Comput. Geotech.* **2020**, *119*, 103382. [CrossRef]
39. Sun, Q.; Dias, D. Uncertainty quantification of tunnel seismic deformations in random soils. *Tunn. Undergr. Space Technol.* **2022**, *128*, 104663. [CrossRef]
40. Wang, C.; Wang, K.; Tang, D.; Hu, B.; Kelata, Y. Spatial random fields-based Bayesian method for calibrating geotechnical parameters with ground surface settlements induced by shield tunneling. *Acta Geotech.* **2022**, *17*, 1503–1519. [CrossRef]
41. Cao, J.; Wang, T.; Zhou, G.; Feng, X.; Zhu, C. Parameter estimation of grouting pressure and surface subsidence on the reliability of shield tunnel excavation under incomplete probability information. *Comput. Geotech.* **2024**, *173*, 106530. [CrossRef]

MDPI AG
Grosspeteranlage 5
4052 Basel
Switzerland
Tel.: +41 61 683 77 34

Applied Sciences Editorial Office
E-mail: applsci@mdpi.com
www.mdpi.com/journal/applsci

www.ingramcontent.com/pod-product-compliance
Lightning Source LLC
Chambersburg PA
CBHW041419210926
51850CB00013B/3315